Butterflies and climate change

Butterflies are particularly sensitive to climate and are important 'bio-indicators' of climatic change. This unique and important book not only explores how butterflies adapt to climatic gradients and weather patterns; it shows how their biogeography and evolution have responded to climate change in the past, and how they are likely to respond in the future as the enhanced greenhouse effect increasingly alters the world's climate.

Roger Dennis begins by explaining the atmospheric systems in which butterflies live, and which impose constraints upon their activity, development and function. He examines how butterflies thermoregulate, despite having very little capacity for generating their own heat. He then covers butterflies' life history strategies, their adaptations to seasonality, and their tolerance of extreme conditions. The discussion takes him into the controversial areas of population dynamics and species diversity, and he presents an important new model to explain how gradients in adult butterfly morphology and colour patterns relate to gradients in climate. Finally he explores further adaptive responses to climatic change, using models to explain past events and to predict the impact on butterfly populations of global warming.

Butterflies and climate change is a major contribution not only to insect ecology, but to our understanding of how anthropogenic climate affects natural populations and ecosystems. The book is readable and accessible to the non-specialist, but fully referenced and with a detailed bibliography.

This book will be of interest to students of ecology, biology, biogeography and climatology; amateur naturalists and entomologists; and non-specialist readers interested in butterflies and global environmental change.

Roger L.H. Dennis is a Fellow of the Royal Entomological Society and the Linnean Society. One of Britain's leading experts in butterfly ecology, he teaches at The Manchester Grammar School.

To Margaret

For, at very least, putting up with so many indigestible soliloquies at the dinner table

Butterflies and climate change

Roger L. H. Dennis

Manchester University Press
Manchester and New York
Distributed exclusively in the USA and Canada by St. Martin's Press

Published by Manchester University Press
Oxford Road, Manchester M13 9PL, UK
and Room 400, 175 Fifth Avenue, New York, NY 10010, USA

Distributed exclusively in the USA and Canada
by St. Martin's Press, Inc., 175 Fifth Avenue, New York,
NY 10010, USA

British Library Cataloguing-in-Publication Data
A catalogue record for this book is available from the British Library

Library of Congress Cataloging-in-Publication Data
Dennis, Roger L. H.
Butterflies and climate change / Roger L. H. Dennis.
p. cm.
Includes bibliographical references and index.
ISBN 0–7190–3505–8 (cloth). — ISBN 0–7190–4033–7 (paper)
1. Butterflies — Behavior — Climatic changes. 2. Climatic changes.
I. Title.
CL562.4.D46 1993
595.78′9045222—dc20 93–9186
CIP

ISBN 0 7190 3505 8 *hardback*
ISBN 0 7190 4033 7 *paperback*

Typeset in Hong Kong
by Graphicraft Typesetters Ltd, Hong Kong

Printed in Great Britain
by Biddles Ltd, Guildford and King's Lynn

Contents

Tables

Figures

Preface

Documentation and explanation of the response and adaptability of biotic systems to agents of weather and climate have long challenged the scientific community. The prospect of unprecedented climatic changes with anthropogenically-induced enhanced greenhouse warming presents new opportunities for research. In many ways, butterflies are ideal organisms for this research – their distinct life-history stages each occupying different habitats and proffering a variety of adaptations, their short generation times, flexible strategies, physical variability, abundance and ubiquity, and their relative ease of identification and study, make them highly suitable for advancing our ecological and evolutionary knowledge.

The primary aim of this short book is to review the relevant literature on the interaction of butterflies and climate agents. I have tracked the subject matter (agents, responses, adaptability etc.) from the individual to the population and to multi-species associations. The frame also scans from the present day, to the past (chapter 4: evolutionary time), to finish with future projections (chapter 5). Limitations of space have made the task more difficult than I first imagined but my hope is that, even where topics have admittedly been given more cursory treatment than I would have liked, readers will find this a useful reference source. As citations are selected to address specific issues and the text does not deal with the mechanics of survey and experiment, it is important to appreciate that it can be little more than this. As such, I would urge students to treat the book as a guide to the research available in the source references.

Some topics in the book are less amenable to 'global' review. These I have illustrated with specific regional cases which include some original material. They occupy more sections in the latter part of the book and should present some welcome relief from the denser reviews in the first two chapters. They allow me to question some research conclusions, and to challenge directly two 'ritornellos' in the literature; thus, to present ideas for theoretical, laboratory and field research. The first tenet is that infraspecific wing phenotype markers are of no direct relevance to individual fitness. In answer to this (chapter 3), a generalized model is offerred for consideration which incorporates climate variability as a prominent factor; case scenarios are fully documented to encourage tests and experimentation. The second tenet is that areas affected by Pleistocene glaciations have been too unstable for macroevolution to have occurred there. A variety of evidence from different sources contests this conclusion. The lack of fossil butterflies thwarts a definitive test and detracts from the undoubted value of butterflies as evolutionary markers. In chapter 4, an appeal is made for better retrodictive modelling for such organisms, and a demonstration given of how much insight can be gained by using simple reconstruction techniques.

Even in a short book, it is probably not difficult to obtain glimpses of an underlying philosophy. Nevertheless, to avoid needless pedantry in flagging the text with qualifiers and cautionary notes, a few points are here made explicit. A pluralistic stance is adopted. I am not averse to 'simple' solutions, but am generally prepared for circumstances to have a multivariate context. I share Gould's (1986) and Rapport's (1991) reservations and objections to notions of optimality, which typically translate into elegant algorithms as in the theory of life-history strategies and population dynamics. The main concern has been that models designed to account for a phenomenon should accord with the sentiments of Lewontin (1972) and the minimal criteria of Leibenstein (1976): namely, they should have an internal logical consistency, make realistic assumptions and, fundamentally, be open to testing, experimentation and falsification. The use of 'systems' modelling admits of unknown factors and provides opportunity for revision, for additional links and parameters, non-linearity, differences in covariances and other modifications – all necessary in the present state of knowledge. I have tried to address Stearns's (1976: 5) conundrum: 'The more certain we are that we know what is causing an evolutionary trend, the less certain we are that the cause is general. And, the more certain we are that a trend is in fact general, the less certain we can be about its cause or causes.' I have attempted this by probing the same issues at extremes of scale and, where possible, by segregating the 'general' (a discussion of the main components) from the 'particular' (specific cases).

For a number of important topics in the book I have found but a dearth of information. There is also uncertainty about many issues for which extensive data are available. As such, many statements are appropriately couched in probabilistic terms. These are areas where further research may usefully be done. Finally, I must caution readers as to the necessarily speculative content of some sections (see section 5.3). Regarding these, my intention has been to encourage new ideas and further work; if it does so, then this book will have fulfilled its task.

Acknowledgements

I am grateful to a number of friends and colleagues for their help in preparing this book, particularly those who have taken the time to read the chapters, thus to sift out my occasional blunders: Keith Bennett (chapter 4), Paul Brakefield (chapter 3), Ernie Pollard (chapters 1 and 2); Andrew Pullin (chapter 2), Tim Shreeve (chapters 1 and 2), Jeremy Thomas and Martin Warren (chapter 5). Needless to say, any views expressed are my own, except where otherwise stated, as are any errors. I would also like to offer my thanks to a number of individuals for additional personal comunications and data: to Peter B. Hardy, Stephen H. Hind, Otakar Kudrna and R. Whyman; to Chris Perkins of Manchester University Geography Department for so kindly providing access to maps; to Elaine Barrow and Paul Brakefield for providing illustrations for the text; and to the many colleagues who have sent reprints over the past two years. In particular, I would like to record my indebtedness to Joel Kingsolver and Paul Brakefield whose excellent research stimulated ideas for section 3.3, and to Bryan P. Beirne whose pioneer research initiated my long standing interest in the historical issues discussed in chapter 4. During this work, I have been fortunate to have had access to the John Rylands University Library, Manchester, and gladly acknowledge their efficiency and assistance. I extend my grateful thanks to W. R. (Bob) Williams and Tim Shreeve for their support over the years, for the enjoyment of co-operative ventures, some of the details of which appear in these pages. George Thomson, with whom I have been in contact about entomological matters longer than anyone else, with his characteristic generosity very kindly set up the 'machine' in which this text evolved and seduced it to work for me. Finally my wife, Margaret, has ensured, as always, that the current task progresses, even when I do not.

R. L. H. D.

1
Atmospheric systems and butterfly biology

Despite the long-term accumulation of climatic data (Meteorological Office, 1980–83; Wallén, 1970) and facilities for research into microclimates (Geiger, 1965), advances in knowledge on insect physiology and behaviour in relation to environmental conditions are the result of development in two areas: firstly, the availability of new techniques with rapid response times (Unwin, 1980); secondly, the realization that mere accumulation of climatic data coupled with physiological studies in static laboratory simulations do not provide adequate experiments (Willmer, 1982). As knowledge of the effect of external agents on butterfly biology is of paramount importance, this chapter first provides a résumé of basic atmospheric patterns and processes for the reader. Subsequently, it reviews the responses of butterflies to meteorological influences and the direct and indirect effects of atmospheric agents on butterfly biology.

1.1 **Determinants and components of weather and climate**

Insects must, if averaged through time, maintain their energy (thermal) and water balance (Willmer, 1982). Although rates of metabolism and water loss are partly functions of internal physiology, both ultimately depend on environmental conditions, the most important of which are temperature and moisture. These correlate with and are influenced by other meteorological variables (i.e., radiation; air movement; precipitation) as well as by feedbacks from different substrates, and affect one another (Geiger, 1965; Rosenberg, 1974). The form that interactions take depends much on space–time scales as well as synoptic meteorology and surface characteristics. This is relevant because biological systems present problems at very different scales. Individual success is determined in the main by microscale processes, as microhabitats can substantially modify screen-registered atmospheric inputs. Nevertheless, as synoptic weather systems can theoretically induce space–time autocorrelation in conditions and responses among aggregated microhabitats, biological systems can be significantly influenced by macroscale meteorology. Germane to scale is the distinction between weather and climate. The weather systems which determine a climate of a locality are governed by many interactions and complex feedback processes involving the underlying surface and atmosphere. Climate described and summarized as average weather is entirely inadequate; variance, frequency, type, magnitude, duration and predictability for meteorological attributes and attribute states critically influence life-systems; thus climate is better regarded as an integration of weather spectra over time at a particular place (Musk, 1988).

A *Climatic gradients and zonal influences*

The relevance of macroscale influences for butterfly biology is evident in the contrasts for global and regional heat and moisture. These two variables have long been recognized to covary with natural vegetation and form the basis of many climatic classifications (Barry & Chorley, 1982; Lockwood, 1985; Henderson-Sellers & Robinson, 1986). The distribution of insolation receipt and diel maximum temperatures is broadly meridional but with tropical rather than equatorial maxima. Corresponding largely with this pattern, the mean range in temperatures also increases from equator to poles (Stevens, 1989; see Fig. 2.7). This pattern depends primarily on the altitude of the sun and length of day which are governed by latitude, global axis tilt and time of year. Thermal contrasts are also affected by the heat capacity and albedo of different surfaces, by air mass movements, ocean currents and air mass stability accompanying pressure systems, all of which reflect on the general circulation in the atmosphere and ultimately on energy transfers in the earth–atmosphere system. Interactions can be complex; for example, the gradients in thermal contrasts from the oceans into interior Eurasia, associated with higher maxima and lower minima, are caused primarily by differential heat capacity of land and sea, but also by prevailing air movements, ocean currents, the relative influence of high and low pressure systems and thus the degree of advection and cloud insulation (see below). Global and regional water balance depends on the precipitation: evapotranspiration ratio. As the primary source for atmospheric water is in the oceans (97.2 per cent), continental interiors have the largest deficit. Moisture levels are more dependent than energy on pressure systems. High moisture levels occur in regions of marked convection (equatorial regions) or surface air mass convergence (intertropical convergence zone, ITCZ; polar front) and low levels in regions dominated by high pressure (subtropical deserts; continental interiors) or areas subject to offshore winds (west coast trade wind belts). As life forms generally require equable distribution of heat and moisture, two types of extreme environmental deficiency can be identified: moisture-deficient temperate and subtropical deserts subject to physical drought, and energy-deficient high latitudes subject to physiological drought. To a lesser degree, these two gradients characterize most climates, even for small regions such as the British Isles (Green, 1964; Hatch, 1973; White & Perry, 1989; see sections 2.2A; 3.2A; 5.3B).

Broad global climatic patterns are substantially disrupted by topography and surface characteristics at a wide range of scales. Systematic trends in climatic variables occur with altitude, and mountain systems can influence synoptic situations (see below). The chief effects of altitude associated with decreased air pressure are decreases in temperature and increased windspeed. Depending on latitude, radiation levels can greatly increase in the thinner atmosphere as can the diel period of direct radiation. Patterns of precipitation and humidity are more complex; prevailing winds determine windward–leeward contrasts, whereas the relative stability of regional air masses can impose a curvilinear (quadratic) pattern on rainfall distribution with altitude. Massenerhebung effects on vegetational zonation in relation to exposure reflect the sheer complexity of mountain climates (Eyre, 1963; Liljequist, 1970).

Aspect and elevation (slope angle) combine to make for enormous differences in radiation receipt which, in conjunction with prevailing winds, influence temperature, humidity and other attributes (i.e., snow cover, ground frost) (Fig. 1.1). Consequently positive and negative relief features develop distinctive diel and seasonal isoline patterns. Small relief features can have surprising effects; for example, hills 30–50 m high can produce 50–80 per cent increases in precipitation compared to surrounding plains during cyclonic episodes (Barry & Chorley, 1982). Surface characteristics produce distinctive climates associated with lakes, forests and urban landscapes. Water bodies and their surrounds record increases in wind speed and rainfall and moderation of temperature extremes. The effect of forests on climate is complicated by woodland height, growth form, architecture, texture, density and reflectivity, the structure of open spaces (e.g., rides, clearings) and position of instrumentation. Whereas radiation levels, wind speed and temperature extremes are greatly reduced and humidity levels increased under the canopy, at forest margins the order for these variables can be reversed. In particular very high temperatures can be experienced within clearings due to the combination of direct and reflected radiation and reduced windspeed (Geiger, 1965). Urban landscapes modify the regional climate by an increased production of heat and hygroscopic nuclei and alteration of surface configuration and roughness. Despite the lower radiation levels, concentric urban heat islands develop and increased condensation (fog; smog) can occur; precipitation is often increased in the vicinity of urban areas whereas windspeeds relate more to the structure of streets and buildings (Chandler, 1965; Landsberg, 1981).

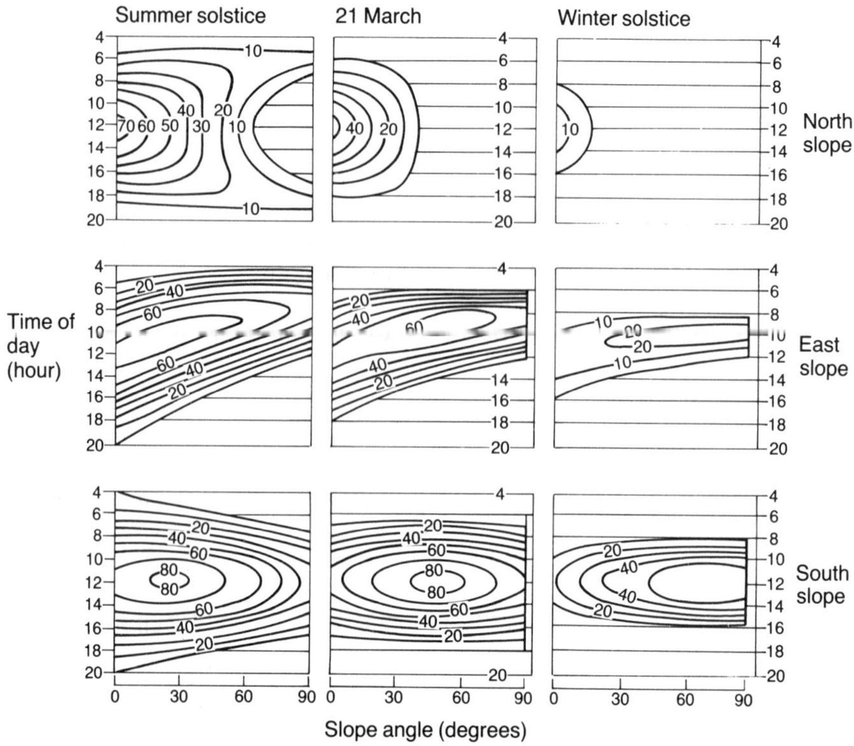

1.1 Direct solar radiation (cal $cm^{-2}h^{-1}$) on cloudless days on north, east and south slopes of all inclinations on three selected days (50°N). (After Geiger, 1965 by courtesy of Harvard University Press.)

At a scale relevant to individual animals, there can be very sharp gradients in temperature and humidity, the steepness and stability of which depends mainly on (i) the nature of the substrate, (ii) proximity to other surfaces and (iii) solar radiation and other meteorological inputs. Horizontal air movement may be very limited close to substrate surfaces, with static 'boundary layers'. Thus, vertical gradients created by the varied heating of, and evaporation from, different substrates may be maintained with minimal mixing for periods of time, though in extent and duration, gradients obviously have both diel and seasonal patterns (Geiger, 1965; Shaw, 1967; Rosenberg, 1974; Oke, 1987). Constancy in hygrothermal conditions over bare soil is proportional to height above and depth below the ground surface, ranges and variances declining far more rapidly with depth than height (Fig. 1.2a). Temperature and humidity gradients are especially steep over the first few millimetres. Ground surface temperature excesses of 10–15°C over screen temperatures (T_{sc}) are typical even in arctic summers (Kevan & Shorthouse, 1970) and in desert environments substrate temperatures (T_s) can reach 80–90°C (Berry & Cloudsey-Thompson, 1960). Thermal conditions with soil depth are not easily predicted; apart from external influences, heat capacity and in turn soil colour, texture and soil moisture levels, as well as seasonal influences such as snow cover, can influence measurements. Ground temperatures of – 2°C have been recorded under 25 cm of snow when T_{sc} is – 30°C (Ylimaki in Ekholm, 1975). Through their own gains and losses of energy and moisture and interactive shading effects on neighbouring substrates, plant surfaces create their own microclimates (Willmer, 1982). Amelioration of microclimatic conditions depends much on depth and density of vegetation (Fig. 1.2b). Boundary layer climates have some similarity to conditions within ground vegetation; higher humidity occurs near the transpiring surface, and temperature levels and gradients may differ between upper and lower leaf surfaces diurnally and with changing meteorological conditions (Willmer, 1982). Surface colour and texture of plants, for instance of tree bark, affect albedo and insulation of living tissue, and in turn can influence the temperatures of microhabitats (Nicolai, 1986).

B *Weather systems*

The characteristics of weather and longer-term climatic patterns are determined by weather systems. Their significance for biological organisms relates to the frequency, duration, timing and suddenness with which organisms are exposed to conditions marginal for activity, development and survival. Weather systems have their basis in air masses and in air mass conjunction. Air masses originate as barotropic structures over stable areas of high pressure; their formation is linked closely with the general atmospheric surface circulation and meridional vertical cells (i.e., Hadley, Ferrel, Polar) and their associated tropopausal jets, whereas their conjunction coincides with surface convergence (ITCZ, polar and arctic fronts). Weather systems differ in numerous ways: in type, location, dimensions, amplitude, periodicity, persistence, seasonality and predictability – thus in degree of complexity of effect. Some parts of the globe have relatively simple weather patterns, e.g., the monotonous weather of equatorial regions and subtropical hot deserts and the seasonal reversals of maritime and continental air as in tropical regions of West Africa under the migration of the ITCZ. Nevertheless, in these regions violent and unpredictable weather phenomena may impinge on biological systems (i.e., thunderstorms, hurricanes, tornadoes).

The most complex weather patterns occur within the prevailing circulation of mid-latitudinal temperate areas influenced by the polar front, a region dominated by the interdigitation of mobile frontal depressions and relatively stable anticyclones. For several reasons the alternation of these pressure systems has enormous significance for biological systems. (i) Strikingly

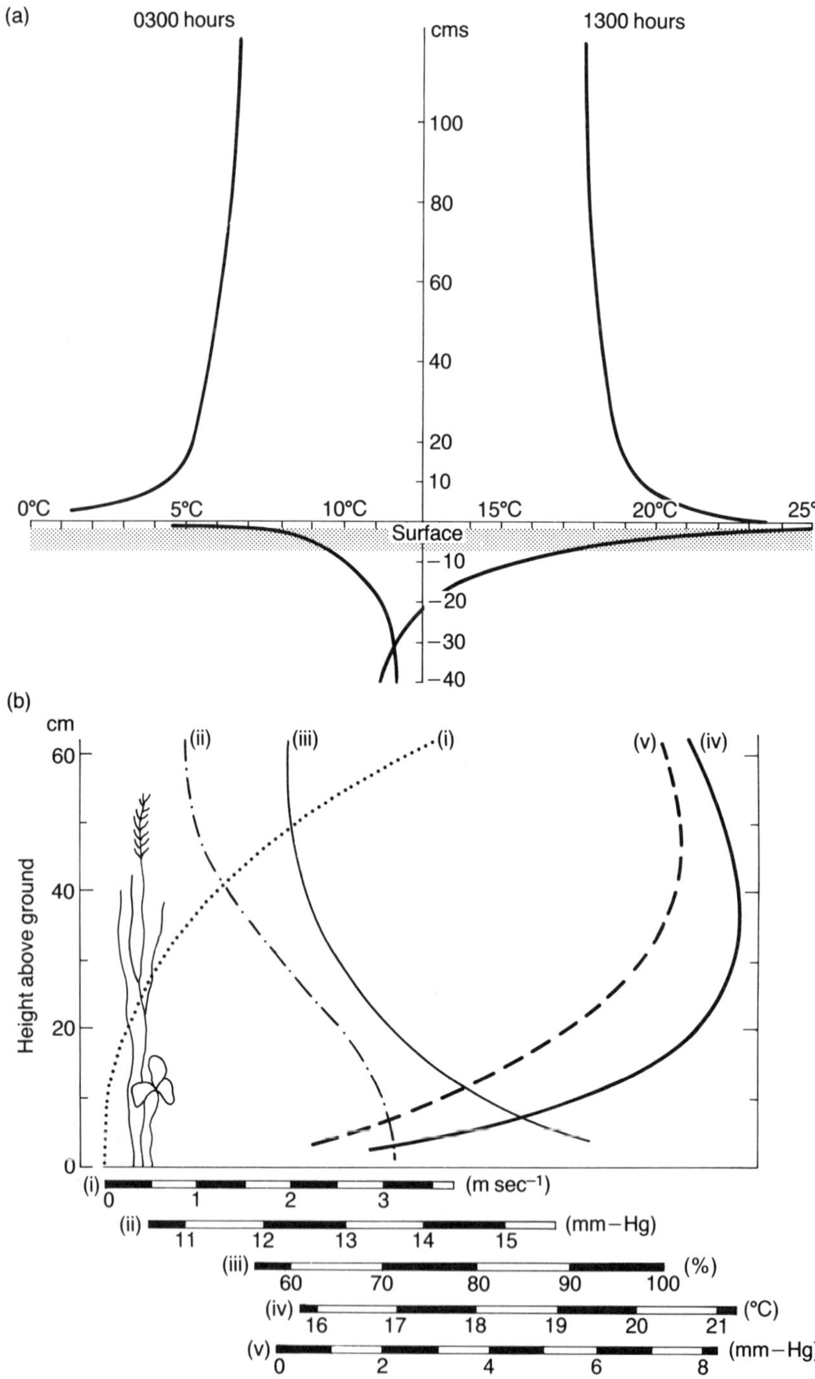
(a)
0300 hours
1300 hours
cms
100
80
60
40
20
10
0°C
5°C
10°C
15°C
20°C
25°C
Surface
Ta
−10
−20
−30
−40
(b)
cm
60
40
20
0
Height above ground
(i)
(ii)
(iii)
(iv)
(v)
(i) 0 1 2 3 (m sec−1)
(ii) 11 12 13 14 15 (mm−Hg)
(iii) 60 70 80 90 100 (%)
(iv) 16 17 18 19 20 21 (°C)
(v) 0 2 4 6 8 (mm−Hg)

different weather types and climatic patterns accompany them (Fig. 1.3; Musk, 1988). (ii) The environments affected by them are generally marginal for organisms in so far as the latter are required to build up much larger excess temperatures than at lower latitudes. (iii) Depressions produce rapid, varied and fluctuating changes in conditions which can enforce immediate inactivity in organisms, or produce conflicting conditions for activity (e.g., steep vertical temperature gradients with high radiation and substrate temperatures but high windspeeds), both of which may influence vulnerability and opportunities for reproduction. (iv) The unpredictability in the alternation and persistence of contrasting pressure systems can produce atypical seasonal weather conditions emitting conflicting signals that may influence individual success, especially in insects with limited tolerances for environmental variation (see section 2.1B).

Longer-term imbalances in synoptic patterns determine climatic differences and changes (see chapter 4). The regional dominance and influence of anticyclones and depressions have a direct impact on European climatic patterns. Stable weather accompanies persistent high pressure systems over eastern Europe and Asia in winter and over the Mediterranean in summer. West coast oceanic areas (e.g., British islands) are particularly influenced by alternations in pressure systems (Riley & Spolton, 1974; Stirling, 1982; Tooley & Sheail, 1985; Gregory, 1988). This contributes to steep lapse rates and to distinctive climatic gradients over relatively small distances (Barry & Chorley, 1982). Mesoscale topography can significantly disrupt or enhance these patterns by generating their own weather systems (Drake & Farrow, 1988). The European mountain belts (e.g., Scandinavia; Alps) can isolate pressure systems, impose climatic boundaries and trigger cyclogenesis in the form of lee depressions (e.g., Gulf of Genoa). Montane wind systems (Föhn or chinook, anabatic and katabatic) also greatly modify hygrothermal conditions as do coastal land and sea breezes. Despite differences in scale all these weather systems are of sufficient magnitude to impinge on biological systems (Geiger, 1965; Pears, 1977).

1.2 Temperatures near the ground, (a) Air and soil temperatures associated with bare ground at Seabrook, New Jersey (39°34′N, 75°13′E) on five bright days in spring. The diurnal range at screen level was 11°C, at 10 cm 15°C and at ground level probably > 20°C. A few centimetres below ground it became negligible. (After Crowe, 1971 by courtesy of the Longman Group.) (b) Temperature, humidity and wind in a Scottish meadow on a sunny June day between 15.00 and 16.00 BST. (i) windspeed (ms^{-1}); (ii) water vapour pressure (mm Hg); (iii) relative humidity (per cent); (iv) temperature (°C); (v) saturation deficit (mm Hg). (From Waterhouse, F.C. in Geiger, 1965 by courtesy of Harvard University Press.)

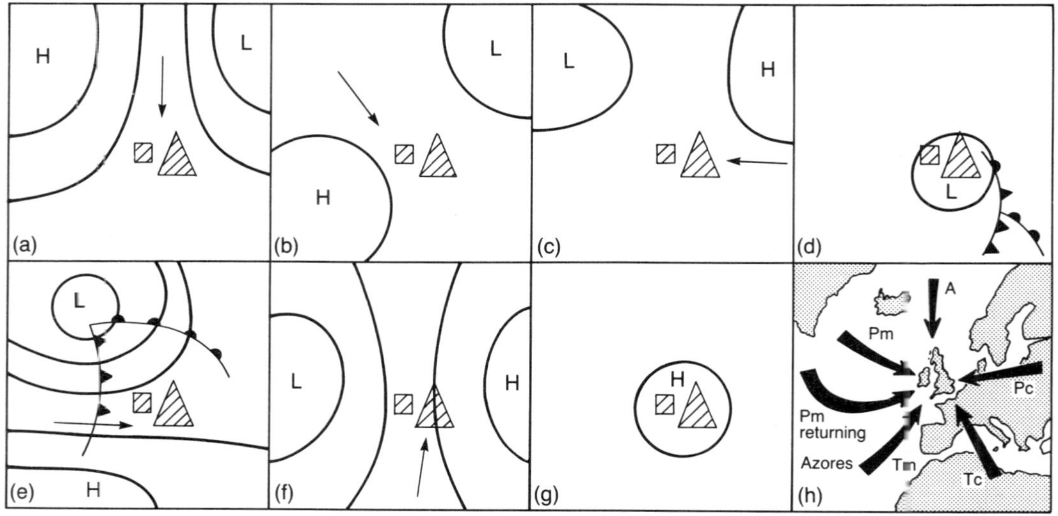

1.3 Schematic maps showing air circulation over the British Isles in relation to anticyclones and depressions. Very different weather conditions result from the varying relative positions of anticyclones and depressions, particularly with the passage of depressions. (a) Northerly; cold winter weather and cool showery summer weather. (b) North-westerly; cooler, showery, changeable conditions with strong winds in winter and summer though southern Britain may have dry bright weather. (c) Easterly: very cold in winter especially in the south and east, fine and dry in west; in summer hot and dry with possible heat waves. (d) Cyclonic: wet and disturbed weather in all seasons with variable wind direction and strength, depending on the sector of the depression affecting a locality at any time. (e) Westerly: unsettled cloudy weather with most rain in the west and north; milder winters and cooler summers. (f) Southerly: milder winters especially in the west; hot and thundery summers. (g) Anticyclones: dry with light winds; hot in summer but cold in winter. (h) Air mass types. H, high pressure; L, low pressure; A, arctic; P, polar; T, tropical; c, continental; m, maritime; cold fronts (triangles) and warm fronts (semicircles) are illustrated in two diagrams. (See Musk, 1988 for further details on air mass stability; by courtesy of Cambridge University Press.)

1.2 Atmospheric influences on butterfly biology

Butterfly populations are constrained by external agents and internal components (Gilbert & Singer, 1975; Gilbert, 1984). Internal components comprise subsets of life-history (viz., stages, instars, sex) and different aspects of butterfly biology (viz., physiology, biochemistry, behaviour etc.). External agents can influence biological systems in a number of complex ways inducing systems responses (e.g., incorporating thresholds, sign, lag time, magnitude, serialization, rate, linearity of response, directness, feedback, multiplicity of effects) and having consequences for individual survival, development and fecundity, thus individual fitness, population parameters and community structure (Calow, 1989). Weather and climate represents but one set of inputs; interaction between extrinsic agents (e.g., climate and hostplant chemistry on thermal tolerance in larvae; Pullin, 1987a; Taylor, 1988) and between intrinsic components (e.g., stage, physiology and egg maturation; Boggs, 1986) compels a multivariate framework of reference in field studies and laboratory experimentation (Calow, 1976, 1989). Furthermore, biological systems can be greatly influenced by spatio-temporal contexts at different scales; an example in the short term is the impact of prior weather conditions on subsequent behaviour such as the lowering of thresholds to ambient temperatures for flight after cool, cloudy periods (e.g., *Coenonympha inornata*, Heinrich, 1986), and in the long term by seasonality on the serialization in the relative investments and expenditure on reproductive or survival potential for different stadia (Calow, 1989).

Feedbacks to stress (*sensu* Underwood, 1989) can be triggered either in ecological time (e.g., immediate responses in differential habitat use or diel activity patterns; Gilbert & Singer, 1975; May, 1979) or in evolutionary time (e.g., ultimate responses as changes in gene frequencies under selection to maintain the insect in the flight activity range, FAR; Watt, 1968), the latter dictating population genetic structure. For biological systems incorporating many interdependent components the ramifications are theoretically extensive (Dennis & Shreeve, 1989; sections 2.1 and 3.3B), although microevolution must often be constrained by intolerance of extreme values especially in environments with fluctuating conditions (Kingsolver & Watt, 1983). The following sections of the chapter first discuss homeostatic mechanisms for maintaining hygrothermal equilibrium, and, subsequently, environmental constraints on survival, growth and fecundity.

A *Mechanisms for maintaining body temperature and water balance*

The subversion of enzyme-regulated biochemical reactions and complex integrated physiological processes by extreme conditions place a premium on the maintenance of optimum body temperature (T_b) and body water levels (W_b). Whereas, with temperature, insects are required to engage mechanisms for heating and cooling, with water balance the problem is generally that of retaining water and replacing losses, as the net gradient of water movements is almost invariably outwards, except in saturated atmospheres (Willmer, 1982). Although requirements for water balance and thermal balance coincide with heat-avoidance behaviour, heat for activity brings them into conflict. In butterflies, particularly short-lived adults, thermal considerations seem to outweigh the needs of water balance; this may simply reflect the lack of extensive field measurements on W_b levels. Because of this and the fact that temperatures affect W_b directly, observations on W_b are integrated into the more extensive details on thermoregulation.

T_b is either regulated (homeothermy) or is subject to environmental conditions (poikilothermy) (May, 1979; Findlay & Findlay, 1983). Compensation for fluctuations in ambient conditions may be behavioural and/ or physiological and heat exchange may be effected endothermically or ectothermically. Endothermy has as its energy source larval and adult feeding (relatively high cost), ectothermy in solar energy (relatively low cost). Heat exchange with the environment includes (i) solar radiation and re-radiation in infra-red from substrates, (ii) convection under the influence of airflow from insect or substrate and (iii) conduction between insect and substrate (Porter & Gates, 1969; Bakken & Gates, 1975; May, 1979). Evaporation from butterflies probably does not effect much cooling (Watt, 1968).

In Lepidoptera, the type of thermoregulatory response is determined primarily by the stage of development, habit (nocturnal or diurnal), structure and size. Ova and pupa are passive poikilotherms, whereas active adults and larvae have some capacity to compensate for ambient conditions. Nocturnal and diurnal Lepidoptera are both dependent on ambient temperature (T_a) and wind velocity (A_v), but diurnal Lepidoptera are additionally influenced by solar radiation (R). Convection from the body surface is the predominant mode of heat loss and solar radiation the major heat input for insects in the sun.

Adult butterflies are homeothermic within a narrow range of T_a. Within this range compensation can occur for low T_a ($T_b > T_a$) or high T_a ($T_b < T_a$). Critical upper and lower limits of T_a occur beyond which T_b approximates T_a. Torpid insects are vulnerable and extreme conditions can be lethal. *Tatochila* butterflies show prolonged survival in cold storage at 2°C. As prior mating significantly reduces survival under these conditions, this suggests that nutrients may be diverted into spermatophores or into somatic maintenance (Shapiro, 1982a). *Papilio polyxenes* can survive cooling to T_a < 0°C, but die at T_a 46°C at lower humidities (Rawlins, 1980). Evidence from many species suggests an inherent upper thermal limit for most multicellular life at c. 45–50°C (Hamilton, 1973). Heat shocks at 45°C significantly decrease survivorship and fecundity in *Colias* spp. (Kingsolver & Watt, 1983). Distinct taxa experience different stress conditions; ambient temperatures at which heat avoidance behaviour is enacted for *Pieris* spp.

are consistently lower than for *Colias* spp. (Kingsolver, 1985a). Adult insects are adapted to activities dependent on flight; typical T_b values for voluntary flight activity range (FAR) lie within 28°C and 38°C (Vielmetter, 1958; Nielsen & Nielsen, 1959; Watt, 1968; Kevan & Shorthouse, 1970; Douglas & Grula, 1978; Rawlins, 1980; Shreeve, 1984; Shelly & Ludwig, 1985). Differences occur for sex (e.g., female *Colias* and *Thymelicus lineola* require higher flight T_b than males) and for taxa at generic level (e.g., *Pieris* and *Colias* differ for FAR, but congeneric taxa are similar) (Watt, 1968; Kingsolver, 1985a; Pivnick & McNeil, 1987). Some species lie outside the typical FAR (e.g., FAR for *Parnassius phoebus*, T_b 17.6–20.1°C, Guppy, 1986; *Danaus plexippus* flies at T_b 10°C, Chaplin & Wells, 1982). Some butterflies have a higher T_b (e.g., *Precis villida*, $T_b < 43$°C at T_a 33°C, Heinrich, 1972). Tropical species such as *Calisto nubila and Mycalesis elia* are apparently able to maintain T_b within a very narrow range because T_a is approximately constant compared to that for temperate species.

Excess temperature ($T_{ex} = T_b - T_a$) varies directly with ambient temperature (T_a) and solar radiation (R), but inversely as the square root of wind speed (A_v) (Digby, 1955; Tsuji *et al.*, 1986). Wickman (1988) found that an increase in A_v of 11.1 times from 0.2 to 2.2 ms^{-1} reduced T_{ex} by half in *Lasiommata megera*. In high latitude species T_{ex} typically exceeds 10°C (Kevan & Shorthouse, 1970; Pivnick & McNeil, 1987) and extends to 17°C (Vielmetter, 1958). Tropical species are closer to optimal T_b and $T_{ex} <$ 10°C (Heinrich, 1972) or < 5°C (Shelly & Ludwig, 1985). The optimal activity range (OAR) of temperate species (e.g., *Colias* spp. T_b for OAR 35–38°C) is closer to typical T_b in tropical butterflies (Watt, 1968). Neither T_{ex} nor FAR in ectothermic Rhopalocera match those of large endothermic fast-flying Heterocera (Sphingidae, Saturniidae) which can respectively have $T_{ex} > 24$°C and FAR 40–45°C (Heinrich, 1974). Relaxation times for equilibrium T_b also depend directly on T_a, R and inversely on A_v. Lag times can be as little as 1.5–4 min in noon insolation at T_a 15°–22°C (Watt, 1968; Douglas & Grula, 1978; Heinrich, 1986) and < 1 min for $T_a > 23$°C (Wasserthal, 1975). Very few studies of T_{ex} have monitored values for R and A_v and equilibrium T_b for different values of T_a. Equally, little research has been carried out on the ability of butterfly adults to maintain equilibrium T_b when $T_a > T_b$ for specific levels of T_a and T_s. *Colias* spp. cease flight and exhibit heat avoidance postures when $T_b > 40$°C (Watt, 1968; Kingsolver & Watt, 1983). *Precis villida* has $T_b < 43$°C in T_a 33°C, a temperature deficiency of 1–9°C below that attained during maximal basking posture for immobile specimens.

Thermoregulatory strategies in butterflies involve the integration of behaviour and structure and endothermic and ectothermic devices. Thermoregulatory activities are not exclusive from other behaviour but are combined, often opportunistically, with mate-location (Dennis & Shreeve, 1988), courtship (Vielmetter, 1958), feeding (Kevan & Shorthouse, 1970), egglaying and roosting (Dennis, 1986a; pers. obs.). Nor is thermoregulation, endothermic or ectothermic, substrate-limited (Guppy, 1986; Heinrich, 1989). Behavioural thermoregulation includes: (i) postural control of solar input, (ii) microhabitat selection and (iii) alteration of diel activity patterns; endothermic behavioural techniques involve: (i) wing whirr and (ii) flight activity.

Basking postures incorporate wing positions: (i) wing surface exposure, (ii) wing angle, (iii) wing overlap; and body positions: (i) body alignment, (ii) anterior–posterior tilt, (iii) height above substrate. Three basic groups of postures have been determined (Fig. 1.4). Lateral basking, invariable exposure of hindwing ventra, is fundamentally distinct from dorsal basking (dorsal wing surfaces exposed) (Clench, 1966). In turn, dorsal baskers are subdivided on the basis of wing angles for attaining maximum body temperatures (T_{bask}), into those which use open wings and absorb solar energy and those that display intermediate wing angles and primarily reflect energy on to the thorax. Additional postures provide fine-tuning adjustments to what in effect is observed to be a shifting series of behavioural responses to deal with a structurally complex temporal environment. However, postural changes are less frequently mapped in relation to changing T_a and R than are changes in activity (Vielmetter, 1958; Rawlins, 1980; Fig. 1.5).

In dorsal absorbance (DA) baskers, maximum basking posture temperatures (T_{bask}) for low T_a are attained with dorsal wing surfaces at 90° to incident solar radiation. For high T_a, T_b is regulated by elevation of the wings (Vielmetter, 1958). The thorax is heated by conduction from the basal 15 per cent of the wings. As T_{bask} is virtually identical in live or dead specimens, little energy is convected in hemolymph between wings and thorax (Wasserthal, 1975, 1983). The body is also heated by air trapped between the wings and the substrate, by conduction from the irradiated wings dorsally and by convection from the substrate. Heat is then conducted through the ventral part of the body. Elevation of the wings influences the area of irradiation and enlarges the air spaces so that convection currents replace heated with cooler air. Variable overlap of the fore- and hindwings alters the mass and thus the heating capacity of the basal wings and the volume of air trapped between them (Vielmetter, 1958). Body alignment (180°) and tilt (head-up) orientate the wings at normal to the sun's rays (Fig. 1.4). When T_a is sufficiently low, appression of wings to the substrate increases energy transfer by conduction (Kevan & Shorthouse, 1970). Heat avoidance typically includes dorsal wing apposition; this can reduce excess temperatures (T_{ex}) by 70 per cent of T_{bask} in *Papilio machaon* (Wasserthal, 1975).

A variant of dorsal absorbance basking termed body basking (Clench, 1966) has been described for species with wide thoraces (e.g., Hesperiidae, such as *Ochlodes venata*). During low T_a, the hindwings are flat but the forewings are maintained aloft and rotated inwards at an angle. The ventral thorax may then be heated from below by the substrate and the trapped

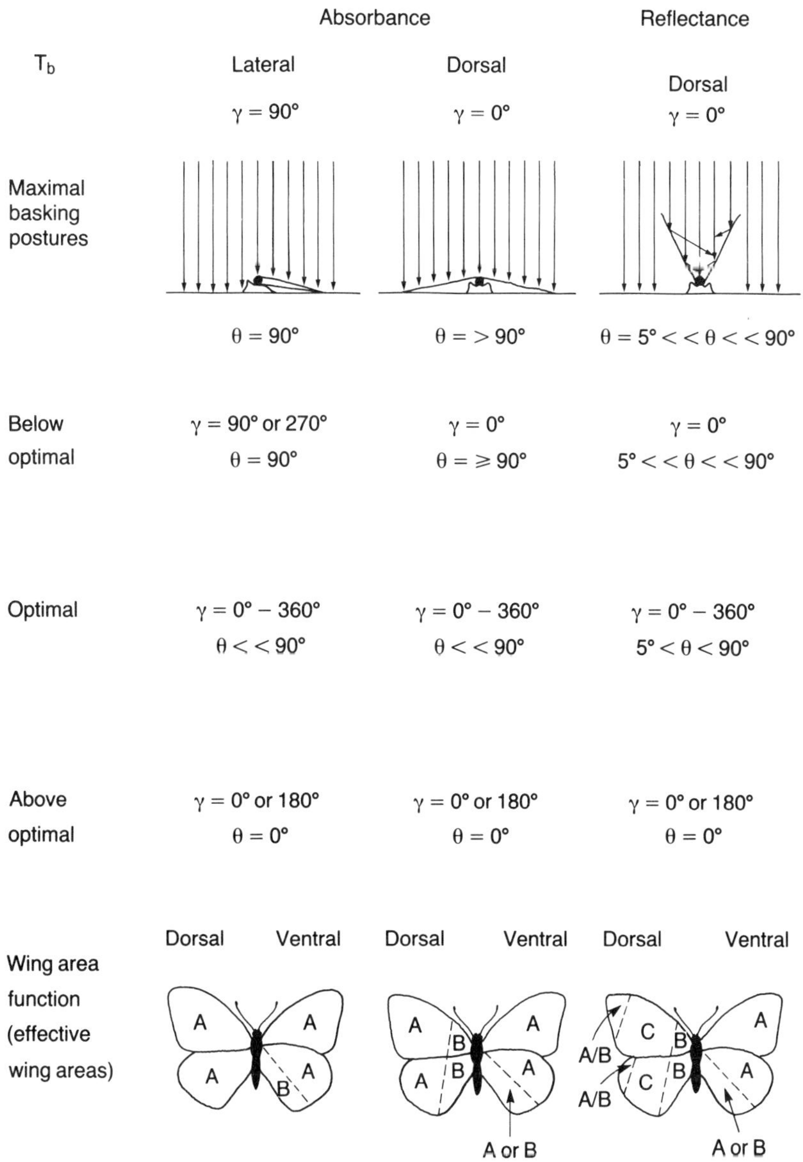

1.4 Basking postures of Papilionidae in relation to body temperatures and the main wing areas of thermal significance to different basking methods. γ, body orientation; θ, wing angle to sun's azimuth; A, neutral; B, absorbing; C, reflective. Wing zones A, increased melanin has no effect on body temperatures; wing zones B, increased melanin increases body temperature; wing zones C, increased melanin decreases body temperature. (Adapted from Shreeve, 1992.)

layer of warm air, whilst the forewing may reduce convective heat loss over the relatively wide body surface which is irradiated directly.

In dorsal reflectance (DR) baskers, the entire wing surfaces act as solar reflectors directing solar radiation on to the body (Kingsolver, 1985a, b). Intermediate rather than the largest wing angles provide the maximum radiation load (Fig. 1.4). There are consistent differences in wing angles used in DR baskers between species (*Pieris* subgenera *Artogeia* and *Pontia* differ) and sex related to melanin deposition on the wings and wing expanse (see below). The wing angle at which intercepted radiation is maximized decreases with increased wing length, and distal wing melanization effectively reduces wing length (Kingsolver, 1985b; 1987a, b). Heat avoidance is attained by wing angles that are either larger or smaller than for maximal T_{bask} but minimal T_b is achieved with dorsal wing apposition. Heinrich (1990) has questioned the importance of reflectance basking in butterflies, but he does not address the direct experimental manipulations of wing reflectivity in *Pieris* carried out by Kingsolver (1987a).

Lateral absorbance (LA) baskers attain maximum T_{bask} by tilting the closed wings so that one ventral hindwing and the body axis is normal to direct insolation (Fig. 1.4). T_b is reduced by decreasing the angle of wing tilt, heat avoidance being accomplished by parallel alignment of the body to irradiation (0° or 180°) with the thorax and abdomen shaded by vertical wing apposition (Watt, 1968; Findlay *et al.*, 1983). LA baskers heat up by absorbance of solar radiation on the basal hindwings; the reduction of the absorbing surface compared to DA baskers is compensated for by an improved heat accumulation capacity of four overlapping wings. There is no evidence that they orientate in relation to wind direction (Kingsolver & Moffat, 1982).

The three ectothermic behavioural strategies (DA, DR and LA baskers) share a number of basic postures (i.e., for heat avoidance) and probably fine-tuning mechanisms such as substrate appression for heat gain (Douglas, 1978; Kevan & Shorthouse, 1970) and stilting for heat avoidance (Findlay *et al.*, 1983). However, significant variation occurs between taxa even within the three categories (e.g., body baskers in DA basking, see above). *Papilio polyxenes* (DA basker) raises its abdomen above open wings when T_a is low (< 14°C) and lowers the abdomen in the shade of the wings (shielding) when $T_a > 22$°C (Fig. 1.5a). These movements are associated with switches in circulation between abdomen and thorax (Rawlins, 1980). *P. polyxenes* also attains maximum irradiation by using its long metathoracic legs to prop the body away from vertical perches. Not all DA baskers align 180° from the sun; Nymphalidae (e.g., *Vanessa atalanta*) often orientate head down on precipitous surfaces; in this way hot air building up under the wings is prevented from escaping upwards by appression of the hindwing margin (Findlay & Findlay, 1983). Guppy (1986) suggests that *Parnassius phoebus* absorbs energy dorsally while in flight; the slow flight speed

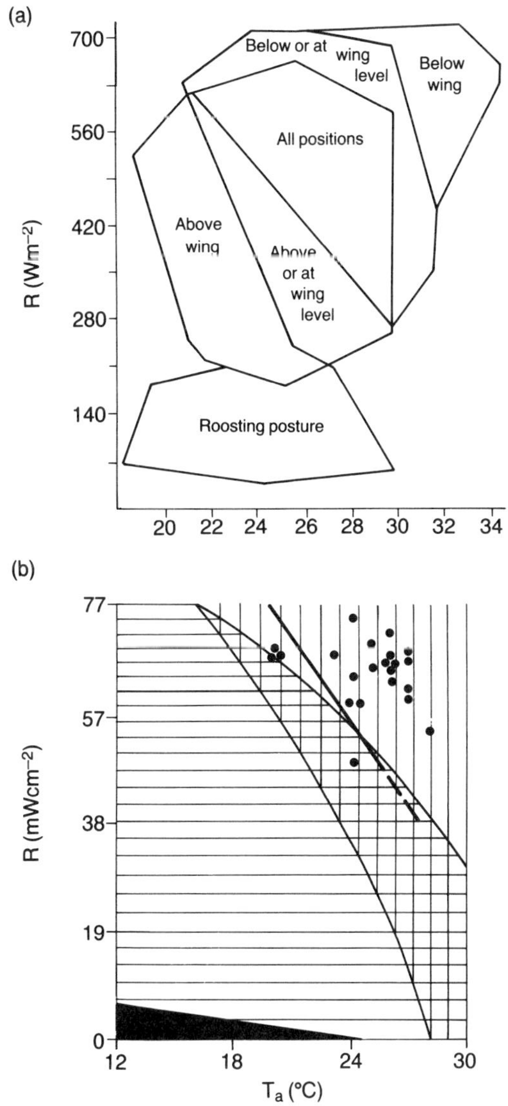

1.5 Modification of behaviour for changing values of temperature (T_a) and solar radiation (R). (a) Abdominal postural changes in *Papilio polyxenes* (Papilionidae). (After Rawlins, 1980 by courtesy of the Ecological Society of America.) (b) Activity in *Heodes virgaureae* (Lycaenidae). Black area, inactive; horizontal shading, basking (actively basking > 50 per cent of time); vertical shading, feeding (> 50 per cent of time sitting); above the thick line, flight activity > 25 per cent (males) or > 5 per cent (females) of time; circles, observation of male courting a female. (After Douwes, 1976 by courtesy of *Oecologia* and Springer-Verlag.)

(< 6 km h^{-1}) and limited wing movement coupled with basal and thoracic pubescence probably do not disturb the boundary layer of still air (see Kingsolver & Koehl, 1985).

The three basic thermoregulatory categories (DA, DR and LA) have clear taxonomic affinity (Fig. 3.2); the different postures influence wing pattern adaptations to other selection pressures (Dennis & Shreeve, 1989; Shreeve & Dennis, 1992; see chapter 3). DR baskers can also absorb energy dorsally and both DR and DA baskers can engage in limited lateral basking (Kingsolver, 1985a, b). It is significant that on the ventral hindwings of *Pieris napi* and *P. callidice* the one melanized ray that does not coincide with the venation lies over the abdomen when the insect is at rest (Shapiro, 1976). LA baskers have the most restricted posture; the wings are rarely opened except for flight or display (Shreeve & Dennis, 1991).

Apparently, not all butterflies show behavioural thermoregulation by DA, LA or DR postural adjustments of wing position, especially in the tropics where T_a is close to FAR (e.g., *Sotriaena medus*, Satyrinae; Heinrich, 1972). However, most subfamilies of Costa Rican butterflies have species that are known to bask; the main exceptions are Ithomiinae and their mimics, Dismorphiinae, but then basking was observed in *Tithorea tarricina* (Ithomiinae) (Srygley & Chai, 1990a). Subsidiary postures such as body shading are used. Alternatively butterflies can change habitat or diel activity patterns. Vertical movements control T_b by exposure to different values of T_a and R. Adjustments in T_b can be effected by shifts in perching and flight duration or frequency (Heinrich, 1972, 1986; Rawlins, 1980), perch elevation (Rawlins, 1980) and flight height (Vielmetter, 1958; Shelly & Ludwig, 1985; Tsuji *et al.*, 1986; Pivnick & McNeil, 1987; Wickman, 1988); selection of different light conditions is perhaps the most important control of inputs from R, T_a and also A_v (see Heinrich, 1972; Shreeve, 1984). Macro-relief in hilltops (Shields, 1967; Scott, 1968; Wickman, 1988), creeks, gullies and depressions (Kevan & Shorthouse, 1970), different aspects (Dennis & Shreeve, 1988; Weiss *et al.*, 1988), micro-topography in hollows and the selection of different substrates (Leigh & Smith, 1959; Kevan & Shorthouse, 1970; Shreeve, 1984) are used for heating or cooling by adjusting for local differences in sun altitude and azimuth, R, T_a, A_v and wind direction (A_d). Females typically adopt warmer microclimates than males (Gilchrist, 1990). Little is known about the interrelationships of thermoregulation with the distribution and activity of insects in space and time, although activity patterns can be significantly influenced by weather conditions (Dennis & Bramley, 1985) and by diel patterns of R and T_a (Shreeve, 1990).

In insects, essentially all endothermic regulation of T_b is by the active flight muscles. In wing whirr (shivering), the antagonistic flight muscles contract primarily against each other. Wing vibrations are of small amplitude and generate heat by positive feedback through temperature of the flight muscles. Shivering is common in nocturnal Lepidoptera (Kammer, 1968;

Adams & Heath, 1964; Heath & Adams, 1967). Large-bodied insects (e.g., Sphingidae) are capable of raising metabolic rates 43 times above resting levels (Krogh, 1948), the rate of change for T_b is similar to that known for small endothermic birds and mammals which regulate the rate of heat production (Heinrich, 1989). Not all butterflies display mechanisms for facultative warming by metabolic heating or muscular activity (e.g., *Colias* spp., Watt, 1968), but shivering has been observed in Papilionidae, Hesperiidae and Nymphalidae (Dennis, 1970; Rawlins, 1980; Findlay & Findlay, 1983). In *Danaus plexippus*, shivering occurs at low T_a 13°C and raises T_b at 1.3°C min^{-1} with T_{ex} 4°C at T_a 15–16°C. This is comparable to T_{ex} in fixed flight at T_a 15°C but lower than the 3.6°C min^{-1} attained by flight at T_a 23°C (Kammer, 1970). Shivering is generally a subsidiary thermoregulatory device in diurnal Lepidoptera. It is expensive and generates less heat than sun basking, but under low R and T_a allows butterflies essential minimal activity to change habitat, regain roosts if dislodged and escape predation (Rawlins & Lederhouse, 1978).

Endothermy in flight is almost obligatory; 80–90 per cent of energy expended in flight degrades to heat (Heinrich, 1974). But heat exchange differs between insects which differ in body size, wing loading (weight/wing area) and speed of flight (Srygley & Chai, 1990a). Large insects with high wing loading (e.g., *Celerio lineata*) not only need to generate sufficient temperatures for continued flight (T_b of flight muscles > 40°C) but require a mechanism for dissipating heat in flight when T_b > 45°C. In *C. lineata* and other Sphingidae, excess heat is shunted to the abdomen via the circulatory system which acts as a radiator (Heinrich, 1970a, b; 1971; 1979). Small insects have rapid heat loss and little accumulation of body heat during flight (T_{ex} < 1°C) but flight is possible at lower T_a and T_b. Small butterflies generate little excess heat in flight and can record deficits (Heinrich, 1972; Shelly & Ludwig, 1985; Shreeve, 1984; Wickman, 1988), so that little danger is posed by overheating. Even in strongly flying *Colias* spp. (wing beat 12–19 Hz), heat generation (c. 100 mW g^{-1} tissue) is sufficient only to raise T_b 1–2°C > T_{bask} at the top of the vegetation and is lower than T_{bask} in the shelter of perch sites (Tsuji *et al.*, 1986).

Overheating in larger butterflies can be avoided by gliding or soaring with reduced wing beat (e.g., *Papilio* spp., Heinrich, 1972; Rawlins, 1980). Myothermic heat production is minimized and convective cooling favoured. In tropical environments *Papilio aegeus* can maintain $T_b = T_a$ = 25°C in the shade by constant flight and wing beating, even when feeding (Heinrich, 1972). Slower wing beats when stationary may alternate convective cooling to different parts of the thorax and abdomen. Wing beat frequencies in butterflies are slower at low T_b (Kammer, 1970) probably owing to the decreased speed of muscle contraction (Neville & Weis-Fogh, 1963) and thus large-winged slow flying butterflies are able to fly at lower T_b; butterflies can be induced to fly at lower T_b than the observed FAR (e.g.,

Parnassius phoebus at 9°C, Guppy, 1986) but their flight is inefficient and they are vulnerable to predators.

Structural modifications for thermoregulation are external, involving (i) size and form of body and wings, (ii) colour and (iii) setae, and internal, affecting adaptations in (i) circulation and (ii) biochemistry. Structural aspects of body and wing form are intimately associated with behavioural postures and movements for ectothermy (i.e., wing patterning) and endothermy (i.e., musculature, shape, size and tensile strength of wings, body–form ratio) but may be compromised by other requirements (see chapter 3). Insects are argued to be well adapted structurally to thermoregulation (Digby, 1955). The approximately spherical shape of the thorax (minimum surface area : volume ratio) minimizes radiative and convective heat loss and the narrow attachment of head and abdomen to thorax minimizes conduction. In many DA baskers (e.g., Papilionidae) the hindwing inner margins are concave and sclerotized and enclose the abdomen to increase heat conduction (Wasserthal, 1975). DA baskers can thus control heat uptake by varying the relative position of wing base and body.

Equilibrium T_{ex} of an insect increases regularly with size, endothermically or exothermically (Digby, 1955; Porter & Gates, 1969; May, 1979), but the rate of heating and time to equilibrium T_b is inversely related to body size (Heinrich, 1972) as is passive cooling in live quiescent individuals (Heinrich, 1974). The transient time constant (τ) reflects the impact of convective and thermal radiative heat flux on T_b in a fluctuating environment; τ is the time it would take a body not at equilibrium temperature to move 1/e (37 per cent) of the way towards thermal equilibrium (Campbell, 1977). As metabolic heating decreases with the cube of linear body dimensions and radiative heating by the square (Digby, 1955), any heating effect of metabolism becomes less important than that of solar basking as size decreases. Although large insects will attain and maintain high T_{ex} fairly easily, the importance of high and well regulated T_b is probably less for many small butterflies, owing in part to their generally lower wing loading (Heinrich, 1974). Males are typically smaller than females and have a reduced wing loading, allowing flight at lower wing beat frequencies and T_b (e.g., *Euphydryas phaeton*. Gilchrist, 1990). Advantages under marginal meteorological conditions have been argued for smaller (Blau, 1981) and larger (Willmer & Unwin, 1981) insects, but this depends on behaviour and structural factors such as wing loading, colour, shape and insulation, as well as the precise nature of microhabitat use and meteorological events (see Kingsolver & Watt, 1983; Pivnick & McNeil, 1987; Gilchrist, 1990). Diverse behavioural repertoires for thermoregulation may be more important than any influence of body mass (Heinrich, 1986); thus, behavioural rather than morphological attributes to avoid overheating are suggested to dominate among the world's largest butterflies (e.g., *Ornithoptera* spp.; *Troides* spp.) (Stone *et al.*, 1988). Size and shape also influence water loss, which is less in larger than smaller insects and less in compact than in elongated forms *ceteris paribus*.

The colour of the body and of the entire wings may influence thermal performance (Kingsolver 1985b; Kingsolver & Wiernasz, 1987; Fig. 1.4). Black (melanism) has high solar absorptivity by comparison with yellows in Coliadinae (pteridines) and Pierinae (sepiapterin) and whites (leucopterin) in Pierinae (cf. 0.7 *versus* < 0.4). Melanin has a reflectivity of < 0.3, below the effective threshold of 0.5; the reflectivity of yellow is 25 per cent as effective as white (Kingsolver, 1988). In dead or alive LA basking *Colias* spp. of the same mass and surface area, dark insects heat significantly faster and attain a steady-state T_{ex} 15 per cent higher than light insects (Watt, 1968), but cool at the same rate. Similarly, tropical *Taenaris myops* with a white body and wings had a lower heating rate than black *Papilio aegeus*, which attained T_{bask} 55°C when irradiated (Heinrich, 1972). Basal melanism increases in many LA, DA and DR baskers with altitude and latitude (Watt, 1968; Guppy, 1986; Kingsolver & Wiernasz, 1987), also as a seasonal polymorphism in spring and autumn broods (Watt, 1969). Dark *Colias nastes* and *C. meadii* fly at lower T_a than lighter ones (Roland, 1982) and dark *Colias* spp. achieve greater reproductive success in alpine and in northern habitats than lighter individuals (Watt, 1968). Spring and autumn melanic forms of LA basking *Nathalis iole* in North America absorb simulated solar radiation more efficiently and to a higher equilibrium T_{bask} of 37°C than summer immaculate forms (T_{bask} 34°C) (Douglas & Grula, 1978). Yet in DA basking *Parnassius phoebus* although basal wing melanization does not affect the flight threshold, it is positively correlated with T_b in suspended dead specimens and with flight duration in conditions of low T_a and low R (Guppy, 1986). Basal melanism development is constrained by critical T_{ex} at high altitude mid-latitude sites owing to the high R levels (100–115 mW cm^{-2}) in fluctuating episodes of high T_a and low A_v, but not in arctic populations experiencing low R (75–85 mW cm^{-2}) (Kingsolver & Wiernasz, 1987).

Not all black butterflies absorb energy at all wave lengths. Some darkly pigmented tropical species are highly reflective in near infrared although relatively non-reflective in the visible spectrum (Heinrich, 1972). Nevertheless, T_{bask} levels in sunlight are critically high (> 45°C) and these butterflies necessarily reduce T_b by selecting shaded habitats and body shading. Colours outside the basal wing region in LA and DA baskers, such as the alba polymorphism in *Colias* spp., have no influence on thermoregulatory performance, but have other consequences (see section 3.3B). However, in DR baskers, distal wing melanism reduces solar reflectivity and T_b (Fig. 1.4; Kingsolver, 1988).

The rough surface of perithoracic and basal wing setae (hairiness, furriness) substantially improves radiation absorption (Krogh, 1955) and reduces convective heat loss, in large insects by c. 50 per cent (Heinrich,

1974). Male *Thymelicus lineola*, with their thicker body fur, have a higher T_{ex} than females, in spite of their smaller size (Pivnick in Gilchrist, 1990). Setae increase in density and thickness along latitudinal and altitudinal gradients (Kevan & Shorthouse, 1970; Wasserthal, 1975; Kingsolver & Moffat, 1982; see chapter 3). Furriness also decreases sensitivity in *Colias* from high elevations to fluctuating R, T_a and A_v caused by their increased melanization and thus higher wing solar absorptivities (Kingsolver & Watt, 1983).

Little research has been done on internal mechanisms for thermoregulation in butterflies. Larger moths cool down in flight by using the abdomen as a radiator and small noctuids have a counter current heat exchanger that allows them to fly at T_a 0°C (Heinrich, 1986; 1989). It is possible that fluid re-cycling in Hesperiidae and Nymphalidae assists cooling (Atkins, 1989; Feltwell, 1990). *Papilio polyxenes* controls heat transfer between the thorax and abdomen (Rawlins, 1980; Fig. 1.5a). A raised abdomen at low T_a cuts off circulation from the abdomen. At high T_a, abdominal contractions increase circulatory exchange between thorax and abdomen; heat is dissipated from a lowered abdomen shaded by the wings. Ligating experiments revealed that hemolymph flow from the abdomen to the thorax during basking increases body temperatures in *Pieris* spp. (Kingsolver, 1985b). However, despite their small size and high temperature requirements, *Colias* spp. apparently do not regulate heat loss by altering hemolymph circulation (Tsuji *et al.*, 1986). Nor do *Colias* populations in cold climates adapt by lowering the limit of FAR (Watt, 1968). Instead, *Colias* are polymorphic for the glycolytic gene locus PGI; the genotypes vary in transient and maximal rates of fuel supply to flight muscles at different temperatures and differ in their ability to initiate and sustain flight at suboptimal T_b (Watt, 1977, 1983, 1985; Watt *et al.*, 1983, 1985). Some genotypes increase mating success by 50 per cent over the least favoured genotypes in a particular environment.

Eggs and pupae are obligatory poikilotherms in which T_b is approximately the same as that of the substrate (T_s). The microclimates of substrates for these stadia vary enormously between taxa, even between individuals, all of which can significantly influence survival (see section 1.2B). Demonstrations that oviposition is specifically adapted to the thermal environment are few and little is known of how egg or pupal structure influences their temperatures. Egg masses of *Euphydryas gillettii* are placed on *Lonicera involucrata* so as to intercept morning sun (south-east orientation) and heat up more rapidly than in other aspects (Williams, 1981). In *Papilio glaucus*, eggs have a western bias in relation to oviposition flights in sunshine, mainly during the latter part of the day, and below 2.5 m in lower windspeeds (A_v) (Grossmueller & Lederhouse, 1985). Cocoons of *Gynaephora* spp. (Lymantriidae) attain elevated T_b and act much as greenhouses (Kevan *et al.*, 1982). They too are orientated north-west to south-east and, it is argued, maximize heat gain from irradiation and meanwhile minimize heat loss from the prevailing winds. Vertical placement in relation to the ground and the degree of cover provide the main controls for R, T_a, A_v and relative humidity. Among butterflies, the least variable microclimates are attained by sub-soil cocoons (e.g., *Hipparchia semele*) and pupae in ants' nests (e.g., *Maculinea arion*) (Ford, 1945), and the most variable by eggs and pupae on bare ground surfaces, the degree of variability depending on duration and seasonal timing of the stage.

Larvae display a number of behavioural and structural thermoregulatory adaptations. Though located on substrates with higher temperatures than T_a, larvae do not necessarily require T_b as high as adults. *Colias eurytheme* have preferred T_bs 25–30°C and *C. eriphyle* even less, much lower than the adults (Sherman & Watt, 1973). When overheated, *Papilio glaucus* larvae raise the anterior portion of their bodies above the leaf surface (Grossmueller & Lederhouse, 1985). Solar basking occurs in a number of Heterocera (e.g., *Choristoneura fumiferana*, Tortricidae: T_{ex} 7°C, Shepherd, 1958; arctiid larvae: T_{ex} 15°C, Krogh, 1955; *Heliothis zea*. Noctuidae: T_{ex} 8°C, Flitters, 1968; *Hemileuca lucina*. Saturnidae: T_{ex} 5°C, Stamp & Bowers 1990). *Gynaephora* spp. orientate to the sun and shelter from the wind attaining T_{ex} 13°C and a 30 per cent heat gain in growing day degrees above 0°C (Kevan *et al.*, 1982). Larvae of *Hyles lineata* (Sphingidae) orient perpendicular to the sun at low T_a and parallel to it at high T_a achieving substantial control of T_b (Casey, 1976). Individual basking of larvae occurs in fifth and sixth instars of *Eurodryas aurinia*, which can raise T_b to 37°C, 15–16°C > T_s of the hostplant and 20°C > T_a at 15 cm. First to fourth instars are gregarious and en masse build up a similar T_{ex} while digesting food (Fig. 1.6). T_b drops to T_s when the larvae leave the cluster to feed (Porter, 1982). Conspecific aggregations, by evaporation or excretion, may contribute to higher humidities if moisture is retained within interstices of the group; water balances in *Pieris* and *Inachis* groups are more easily achieved than in the same animals when solitary (Willmer, 1980; Fig. 1.7).

Larvae can adjust their microhabitats during diel activity by altering the height above the ground (e.g., *Hyles lineata*; Casey, 1976) or selecting exposed or shaded sites (Sullivan & Wellington, 1953; Capinera *et al.*, 1980; Grossmueller & Lederhouse, 1985). Post-diapause *Eurodryas aurinia* adopt substrates such as dead oak leaves, dead grass or molehills as to their properties for solar radiation absorbance (Porter, 1982). When overheated they move to less exposed sites as do *Colias* spp. (Sherman & Watt, 1973).

Thermoregulation and water balance in larvae are affected by size, colour, setae and other structural features, such as silken webs (May 1979; Casey & Hegel, 1981). Smaller third instar *Pieris* larvae lose water more rapidly than the fifth instars, though differences may occur in integumen permeability (Willmer, 1980). Larger insects can maintain higher T_b in sunlight than small ones; this together with their black colouration may facilitate

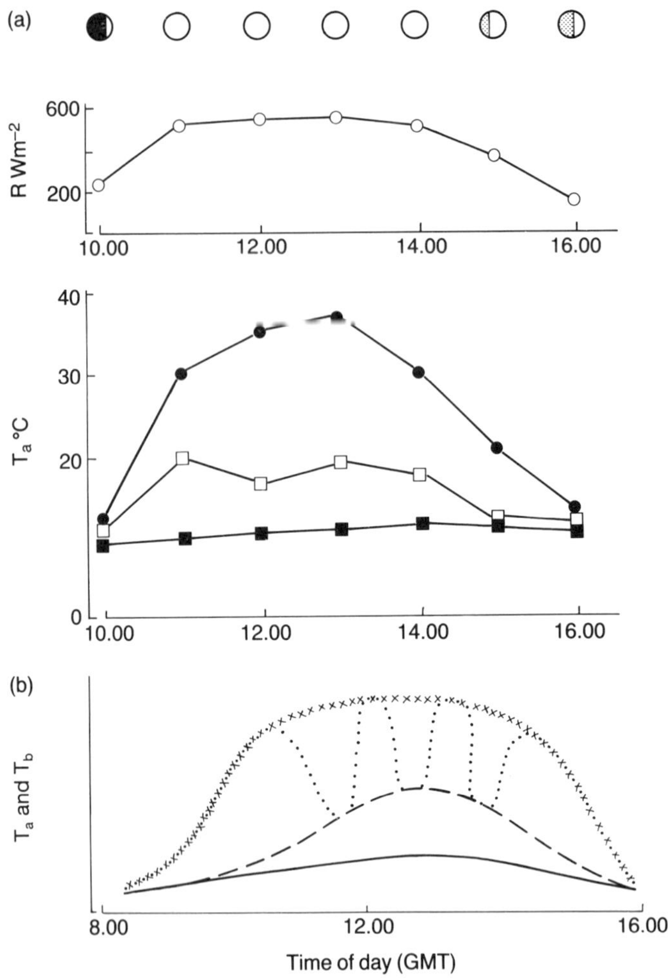

1.6 Thermoregulation in gregarious *Eurodryas aurinia* (Nymphalidae) larvae. (a) The relationship between fourth instar larval cluster temperature and environmental temperatures during the course of a sunny day. Black square, air temperature at 15 cm; white square, surface temperature of hostplant; small black circle, larval cluster temperature; small white circle, solar radiation intensity. The cloud cover is indicated as absent (large white circle), hazy (large grey circle) or overcast (large black circle). (b) A diagrammatic representation of the feeding strategy of fourth instar larvae through a clear day in which solar radiation reaches a peak at midday. . . . . . , individual larval temperature as measured by a hypodermic probe; xxxxx, basking cluster temperature; _____, T_a at 15 cm; - - -, T_a at vegetation level. Individuals commute between the foodplant to eat and the cluster to heat up and to digest food. (From Porter, 1982 by courtesy of *Oikos*.)

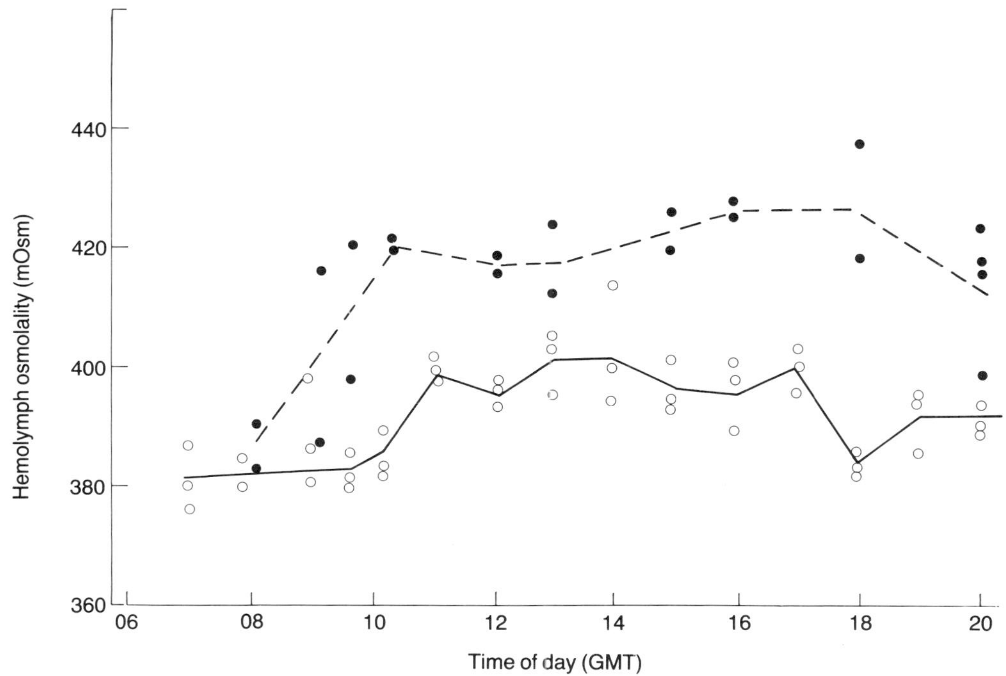

1.7 The effects of the environment on hemolymph concentration in solitary and gregarious third instar larvae of *Pieris brassicae*. Hemolymph osmolality is shown as it varied during a day of decreasing and subsequently rising relative humidity, the increments being larger for solitary specimens of *P. brassicae*. Black circle, solitary and segregated experimentally; white circle, gregarious. (From Willmer, 1982 by courtesy of Academic Press Ltd.)

more solitary behaviour in late instars of Nymphalids, such as *Eurodryas aurinia* and *Euphydryas editha* (Porter, 1982; see Emmet & Heath, 1989; Weiss *et al.*, 1988). The thermal environment of silken webs or tents depends much on cohort size and age. Young *Malacosoma americana* and *M. pluviale* (Lasiocampidae) use tents to escape overheating; conversely, the larger tents of older larvae absorb radiation and reduce convection so that the interior can be many degrees in excess of T_a (Sullivan & Wellington, 1953). This can contribute to the excess in T_b beyond predicted values of T_{ex} for solitary larvae on the upper surface of tents (Joos *et al.*, 1988; Casey, 1976). The webs of Nymphalidae undoubtedly function in much the same manner, but require study. Solitary occupancy of leaves curled with silk by *Papilio glaucus* larvae can produce T_{ex} 3.5°C, corresponding to the temperature of the air pocket in the curled leaf, which probably serves as a parabolic reflector of radiation and as barrier to convective heat loss (Grossmueller & Lederhouse, 1985). It may also maintain a high relative humidity and benefit W_b balance (Willmer, 1982).

Although requirements for water balance and thermal balance coincide for heat avoidance behaviour, butterflies have developed a number of specializations during different stages which either specifically reduce water loss (W_b) or contribute to water retention but may primarily influence other aspects of individual success. These specializations include: (i) modifications in phenology and development (e.g., diapause; reduction in time of vulnerable stages; very rapid growth rates); (ii) selection of cool moist microhabitats (e.g., larvae of *Meganthymus* and *Agathymus* (Hesperioidea) develop inside agave and yucca plants); (iii) temporal modifications in activity (e.g., crepuscular or dawn flight as in *Euphilotes* spp., Lycaenidae); (iv) morphological adaptations (e.g., oral and anal 'plugs' in diapausing *Euphydryas* spp.). Such adaptations are as varied as thermoregulatory devices (Sbordoni & Forestiero, 1984) and are complemented by attributes which enhance uptake of moisture (e.g., short, wide probosces) and the timing of active periods with moisture availability (e.g., emergence triggered by rains).

B *Ambient conditions and butterfly biology: tolerances and constraints*

Weather and climate are significant modifiers of activity in small ectotherms and it is activity, budgeted in expenditure of time and resources, that controls reproduction and determines the ability to survive (see section 2.1A). Resources for adults emanate mainly from (i) larval feeding, but also from (ii) adult feeding, (iii) parental contributions in ova and, for females, (iv) male contributions in spermatophore and mating plugs. These resources are distributed among somatic maintenance and investment in reproduction (Boggs, 1986). Unfortunately, the very complexity of full life-cycle budgeting has resulted in partial, often anecdotal, appraisal of the impact of external agents on butterfly biology (Stearns, 1989). A repercussion is the indeterminate significance of indirect, but theoretically plausible, contributions to reproductive success and survival.

In adults, reproductive potential (i.e., mate-location; courtship, host-seeking, oviposition etc.) and survival (i.e., predator escape, roost-site selection) depend on flight. The voluntary flight activity range (FAR) relates to T_b, which in turn depends on flight space – the combined ranges of R, T_a and A_v, in which flight can or does occur in a species (Kingsolver, 1985c). Few calculations of flight space have been attempted (e.g., *Argynnis paphia*, Vielmetter, 1958; *Colias philodice*, Leigh & Smith, 1959; *Papilio polyxenes*, Rawlins, 1980; *Colias nastes*, Roland, 1982; *Pieris virginiensis*, Cappuccino & Kareiva, 1985); as differences can be substantial (e.g., *P. polyxenes*, T_a 19–30°C; C. *nastes*, T_a 6–20°C) despite similar requirements of T_b, this suggests species-specific distinctions in thermal characteristics and mechanisms. Kingsolver (*loc. cit.*) has expressed the requirement for the development of models linking thermoregulation to flight space as empirical investigations, although essential, may represent but a subset of the entire space; such modelling reveals a 20 per cent difference in the flight spaces of *Colias meadii* and *C. philodice eriphyle* determined by distinctions in hindwing ventral melanization. Weather, flight space and lifespan allows the calculation of potential flight time; as massive differences can occur within populations between years and between different populations, flight time is a fundamental measurement for population ecology (Kingsolver, *loc. cit.*).

Weather also affects other aspects of flight: speed and efficiency, duration of individual flights (Heinrich, 1986), frequency of flights, timing and spatial bias in habitats (Dennis & Bramley, 1985, and height above ground (Baker, 1984; Wickman, 1988). Weather affects activity in the tropics as well as in the temperate zone (Srygley & Chai, 1990a). High as well as low T_a can terminate flight (e.g., *Colias* spp., Kingsolver & Watt, 1983; *Phulia nymphula*, Courtney, 1986; *Lasiommata megera*, Wickman, 1988) and males may have a lower FAR than females (Kingsolver, 1983; Pivnick & McNeil, 1987). The rapid response of T_b to meteorological pulses, such as intermittent cloud cover, implies potentially extensive repercussions for butterfly biology. This is controlled to a large extent by the dependency of flight characteristics on resource distribution, population density and predation intensity. For example significantly longer flight duration is required in Colorado Rockies *Colias* populations at mid and high altitudes where hostplants occur at low density than among alfalfa fields at lower elevations (Kingsolver, 1983).

Both flight height and speed of butterflies are directly influenced by wind during migration and more trivial movements (*sensu* Baker, 1978). Most butterfly activity is restricted to that part of the planetary boundary layer in which the air speed of the insect exceeds wind speed. In this zone, an insect is capable of flying in any fixed compass bearing (Drake & Farrow,

1988). Several species (e.g., *Danaus plexippus, Ascia monuste*) have been observed flying low against headwinds but to attain high altitudes, as much as 1–1.5 km, when the wind direction is favourable (Urquhart, 1960; Shields, 1974; Schmidt-Koenig in Baker, 1984; see section 4.3C). *Danaus plexippus* and *Nymphalis antiopa* use thermals and the lift produced by upslope winds to soar and glide, and thus to migrate over long distances (Gibo & Pallet, 1979; Gibo, 1981). Similar soaring behaviour has been noted at Chesil beach in Dorset in *Aglais urticae*, which used a sea breeze to attain height and then glided inland (Dennis, pers. obs.). The prospect exists of butterflies utilizing highly predictable terrain-induced air motions (e.g., anabatic and katabatic winds) to 'invest' in new regions with minimal effect (Drake & Farrow, 1988). This may explain the ability of allegedly sedentary species to cross unsuitable biotopes on valley floors and colonize appropriate habitats on opposite slopes as in the Alps (McManus, 1988).

Apart from floral nectar, sources for adult feeding include pollen, honeydew, aqueous deposits, solids, animal and plant excretions (Horton, 1977; Boppré, 1984; Atkins, 1989; Shreeve, 1992). These supplement larval diet with energy and nutrients directly as water, amino acids, carbohydrates, sugars, trace elements, lipids and toxins (Watt *et al.*, 1974; Baker & Baker, 1973; 1975; Marshall, 1982; Shreeve, 1992) and indirectly as proteins, lipids and salts in male spermatophores and pheromones (Arms *et al.*, 1974; Marshall, 1980; 1985). The impact of weather and climate on adult feeding depends on the extent to which 'nectar' sources are limiting for adult butterflies (Courtney, 1986: 55). This in turn depends on the requirements and availability of resources and time constraints placed on activities.

The amount of time allocated to adult feeding varies from < 2 per cent of active time (e.g., *Anthocharis cardamines*, Wiklund & Åhrberg, 1978; *Calisto nubila*, Shelly & Ludwig, 1985; *Coenonympha inornata*, Heinrich, 1986) to > 15 per cent of active time (e.g., *Colias* spp., Kingsolver, 1983; *Lasiommata megera*, Dennis, 1982–83). *Thymelicus lineola* can spend 65–100 per cent of their time feeding in the mid-part of the day (Pivnick & McNeil, 1985). Where supply is unlimited, feeding time relates to viscosity of nectar; as the energy content of nectar increases linearly with sugar concentration, while viscosity increases exponentially, increases in viscosity limit the rate of uptake at higher concentrations (May, 1985; Pivnick & McNeil, 1985). Although maximum rates of energy uptake are now known to be generally higher than previously (35–40 per cent sucrose versus 20–25 per cent sucrose; cf. Murphy, 1982, Kevan & Baker, 1983 with May, 1985), selection on feeding activity for sucrose concentration seems less probable as the graph of energy intake *versus* sucrose concentration is more platykurtic than earlier simulations suggested (cf. Kingsolver & Daniel, 1979 with May, 1985).

Although fecundity is generally believed not to be enhanced by additional amino acids acquired by adults (Moore & Singer, 1987; Wickman &

Karlsson, 1987; Karlsson, 1987; Karlsson & Wickman, 1989) energy reserves (e.g., carbohydrates) from adult feeding are known to make a significant contribution to somatic maintenance, reproduction, offspring survival and perhaps even defence (Stern & Smith, 1960; Baker & Baker, 1973; Boggs *et al.*, 1981; Chaplin & Wells, 1982; Murphy, 1983; Murphy *et al.*, 1983; Boppré, 1984). The relative contribution compared to larval resources relates much to the degree of egg maturation and adult longevity (Boggs, 1986; Garcia-Barros, 1988); even in species such as *Euphydryas editha*, with > 40 per cent of its egg mass fully developed at eclosion, nectar can play an essential role in increasing production of late egg masses in years with greater than normal rainfall, increasing population size and reducing the probability of population extinction in dry years (Murphy *et al.*, 1983). Adult feeding is critical for accumulation of abdominal lipid in overwintering *Aglais urticae* and *Inachis io* (Pullin, 1987b); this depends substantially on weather conditions preceding hibernation; lipid reserves in *Danaus plexippus*, replenished by overwinter nectaring, also depend on atmospheric conditions (Chaplin & Wells, 1982; Calvert & Brower, 1986).

For many species nectar resource distribution may be limiting. This would depend on (i) accessibility based on corolla length and proboscis extension, (ii) nectar quality and quantity (i.e., nectar species, flower age, time of day and exploitation by competitors), (iii) comparative nectar source geography and (iv) temporal availability. Weather and seasonal climate can influence a number of these parameters. Nectar concentration is very sensitive to both relative humidity and state of depletion (Pivnick & McNeil, 1985). Duration of feeding increases with sucrose concentration, but generally, as nectar becomes concentrated by evaporation during the day, duration of feeding is partly offset by faster feeding rates at higher temperatures (Pivnick & McNeil, *loc cit.*). Temperatures probably have no effect on meal size (Pivnick & McNeil, *loc cit.*). Lower T_a is associated with lower feeding rates but dilute nectar may exist in larger quotas and butterflies are often able to feed at temperatures marginal for vigorous flight (Chaplin & Wells, 1982). *Parnassius phoebus* can crawl around to feed at T_a 6°C (Guppy, 1986). Feeding can also be carried out in conjunction with basking and opportunistic mate-location (*Maniola jurtina*, Dennis, pers. obs.). The faster feeding rates and shorter durations at high T_a may be offset by lower nectar quotas and longer visit and search times. Other weather conditions, such as rainfall, can affect adult feeding directly (Boggs *et al.*, 1981). As females feed for longer durations than males and ingest larger meals, they are more subject to weather and climatically-induced time constraints; female longevity and fecundity are more severely reduced by lower feeding activity in very cool, cloudy or rainy weather (Pivnick & McNeil, 1985).

Despite the variety of butterfly mate-locating behaviours and the potential continuum of activities, for instance in perching, patrolling, defence, male aggregations and fidelity to sites, categorization of behaviour revolves primarily around the perch-patrol (sit *versus* search) spectrum associated with individual apparency and predictability (Dennis & Shreeve, 1988). At one extreme are invariant male patrollers; at the other, apparently all territorial males, at times, also patrol habitats. Separating these extremes are potentially substantial differences in energy consumption and required conditions. Virtual female monogamy, low density of receptive females and low mating frequencies are typical but not universal among butterfly species (Heinrich, 1986). Unlike nectar feeding, patrolling depends more on duration than frequency of flight, which is energetically expensive and is limited by T_a. Thus, it is significant that males can fly at lower T_b than females (Pivnick & McNeil, 1987). Perching permits thermoregulation by males in a warmer microclimate and females flying to mating sites are less dependent on flight duration than flight frequency. Moreover, energy consumption by perching males, in the form of increased interaction and courtship encounters, may well correlate with weather conditions permitting flight activity (Dennis & Williams, 1987). Territorial perchers can adjust T_b by switching to flight (Heinrich, 1972) or by selecting warmer or cooler microclimates within territories (*Lasiommata megera*; Dennis, unpub. data), although this may reduce visual range and overheating can force some species to abandon territories (Lederhouse, 1982). Invariant patrollers are less flexible in adjusting to marginal conditions, as there are no obvious sites to await males (Dennis & Shreeve, 1988); furthermore, patrolling height, which influences apparency among conspecifics, is adversely influenced by windspeed (Wickman, 1988). It is possible that courtship and mating can take place in lower T_a than mate-location especially where butterflies aggregate communally (Willmot, 1985). Some species actually mate when going to roost (Mikkola, 1976).

Egg development – numbers, size and weight – depends primarily on hostplant quality and larval resources, and is usually fixed at the end of metamorphosis (Jones *et al.*, 1982; Boggs, 1986; Karlsson, 1987). This can be affected by weather and climate during larval feeding (see below). Egg maturation is T_a dependent (Stern & Smith, 1960; Wigglesworth, 1972) and as females bask more than males, this could be an adaptation to enhance egg development (Shreeve, 1984) especially as development is retarded in cool, cloudy and rainy weather (Pivnick & McNeil, 1987). A further indication is the negative correlation in number of oocytes in the ovaries with elevation in *Colias* spp. (Springer & Boggs, 1987). Some female Lepidoptera are known to resorb eggs during periods of nutritional stress (Boggs & Watt, 1981) and egg size tends to decrease with adult age (Jones *et al.*, 1982; Karlsson, 1987, but see Moore & Singer, 1987), although egg size may not necessarily affect survival (Wiklund & Persson, 1983; Karlsson & Wiklund, 1985).

Egg productivity and fecundity is restricted by flight time. Oviposition rate and adult survivorship, thus overall fecundity, is adversely influenced by cool, cloudy weather (Dempster, 1967; Pivnick & McNeil, 1987) and by overheating (Takata, 1961; Kingsolver, 1985c) and is correlated with R and T_a (Gossard & Jones, 1977; Pollard, 1979; Courtney & Duggan, 1983; Fig. 1.8). Optimum ranges of T_a occur for fecundity (Taylor, 1988); it is possible that as T_a rises, daily oviposition rate is counterbalanced by adult longevity (Gilbert & Singer, 1975; Boggs, 1986). The level and timing of egg maturation at eclosion can influence fecundity as time may need to be invested in adult feeding, an additional time constraint vulnerable to weather effects. *Thymelicus lineola*, which emerges with no mature eggs, can increase egg production 27 times by adult feeding (Pivnick & McNeil, 1987) and *Colias* spp. realize only 3–5 per cent of potential reproductive output in the absence of adult feeding (Watt *et al.*, 1974); adult feeding may also be important via male contributions (Adler & Pearson, 1982; Boggs, 1986). Fecundity is often much less, in the case of *Colias* spp. 20–50 per cent less, than maximum potential (Tabashnik, 1980; Kingsolver, 1983) and egg shortfall is a prominent key factor in population dynamics (Dempster, 1983), although most studies, thus far, have been restricted to marginal climates (Courtney, 1986). Egg production excess seems to be maladaptive, but may be influenced by the stochastics of reproductive success in seasons with favourable weather conditions (Courtney, 1986).

Butterflies are most vulnerable during long periods of enforced inactivity and roosting (section 1.2C). Microhabitats selected by many species (e.g., crevices in rocks and dense ground vegetation) typically have low wind speed (A_v) and are well insulated. Most butterflies (e.g., *Polyommatus icarus* and *Lasiommata megera*) select roosts above the ground; overnight temperatures are typically higher above the ground (Fig 1.2) permitting higher T_bs and such elevated positions may allow immediate evasion of nocturnal predators (Dennis, 1986a). *L. megera* adopts sheltered situations during windy weather but in calm cloud-free warm conditions chooses sites exposed to sunlight and even side basks whilst roosting (Dennis, 1986a). Roosting in the tree canopy (e.g., *Apatura iris*; Emmet & Heath, 1989) may have the benefit of re-radiated energy from the ground after dusk and the first sun's rays at dawn. Grounded overwintering *Danaus plexippus* in Mexico have been shown to increase their survival significantly by crawling upwards one metre. They benefit by avoiding night-time low T_a and increased moisture near the ground (Calvert & Cohen, 1983). Insects freeze internally at higher T_a when their surfaces are covered with moisture than when dry (Salt, 1936; see section 2.1B). On cold nights grounded *D. plexippus* become moribund and below T_a 4°C can no longer crawl. In Mexico, active adults in overwintering colonies of *D. plexippus* respond rapidly and repeatedly to periodic cloud shading (Calvert *et al.*, 1992). Large numbers basking in sunshine within their colonies take flight during cloud shading before reforming clusters on conifer boughs. Similarly, monarchs in streaming flights down arroyos to water sources reverse direction and return

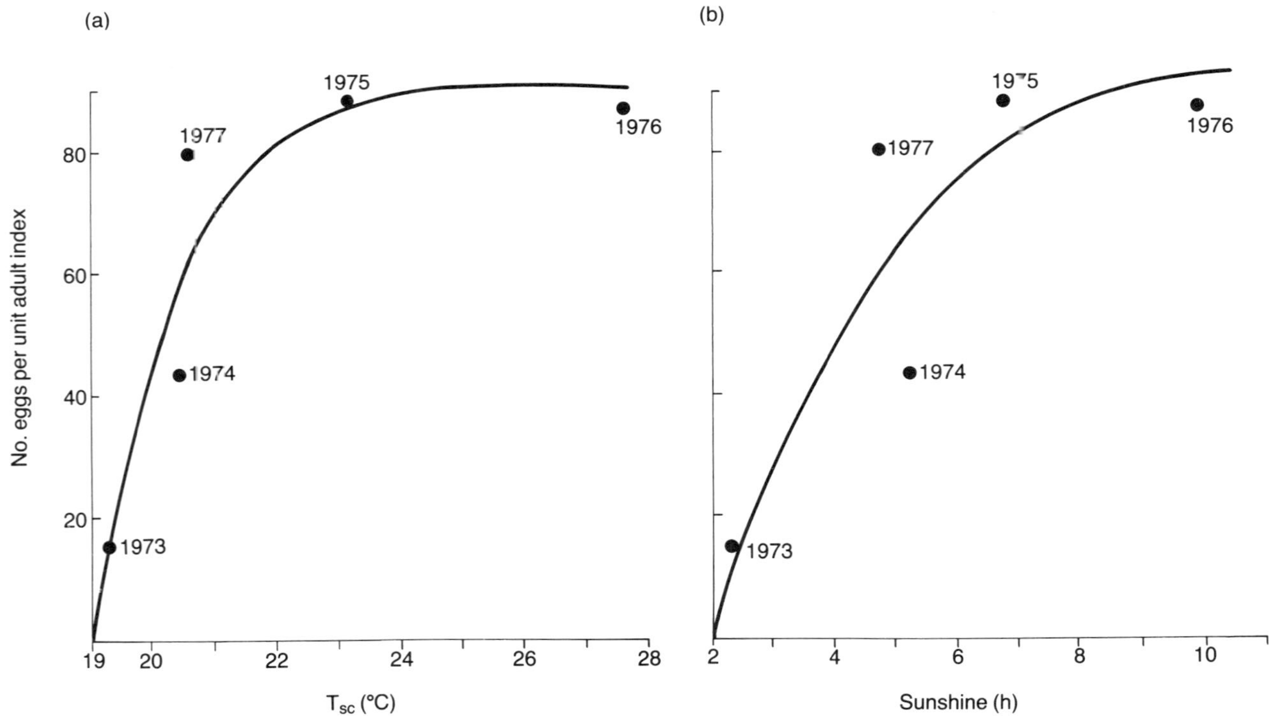

1.8 Relationship between egglaying in *Ladoga camilla* (Nymphalidae), as measured by the number of eggs per unit index (counts on transect) and climate. (a) Mean daily maximum temperature. (b) Mean daily sunshine hours. Curves drawn by eye. (From Pollard, 1979 by courtesy of Blackwell Scientific Publications Ltd.)

to their colonies during cloud shading. Calvert, Brower and Lawton hypothesize that both these behaviours are physiologically triggered by rapidly lowering T_b. The behaviours are interpreted as an adaptation to unpredictable duration of cloud cover which would otherwise strand butterflies in marginal microclimates and possibly subject them to greater bird predation. Within colonies, the adaptive significance of the cloud-induced behaviour may be that it allows individual butterflies to select positions on boughs that are better insulated by the forest canopy against radiant heat loss to the open sky at night.

Weather and climate can profoundly affect the timing and location of behaviour, though in butterfly adults not to the extent of varying activity from night to day seasonally in relation to T_a; even so, nocturnal overseas migration has been recorded (Williams, 1958; Shields, 1974). Light intensity, related to dusk or dense cloud cover, inhibits activity even if T_a is favourable (Douwes, 1976; Saunders, 1982; Rawlins & Lederhouse, 1978). Daily weather conditions influence the rhythm and timing of different activities (Pivnick & McNeil, 1987; Fig. 1.9) and, if outside the flight space, then preclude activity altogether, occasionally for several days (Kevan & Shorthouse, 1970). Few observations have been undertaken on the potential influence of variable weather conditions during different stages (Kingsolver, 1985c), nor on ways in which butterflies may compensate for previous adverse weather conditions (Heinrich, 1986). Within-habitat bias is effected mainly by weather conditions on flight activity or seasonal climate on resource quality. This can affect demographic structure, especially population density, which in turn can influence reproductive success (e.g., through conspecific competition) and survival (e.g., facilitating predator search images).

Nectar source availability in space and time influences feeding mainly via transit costs – the amount of movement necessitated between nectar sites. As a result, dense concentrations of suboptimal sources may be more profitable than low density optimal sources, especially where time constraints are imposed by weather conditions (May, 1985). Seasonal climate, particularly drought or severe frosts can decrease resource availability, increase transit time and influence survival and reproduction (Murphy, 1983). Within-habitat variations in nectar resources associated with topography (i.e., aspect, soil moisture) or vegetation structure (i.e., shade) can result in spatial and temporal bias in feeding, in turn influencing the location of other activities and population structure. Gilbert & Singer (1973) found that the spatial association between larval and adult resources played a key role in movements of *Euphydryas editha*, and in a Central Sierran habitat for *E. chalcedona*, with widespread larval resources but a single concentration of nectar, adults were found more often near nectar and there was a distance decay in the distribution in the number of offspring from nectar (Murphy, 1983). Although seasonal climate and diel weather can make for different rewards at different times and in different situations,

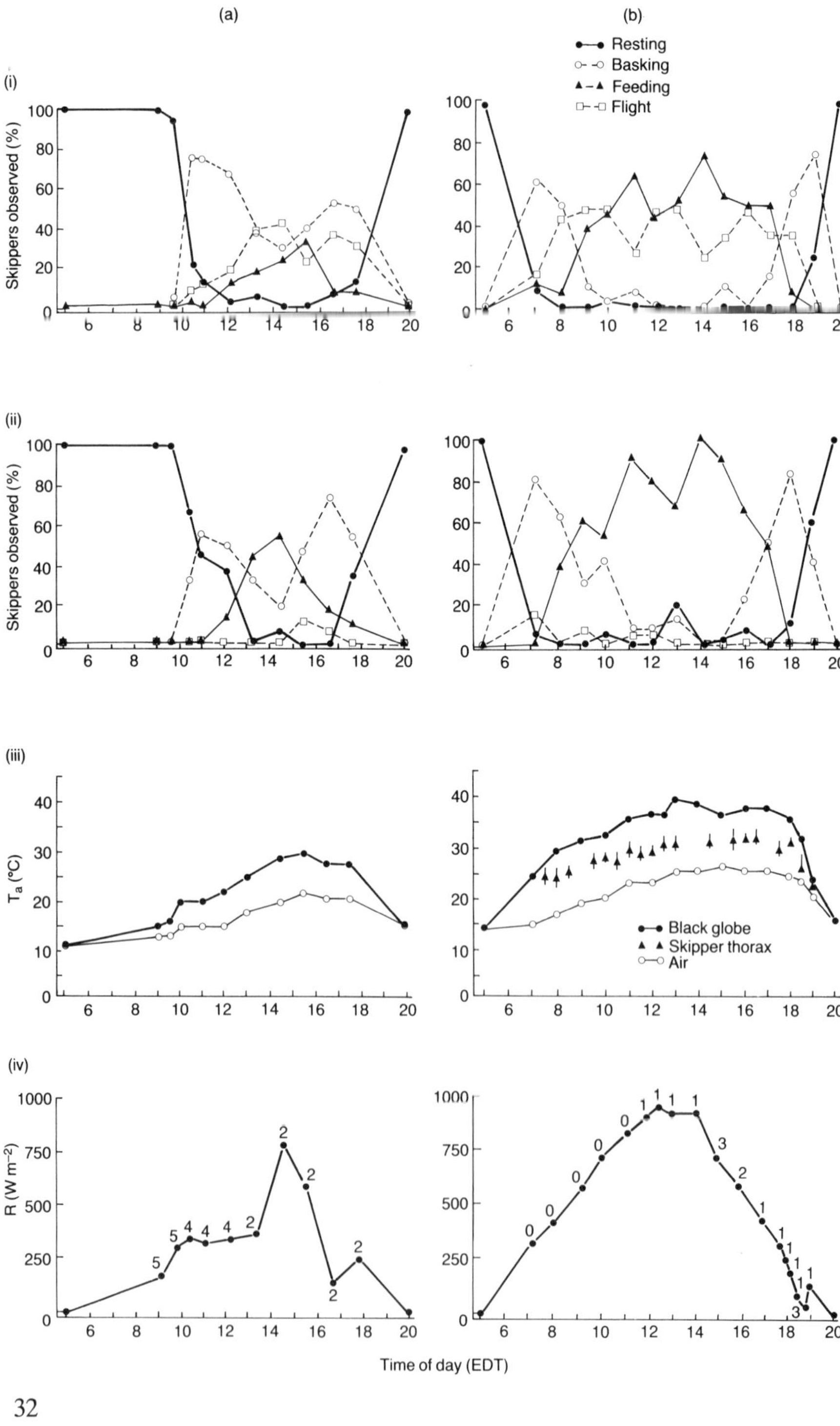
(a)
(b)
Resting
Basking
Feeding
Flight
(i)
(ii)
(iii)
(iv)
Skippers observed (%)
T_a (°C)
R (W m^{-2})
Black globe
Skipper thorax
Air
Time of day (EDT)

and learning may exist for floral preferences (Wiklund & Åhrberg, 1978), butterfly adults are generally opportunistic users of nectar (Courtney, 1986), and patterns of feeding may relate more to thermal costs and constraints of other activities (Willmer, 1983).

The mode of mate-location influences diel and seasonal activity patterns and habitat use. Territorial species can adjust their mate-location in relation to weather, climate and population structure. Territoriality declines and patrolling increases with a rise in T_a in *Pararge aegeria, Lasiommata megera* and *Coenonympha pamphilus* and increases seasonally in *P. aegeria* (Dennis, 1982–83; Wickman & Wiklund, 1983; Shreeve, 1984; 1987; Wickman, 1985a, 1988). *Ochlodes venata* which typically patrols am and is territorial pm, can adjust its activity on cloudy days to patrol in the afternoons, probably in relation to receptive female density (Dennis & Williams, 1987). In territorial species, shifts in behaviour are accompanied by marked changes in habitat use diurnally and seasonally which can greatly influence demographic structure, especially population density (Shreeve, 1984; Dennis & Williams, *loc. cit.*). Invariant patrollers can compensate for changes in T_a, R and A_v only by adjusting flyways, as mate aquisition during enforced rest must be entirely fortuitous (Dennis, 1986b).

High T_a and low A_v is associated with hilltopping; such altitudinal bias is independent of the perch–patrol continuum (Shields, 1967; Scott, 1968). Hilltopping can influence mating success (Wickman, 1988), especially for species at low density, but can be maladaptive at high density (Ehrlich & Wheye, 1986). As more relatively larger species hilltop and larger rather than smaller species can reach injuriously high T_b (Scott, 1968), this behaviour probably evolved in response to heat stress. Wickman (1988) observed that the distribution of male *Lasiommata megera* followed the sun's azimuth from east to west, but was also influenced by wind direction, with males preferring the lee side of hilltops. The selection of microrelief features was found to be influenced by the windspeed gradient and the need for T_as suitable for both territorial flights and perching.

1.9 Daily pattern of activity in *Thymelicus lineola* on (a) 17 July 1982 and (b) 23 July 1982. (i) males; (ii) females; (iii) black globe, air and *T. lineola* thoracic temperatures (mean ± SD); (iv) solar radiation (diffuse and direct) and cloud index (0, clear sky; 1, < 10 per cent cloud; 2, 10–75 per cent cloud; 3, thin overcast; 4, 75–95 per cent cloud; 5, solid cloud; 6, sombre). Courting activity of the males is not presented but was included in the calculation of percentages. The sun was visible above the surrounding landscape in clear weather between 05.45 and 19.30 h EDT. By comparison with 23 July, an 'ideal' day, the early morning of 17 July was very cool and cloudy. Consequently, there was no activity until 10.00 h, when $T_{bg} > 20°C$ as solar radiation increased. (From Pivnick & McNeil, 1987 by courtesy of Blackwell Scientific Publications Ltd.)

Bias for oviposition within habitats can occur directly, as a result of A_v, A_d, sunshine and shade (Wiklund & Åhrberg, 1978; Pivnick & McNeil, 1987) or indirectly as result of climatic influences on hostplants (Singer, 1972; Weiss *et al.*, 1988), nectar source quality (Murphy, 1983) and seasonal availability (Dennis, 1983a). Climate may also be partly responsible for microhabitat bias in oviposition, determining modal egg orientation on hosts (Williams, 1981), 'edge' and 'recess' effects (Dennis, 1983b, 1984; Porter, 1992). Atmospheric inputs may also combine with various resource attributes in the habitat to affect fecundity (e.g., abundance of oviposition sites; larval and hostplant quality and spacing; nectar resource quality and spacing; egg avoidance behaviour and male harassment) (Boggs & Watt, 1981; Porter, 1992). The impact is generally on flight time via resource quality and spacing (Shreeve, 1984; Myers, 1985; Courtney, 1986). Selection should generally determine a balance between levels of fecundity and offspring survival, much as in egg avoidance (Thomas, 1984), the evolution of egg batch size (Courtney, 1984) and of hostplant range (Wiklund, 1981). Suppression of egg avoidance, batch size and polyphagy relate not only to host quality or dimensions but also to time constraints imposed by adverse weather and the need to maximize reproductive potential. This can result in apparently maladaptive behaviour in which most eggs are placed on least suitable hosts (e.g., *Anthocharis cardamines*; Courtney, 1982a, b) or individual plants become overloaded (Dennis, 1982).

Eggs and pupae develop faster at higher T_a (Pollard, 1979; Weiss *et al.*, 1988), though tolerances of extreme values of T_a clearly vary widely for eggs and pupae of different species, particularly between diapause and non-diapause groups (Beck, 1980; section 2.1B). Freezing temperatures can cause high mortality in pupae (e.g., *Danaus plexippus*, Urquhart, 1960) and as high T_a results in desiccation, there are probably optimum values of T_a and relative humidity (H_r) for egg development and survival (Taylor, 1988). Eggs of some species (e.g., *Lasiommata* spp. of the Pararginae) are resistant to desiccation and adapted to low-humidity microhabitats (Karlsson & Wiklund, 1985). Butterfly eggs, and pupae to a lesser extent, are small enough to fit into the boundary layer enveloping plant surfaces; precise selection of substrate ensures very different values, means and extremes of T_a, A_v and H_r in relation to exposure, aspect and insulation (Porter, 1992). Thus, during oviposition climate may occasionally take precedence over host species (e.g., *Pararge aegeria*, Shreeve, 1986a).

Larval growth rates increase and developmental times decrease with T_a (Shreeve, 1986b; Joos *et al.*, 1988; Casey *et al.*, 1988; Weiss *et al.*, 1988) or with accumulated thermal units (Scriber & Slansky, 1981; Pullin, 1986a), to near lethal levels in some species (May, 1979), but survival generally has a distinct optimum range at lower T_a (Bursell, 1974) and larvae thermoregulate to optimize growth and development. Many insects are able to continue their development in fluctuating temperatures with means below

T_0min (Behrens *et al.*, 1983). Insolated *Pieris rapae crucivora* larvae develop more rapidly than *P. napi nesis* or *P. melete* in shade, producing six rather than three broods (Ohsaki, 1982). Males develop more rapidly than females (Shreeve, 1986b). Modal developmental T_0min is c. 11.5°C for some 85 species (Utida, 1957; Scriber & Lederhouse, 1983) but can be as low as 6°C (Lees, 1962b) and 8.6°C (Pullin, 1986a). Optimum T_b is c. 35°C for *Eurodryas aurinia* (Porter, 1982) but this represents T_0max for many species (Guppy, 1969; Sherman & Watt, 1973; Rawlins & Lederhouse, 1981; Scriber & Lederhouse, *loc. cit.*). Light has several important influences on feeding; via thermoregulation, diel feeding cycles and diapause-induction (section 2.1B). Other weather events which affect development directly are rain (e.g., larvae may be dislodged and drown such that rain can be a major mortality factor for eggs and larvae in *Pieris rapae*; Harcourt in Courtney, 1986), frost, typically during spring or autumn (Dolinger *et al.*, 1973; Chew, 1975; Shreeve, 1986b) and even snow (Ehrlich *et al.*, 1972). Trade-offs for different mortality factors can occur. Thus *Glaucopsyche* butterflies fly early in the spring to take advantage of hostplants low in alkaloids but as a result are subject to an increased likelihood of mortality from frosts (Dolinger *et al.*, 1973). Diapausing larvae of high latitude species are adapted to withstand extreme cold ($T_a < -20$°C) (see section 2.1B); many species especially of marshland habitats survive immersion in water (e.g., *Lycaena dispar*, Ford, 1945), even when this is frozen (e.g., *Coenonympha tullia*, Melling, 1984).

The compartmentalization of biomass (B) allows an account to be made of the relative influence of abiotic conditions and nutrition on consumption (RCR, CR) and digestion (AD, ECD, ECI, NAR, NCR, NUE). Above or below an optimal T_a, most performance values tend to decline, although approximate digestibility (AD) generally shows little change. Variable or cycling temperature regimes frequently stimulate performance relative to constant T_a (Matheson & Decker, 1965; Scriber & Slansky, 1981) but exceptions occur (Guppy, 1969; Poston *et al.*, 1977). The most commonly reported cause for increased growth rates has been increased feeding rates (Sherman & Watt, 1973; Mathavan & Pandian, 1975; Scriber & Lederhouse, 1983) whereas digestive efficiencies seem to be relatively independent of T_a in a number of species (Scriber & Slansky, 1981). In *Papilio glaucus*, and *Malacosoma* spp (Heterocera), increased growth rates with higher T_a were paralleled by food processing rates and consumption rates but not necessarily by feeding duration and frequency (Scriber & Lederhouse, *loc cit.*, Casey *et al.*, 1988; Joos *et al.*, 1988), implying as in *Eurodryas aurinia* a positive feedback of food processing rates on subsequent meal size during inactive periods (Porter, 1982).

Growth performances can be influenced indirectly by weather and climate on hostplant quality. Growth rates and development time are significantly affected by plant water content and nitrogen levels (Feeny, 1975;

Slansky & Feeny, 1977; Scriber & Slansky, 1981) but the extent to which nutritional indices are related to differences in total N_2 and water content in food is uncertain (Pullin, 1987b). In some species (e.g., *Pieris brassicae*) low N_2 concentration in the diet is associated with high larval mortality in the field (Gorovaya & Samersov, 1976). As leaf water and N_2 per cent of forbs, grasses and trees declines with age, seasonal weather (temperatures, rainfall) can adversely affect larval growth, especially if moisture plays a role in N_2 availability. Larvae on trees are most likely to be affected as this is a poor substrate compared to forb leaves, mainly owing to low AD, and the microclimate is often cooler (Scriber & Slansky, 1981). Even when regrowth foliage is available for larvae, this too can be influenced by previous and current seasonal weather conditions (Pullin, 1986b, 1987a). Climatic influences, in conjunction with topography and light environment, on hostplant quality and availability also have implications for within-habitat bias, with concomitant repercussions for larval density, predation, parasitization, hostplant overloading, exploitative competition and starvation. Nutritional imbalances often entail increased catabolism and excretion and this frequently slows growth and reduces conversion efficiency (House, 1974); but interactions between temperature and nutrition can be complex (Taylor, 1988) and in some cases larvae are able to maintain homeostasis through behavioural, physiological and metabolic compromises, often involving compensations via consumption rates and digestive efficiency (Mattson, 1980; Miles *et al.*, 1982). The outcome is greatly complicated by larval age, particularly the tolerances, detoxification capabilities and behavioural repertoire of different instars (Goettel & Philogene, 1978); early instars are fragile, relatively immobile animals and particularly vulnerable to differences in host quality (Courtney, 1986).

Growth performance influences individual success directly via development time, total biomass and energy and nutrient reserves. Reduced development time affects survival (section 1.2C), and annual reproductive capacity (section 2.1A), whereas pupal weight may influence fecundity directly (Cook, 1961; Hough & Pimentel, 1978) or indirectly via activity (Wickman & Wiklund, 1983; Shreeve, 1986b) or longevity. The expected negative correlation between pupal weight and growth rates from cross-taxonomic comparisons (Begon & Mortimer, 1981; Calow, 1989) depends at conspecific level on other factors such as food status (Mathavan & Muthukrishnan, 1976), concordance in feeding indices (Pullin, 1987a) and diapause-induction (Scriber & Slansky, 1981); some species (e.g., *Pieris rapae*) are capable of attaining the same pupal weight for different growth rates on different hosts (Rahman, 1969; Chew, 1975).

C *Weather, climate and enemies*

Mortality from enemies (predators, parasitoids and microparasites) differs according to age, because of stage-idiographic behaviour and activity, but also because the spectrum of enemies varies with insect size; eggs and early instars are affected more by invertebrates and later stages more by vertebrates (Dempster, 1967, 1971; Thomas, 1974; Pollard, 1979). Weather and climate determine the vulnerability of butterflies to enemies by influencing their activity, spatially and temporally, their degree of apparency and physiological functioning; thus, their exposure to enemies and their ability to monitor, evade and engage secondary defences against attacks. Environmental conditions which induce inactivity in mobile stages (larvae, adults), especially low T_a, correlate with vulnerability to successful attacks; small parasitoids can be active at low T_b (Shreeve, 1986b) and vertebrate predators can function outside a butterfly's activity range.

Adult butterflies are probably most vulnerable when roosting and when adverse weather enforces extended periods of diurnal resting, as in *Papilio polyxenes* (Rawlins & Lederhouse, 1978; Lederhouse, 1983; Bowers *et al.*, 1985); risks may be proportional to the increased period on roosts or may be raised owing to increased apparency in roosts (Lederhouse, 1983) and a change of predator behaviour during daylight. Predation has been shown to be inversely related to T_a in winter roosts of *Danaus plexippus* (15k butterflies killed d^{-1}, 1978–79); low T_b restricts their ability to bask, display aposematic coloration and regain roost perches (Calvert & Brower, 1986). Wing damage evidence in *Euphydryas chalcedona* also suggests that increased bird predation is correlated with cool, cloudy weather (Bowers *et al.*, 1985). Unusually high levels of bird predation on this butterfly occurred in the cold weather of 1980 compared to that of 1979 (Bowers *et al.*, 1985). Low light intensity during cloudy conditions may adversely influence the ability of adults to monitor predator attacks, and may explain abandonment of flight in conditions otherwise adequate for flight (Douwes, 1976; Rawlins & Lederhouse, 1978; Pivnick & McNeil, 1987).

Flying butterflies are not free from predation (Carpenter, 1939; Wourms & Wasserman, 1985). At Corcovado National Park, Costa Rica, evasion among some 55 brightly coloured tropical species falls into two 'adaptive suites' based on their palatability to *Galbula ruficauda melanogenia* (Chai & Srygley, 1990; Srygley & Chai, 1990a). Palatable and unpalatable species are distinct in morphology, flight, habitats, activity periods and thermal ecology (see section 3.2). Srygley and Chai (1990a) concluded that the higher thoracic temperature of palatable butterflies indicates that predation is a strong selective force on thermal physiology of butterflies. The average T_b of the thorax for 17 palatable species is 4°C higher than the average for 24 unpalatable species (Fig. 1.10). The former fly faster and more erratically, are more frequently pursued by birds but escape more often from attacks than the latter. For palatable species, thermal specialization results in more frequent basking behaviour and the restriction to open, sunny habitats during the midday period. The significantly higher thermal excess of palatable species within the same habitat results in larger

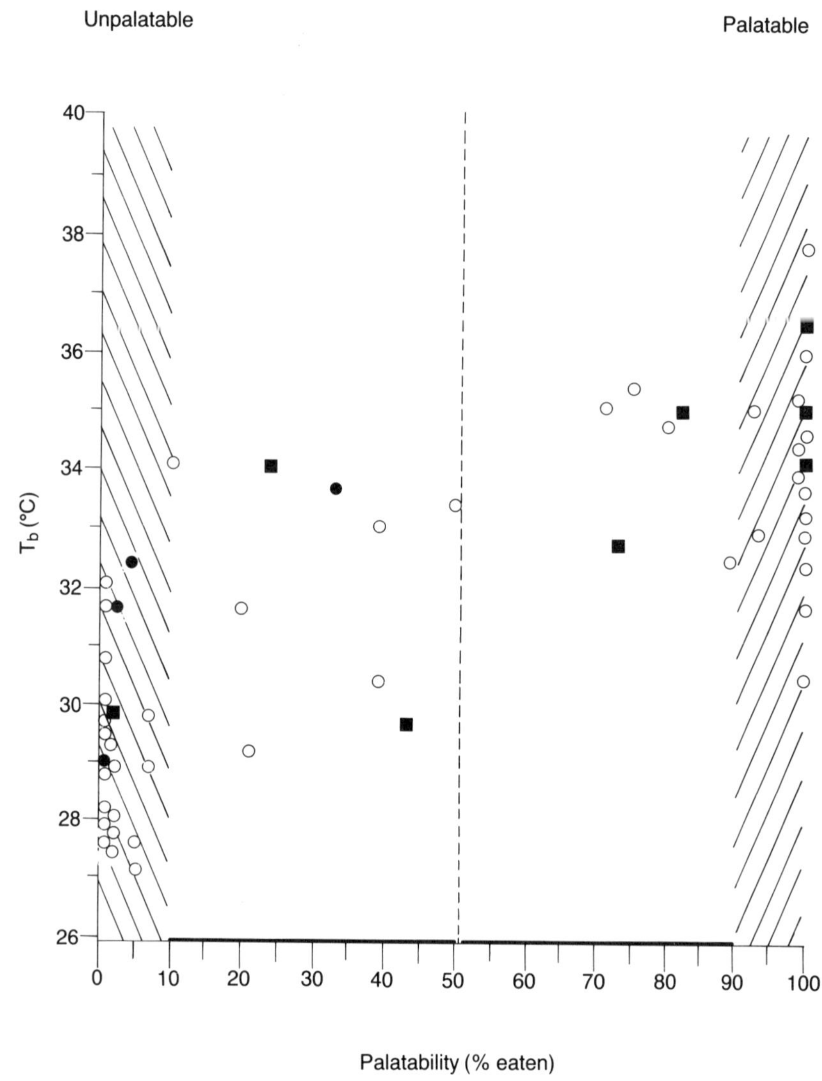

1.10 Scattergram of palatability of tropical butterflies in Corcovado National Park, Costa Rica to *Galbula ruficauda melanogenia* and their thoracic temperatures during flight activity. Each point represents the mean result for one species. Shading, species in the palatable or unpalatable classes. The mean difference in T_b for the palatability groups (4°C) is highly significant (Student's t-test, $t = 6.54$, $p < 0.001$); at least 2°C of this difference may be attributable to differences in the thermal environment (T_a); mean values for T_{ex} and T_a on 55 species are positively correlated ($r_p = 0.3$, $p < 0.05$). White dot, Nymphalidae; black dot, Papilionidae; black square, Pieridae. (From Srygley & Chai, 1990b; © 1990 The University of Chicago; by permission.)

part from the greater net influx of radiative energy as R tends to be higher when and where palatable butterflies are active. At least two other reasons may account for the higher T_{ex} in palatable species: (i) the greater production of metabolic heat from a more powerful wing beat and powerful thoracic musculature; (ii) the possible retention of more metabolic heat due to a wider thorax which may influence surface area : volume ratio and a shorter abdomen insulated by the hindwings. In contrast, high flight speed and high temperatures, thus warm microhabitats do not have the same advantage for unpalatable species and their mimics (e.g., Ithomiinae and Dismorphiinae), as slow flight facilitates aposematic signals (i.e., flight behaviour, wing and body morphology and colour) to predators, recognition and rejection. These strategies imply the existence of trade-offs between increased flight performance, energy economy and nutriment investment in the evolution of flight physiology (Srygley & Chai, 1990a; see section 2.1A).

Measurements of thoracic temperatures of two species of unpalatable butterflies (i.e., *Tirumala hamata* and *Euploea tulliolus*, Danainae) in Papua New Guinea (Dudley, 1991) urge caution in the interpretation of thermal differences between palatable and unpalatable species. Both sexes flew under conditions of intense solar irradiance, had a mean T_{ex} of 5.5°C, yet T_{ex} was not positively correlated with solar irradiance, body mass or wing loading. Evidently, equilibrium temperature of flying butterflies is influenced by a number of factors (e.g., choice of microhabitat, flight metabolism, convective cooling); interactions between these factors can be complex and association between palatability and thermal biology in butterflies thus has a variety of possible outcomes (Dudley, 1991).

Resource use (e.g., nectaring; single as opposed to batch egglaying) during low T_a, R and high A_v may be less efficient outside optimal T_b for flight activity and expose butterfly adults to increased predation (Pivnick & McNeil, 1985) despite opportunistic basking. However, as butterflies are most apparent when basking, there must be powerful selection for shortened basking times. It is probably not coincidental that wide field of view at mate-location perches, which allow detection of mate incursions into territories, also permits monitoring of predator attacks. The smaller wing angles used by pierid DR baskers during dawn warm-up may imply a compromise between apparency and thermoregulation (Kingsolver, 1985a, b). Risk of predation may also be increased by weather or climate-induced shifts in habitat or microhabitat use. For example, the enforced movements of larvae from their usual backgrounds, caused by the effects of drought on host quality, and habitat bias in resource use, caused by seasonal influences on resource quality in varied topography, increase apparency in the former case and in the latter raise prey density and may enhance olfactory cues for parasitoids (Laing & Levin, 1982; Kitano, 1978; Sato, 1979) or search images of predators. Seasonal climate can lead to the adoption of different habitats with different predator regimes as in the case of Mediterranean *Maniola jurtina* which adopts woodland cover during midsummer 'aestivation'. The effects of habitat shifts are not easily predicted. Enemy-escape can occur when partial spatial incongruity exists between host and prey in habitat preferences (e.g., *Anthocharis cardamines* and parasitoid *Apanteles saltator*; Courtney, 1986). The full consequences of habitat shifts are often unexplained and can involve compromises. In *Papilio glaucus*, a western bias in oviposition on host trees involves a trade-off between growth rates and risk from increased mortality; evidently the cost of higher mortality rates inferred from increased predator activity in warmer locations does not exceed the liability of retarded growth rates for other aspects (Grossmueller & Lederhouse, 1985).

Increased development time with low T_a exposes stadia to predation and parasitization for longer periods (Thomas, 1974; Pollard, 1979); this may be only partly balanced by any increase in size during larval feeding and possibly increased fecundity (but see Leather, 1988). Protandry may subject females to several days of additional mortality as immatures compared to males. Cold weather also often increases food requirements of warm-blooded predators (Gilbert & Singer, 1975) and can lead to increased predation, especially where attainment of a relatively well-protected winter diapause stage is delayed (Shreeve, 1986b). In comparison to aposematic species, certain cryptically-coloured larvae tend to feed less often during the day and exhibit slower growth; this may extend development by several weeks and their exposure to a different suite of predators (Scriber & Slansky, 1981).

Although life-cycles of specialist parasitoids are closely tuned to their hosts, they are susceptible to density-independent weather-mediated overwinter mortality (Courtney, 1986), and weather can influence the synchronization of host and prey. Porter (1983) observed that during cool, but sunny, weather, asynchrony in development between *Eurodryas aurinia* and its parasitoid *Apanteles bignellii* can reduce parasitization levels, because *E. aurinia* larvae are influenced by R and T_a, whereas the braconid cocoons which depend on T_a alone emerge too late to impregnate the larvae. A similar observation has been made for *Euphydryas maturna* in central Sweden (Eliasson, 1991). At this location, it is suggested that snow cover protects larvae against parasitization by preventing them from becoming active too early in spring, when air temperatures may be high but radiation low. Variable development rates may also facilitate parasitism-escape as in *Pararge aegeria* during autumn in relation to attacks by *Apachthis resinator* (Shreeve, 1986b).

Susceptibility of early stadia to bacterial, microsporidial and viral infections may relate to 'stress' triggered by hot weather (e.g., African *Colias electo*; Polson & Tripconey, 1970) or cold weather (e.g., *Pieris rapae*, Biever & Wilkinson, 1978) although any relationship is complicated by nutritional factors (Courtney, 1986). There is a danger of increased pathogenic attacks at high relative humidity (Willmer, 1982). Death of diapausing larvae, probably caused by fungi, during dull, humid weather, constitutes a major mortality factor in Dutch *Lycaena dispar* (Bink, 1972).

2

Climate, butterfly populations and distributions

Populations persist when natality and immigration exceed mortality and emigration. This chapter investigates the influence of atmospheric factors on population fluctuations and by extension the chain of causation underlying distributions and species' abundance patterns. As adaptation forms an essential part of a population's ability to persist in a specific habitat, attention is first focused on life-history strategies and trade-offs linked to survivorship and fitness traits.

2.1 **Life history strategies**

The constraints imposed by internal biological state variables and external environmental parameters preclude an organism from simultaneously maximizing all aspects of reproductive performance – the Darwinian demon (Law, 1979). Instead, life-history strategies have evolved that apportion more or less nutriment to gonads or soma (Karlsson & Wickman, 1989), combinations of traits and fecundity and survivorship schedules most suited to maximize fitness (Stearns, 1976, 1977, 1989; Boggs, 1981; Begon & Mortimer, 1981; Partridge & Harvey, 1988). Emphasis has been placed on two themes: (i) the determination of systems of co-evolved adaptive traits in reproduction, growth, development and survivorship, and (ii) mechanisms for survival of adversity and synchronization of life-cyle events with a strong phenological component; the former may necessarily incorporate the latter, and both are substantially influenced by climate.

The result of cross-taxonomic comparisons has been the formulation of a supposedly self-reinforcing system in which suites of characteristics are divided into r-selected and K-selected 'strategies', as to parameters in the logistic equation* (Stearns, 1976; Horn, 1978), corresponding with the type of environments purportedly producing them, r-selecting (unpredictable and ephemeral) and K-selecting (predictably seasonal or constant) (MacArthur & Wilson, 1967; Pianka, 1972; Southwood, 1977). In part, the habitat categories (Southwood, 1977) specifying degrees of 'favourableness' underlying the r and K paradign are potentially climate-driven (Dobzhansky, 1950), but how generally applicable are the combination of life-cycle traits to the Lepidoptera? A review of organisms from 35 studies revealed that only 18 conformed to the model (Stearns, 1976).

Although repetitive elements of r- and K-ness may re-emerge in studies, an expectation of strategies as co-evolved assemblages of traits that must be found together is misleading for the following reasons: (i) the covariances of phenotypes and certainly of genotypes are typically well below unity (Roff, 1990); (ii) phenotypic plasticity and threshold responses in phenetic traits occur frequently and intermediate structure (physiology) often has a modulating capacity on trade-offs in phenotypes (Shapiro, 1976; Brakefield, 1987b; Partridge & Harvey, 1988; Stearns, 1989); (iii) behavioural outlets, such as microhabitat choice for thermoregulation, are often available to organisms; (iv) there is a multiplicity of potential interrelationships among biological state variables promoting constraints and trade-offs and thus complexity in their manifestation (see Brakefield, 1987b); and (v) the equally varied and often disruptive nature of extrinsic parameters can select for different traits to the extent that they are uncoupled. Integration of traits within taxonomic groups thus has no obvious outcome; pathways leading to and from juvenile and adult size, two of the key attributes distinguishing r-ness and K-ness, are often numerous, shared, imprecise, conflicting and indirect (See Blau, 1981; Fig. 2.1). Although adult size and fecundity are often positively related in Lepidoptera (Karlsson & Wickman, unpub. manuscript), this relationship is not absolute (Springer & Boggs, 1987). In a classification of 145 butterfly species on 12 life-history characteristics (Bink & Siepel, 1986; Bink, 1989), two additional categories have been added to r and K strategists, 'stress tolerators' and 'fixtures' after an analogy from botany (Grime, 1979), by further subdividing Southwood's habitat space; yet both may represent subsets of K strategists. Even so, the r and K division is useful. There is fairly general agreement that, with certain simplifying assumptions (i.e., stable environment, frequency-independent weak selection, demographic ergodicity), the optimal life-cycle maximizes

* r and K are terms in the logistic equation which describes the actual rate of increase per individual of a population given by: $\delta N/\delta t = rN(K - N/K)$, where N is the population size, r is the intrinsic rate of increase (the maximum rate of increase = birthrate – death rate under fixed conditions of temperature and humidity) and K defines the carrying capacity determined by resources. r-selected individuals attempt to maximize their fitness by improving their ability to reproduce rapidly in an uncrowded environment (i.e., increasing r), whilst K-selected individuals maximize fitness by increasing their contribution to a population that remains at its carrying capacity (see Varley *et al.*, 1973; Begon & Mortimer, 1981).

the intrinsic rate of natural increase, r. Thus quantum increases in the life-span have prompted Shapiro (1975a, 1980a) to treat univoltive and biennial regimes as K-type compared to multivoltinism (r-type) in Lepidoptera.

Sympatric populations displaying r- and K-ness suggest that there are potential trade-offs between life history traits; their investigation provides an effective tool for determining structure in biological systems. Parameter r or, more appropriately for seasonal environments, r_a the annual rate of increase, provides a suitable measure of fitness for modelling trade-offs in Lepidoptera life histories (Roff, 1980; 1983; Smith *et al.*, 1987). In the absence of trade-offs, selection would drive all traits correlated with fitness to limits imposed by phylogeny and design. Strong experimental evidence for the existence of absolute trade-off constraints is provided by the maintenance of many traits within limits (Stearns, 1989) and the demonstration of negative genetic correlations between fitness characters (Reznick, 1985). Although the heritabilities of life-cycle traits may be relatively low (h^2, 0.30 ± 0.05), there is still sufficient variation to make possible quite rapid response to selection (Roff, 1990). Populations benefit through dynamic homeostasis and the retention of additive genetic variance in components of fitness (Smith *et al.*, 1987). Interpretation of trade-offs and certainly of complete life-history systems may be more refractory; multivariate influences on key variables (e.g., size) and the potential for interactions between trade-off circuits suggest a multiplicity of plausible adaptive hypotheses and some non-adaptive ones too (cf. Smith *et al.*, 1987; Gilbert, 1984a, b, c, 1986). Appropriate experimental designs and cautionary tales are discussed in Stearns (1976, 1977, 1989), Partridge & Harvey (1988) and Rapport (1991).

A *Trade-offs in reproduction, growth, development and survivorship*

The principle life-history trade-off amongst Lepidoptera is between fecundity and pre-reproductive mortality as influenced by differential growth rates and development times (Roff, 1980; Sibly & Calow, 1989; Fig. 2.1). Natural selection will generally favour gene-determined traits that maximize survivorship and fecundity and minimize development time (Schaffer, 1974; Calow, 1989). But larger adult insects which have higher fecundity (Tyndale-Biscoe & Hughes, 1968; Clifford & Boerger, 1974; Hough & Pimentel, 1978; but see Gilbert, 1984c and Leather, 1988) are usually the biproduct of extended development (Robertson, 1960a, b; Masaki, 1973; Kamm, 1972) and development time may be correlated positively with mortality (Lewontin, 1965). When development time is affected by instar numbers or switches in voltinism (e.g., *Pararge aegeria*; Shreeve, 1986b; Nylin *et al.*, 1989; *Coenonympha pamphilus*, Wickman *et al.*, 1990), the trade-off is between r of current and subsequent generations (Shapiro, 1975a; 1980a). With correlation of size and development time, 'saw-tooth' clines may be generated at the point of transition in voltinism (Fig. 3.8; Masaki,

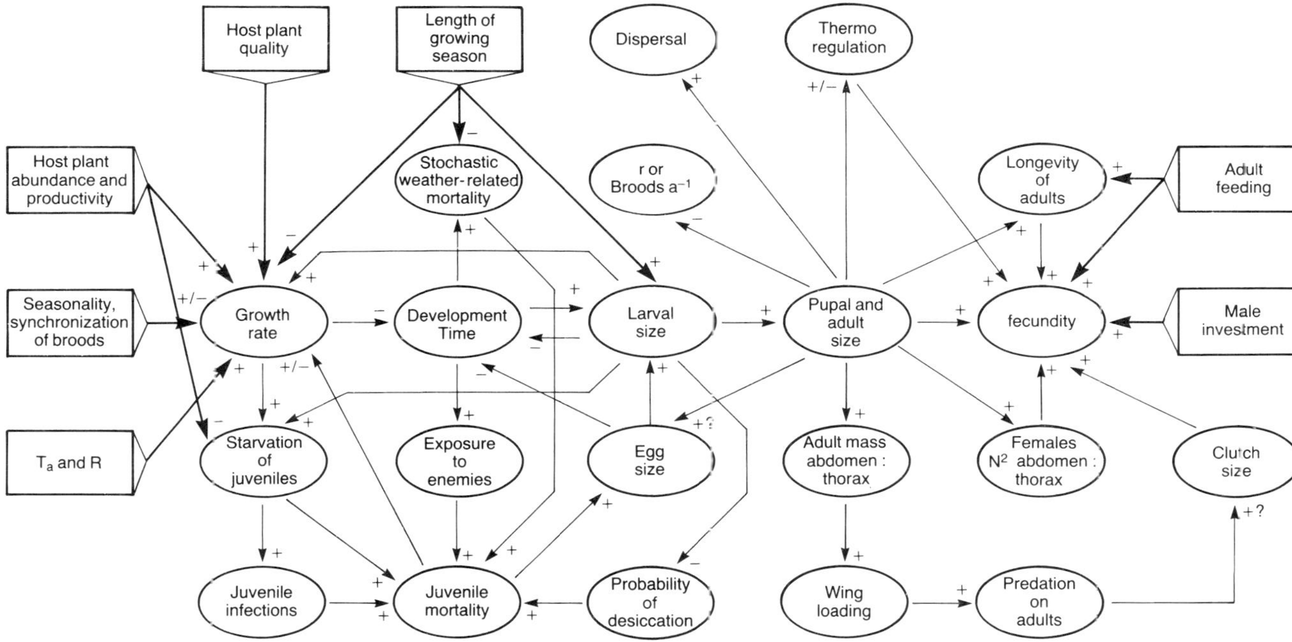

2.1 Links and feedbacks in some basic insect life-history traits. Boxes, environmental variables and basic inputs; ellipses, biological attributes. Signs (+, –) give the generally observed polarity of correlations, though some positively covarying environmental variables (e.g., T_a and length of growing season) can have a contradictory outcome (e.g., for adult size) for different species depending on the interaction of other life-history attributes and the degree to which life-history traits are shielded by phenotypic plasticity. (From Lewontin, 1965; Pollard, 1979; Roff, 1980; Blau, 1981; Gaston & Lawton, 1988; Leather, 1988; Calow, 1989; Sibly & Calow, 1989; Wickman *et al.*, 1990; Nylin & Svard, 1991; Nylin, 1992.)

1978; Roff, 1980, 1983; but see Ritland & Scriber, 1985). However, as any one variable in this relationship from growth rate to fecundity may be involved in other trade-offs, including changes in behaviour, the outcome is uncertain, particularly as total fecundity depends on adult longevity (see Pullin, 1986b). One evolutionary outlet is an increase in egg size with a concomitant reduction in development time (Sibly & Monk, 1987; Wickman *et al.*, 1990). In resource-limited environments, a trade-off may develop between high growth rates, causing starvation and infections, and dormancy stage weight correlated with overwintering mortality (Wickman *et al.*, 1990). In temperate butterflies, large adults allocate mass (both sexes) and nitrogen (females only) allometrically to the abdomen. This can produce proportional increases in fecundity and oviposition rate, but the isometrically incremented, and thus disadvantaged, thorax musculature and wing expanse may have adverse consequences for wing loading and perhaps for dispersal, predator-escape and flight time, the last of which can be equated with oviposition rate (Wickman & Karlsson, 1989; Karlsson & Wickman, 1989, 1990). Srygley and Chai (1990b) find that palatability is correlated positively with thoracic mass but negatively with abdominal mass among lineages and positively with flight speed within lineages of neotropical butterflies; differences between sexes for sex-limited mimics correspond with these observations. Hibernal diapausing adults of *Polygonia c-album* allocate more resources to the thorax compared with directly reproducing individuals, following the prediction of Boggs (1981) but as they are larger, resource allocation to the abdomen does not differ between the two pathways (Karlsson & Wickman, 1989). Longevity, dispersal, apparency to predators and thermoregulation are also suggested to be influenced by adult size (Blau, 1981) with inferences for fecundity, clutch size and oviposition rate depending substantially on environmental parameters which may be influenced more or less by weather and climate (sections 1.2B and 3.3C). Thus, reversals in associated traits are not altogether unexpected although their interpretation may be uncertain. In *Papilio polyxenes*, first and second brood larvae have similar development times, but second brood larvae grow slowly under short photophase to produce small autumn pupae and small spring adults having lower fecundity levels. The differences in adult size between the two broods have been related to seasonal thermal strategies (Blau, 1981), but may also be accounted for by the effect of seasonal resource distributions and host quality on dispersal and growth rate.

To date, investigations of more subtle trade-offs are not far advanced. Not only are the costs uncertain, but so too are the ramifications of probable benefits of investments in adult feeding (see section 1.2B), male ejaculates (cf., Rutowski *et al.*, 1987; Watanabe, 1988 and Pivnick & McNeil, 1988 *versus* Greenfield, 1982; Jones *et al.*, 1986; Svard & Wiklund, 1988), polyandry versus monandry (Svard & Wiklund, 1989), egg size (Wickman *et al.*, 1990), changes in egg size with age (cf., Moore & Singer, 1987 *versus* Murphy *et al.*, 1983; Chew & Robbins, 1984), batch size (Courtney, 1984) and delay in gonadal and egg maturation (Scali, 1971; Garcia-Barros, 1987, 1988; Porter, 1992) despite plausible hypotheses. Moreover, although genetic variance for traits, platykurty in trade-off curves (Sibly & Calow, 1989; Smith *et al.*, 1987), bet-hedging in partial broods and photoperiodically controlled development rates (e.g., *Coenonympha pamphilus*, Wickman *et al.*, 1990; *Pararge aegeria*, Nylin *et al.*, 1989) suggest fine-tuning in life-history strategies, ambiguous or contrary evidence occurs for the same or other traits (i.e., carry-over diapause, Shapiro, 1980a; egg size, Wickman *et al.*, 1990; phenology, Nishida, 1955). Among weather and climate agents that influence trade-offs are drought (Smith *et al.*, 1987; Garcia-Barros, *ibid.*), length of season (Shapiro, 1975a; Pullin, 1987a), summer temperatures and insolation (Nylin *et al.*, 1989), variability in spring (Shapiro, 1979) and autumn weather (Shreeve, 1986b) and conflicting conditions of winter temperatures and summer drought (Shapiro, 1984a).

B *Adaptations to seasonality*

Very few parts of the world lack any seasonality – mainly at equatorial latitudes. At least three aspects of seasonal changes produce different responses to cyclic and acyclic environmental fluctuations: their magnitude, predictability and duration (Tauber *et al.*, 1986). Adaptations are basically of two kinds: (i) to adversity in conditions; (ii) the timing and synchronization of life-cycle events with resources and conditions. Because of the latter, in particular, there is no simple dichotomy into active and inactive periods in the year which correspond loosely with overall levels of seasonal autotroph productivity (see Nylin *et al.*, 1989). Sympatrically distinct organisms have different life-cycle strategies if only because of distinctions in phylogenetic constraints – therefore in vulnerability to external influences – and in resource requirements. Adaptations to seasonal adversity mainly include: diapause, quiescence, dormancy, seasonal polyphenism in morphology and changes in behaviour and habitat choice; to synchronization with seasonality: diapause, voltinism, stage longevity, growth and development rates (Wolda, 1988). As diapause-mediated adaptations require predictable cues, Tauber *et al.* (1986) distinguish between adaptations to seasonally unpredictable more localized exigencies (non-diapause quiescence, migration, polyphenism) and those to seasonally predictable habitat deterioration (dormancy, seasonal migration and polyphenism) that are usually, but not invariably, diapause-mediated (Shapiro, 1980c; Hoffman, 1978).

Seasonal adversity, *sensu lat.*, is regionally highly variable and results in very different phases of relative activity and dormancy in temperate and tropical insects (Masaki, 1980; Denlinger, 1986; Table 2.1). The distinctive stages of holometabolous organisms largely dictate phenology and

Table 2.1 Seasonality in insects

(a) Seasonal deficits in energy and moisture for insects within climatic zones

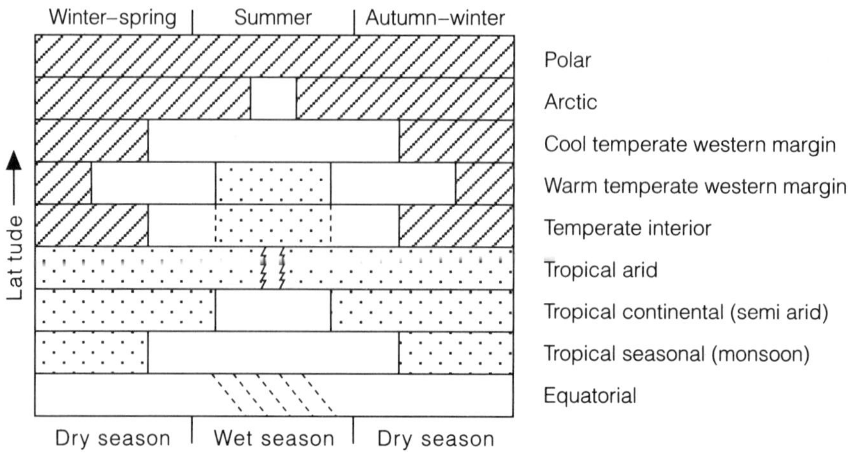

(b) Classification of seasonal (1, 2) and aseasonal patterns (After Wolda, 1988, modified and reproduced by permission of Annual Reviews Inc)

1 Phenomenon does not occur during the entire year
  A One clear seasonal maximum
  B More than one seasonal maximum
  C A broad seasonal maximum
  D Clear maximum, at predictable supra-annual intervals, related to seasons
2 Phenomenon occurs around the year
  A One sharp seasonal maximum
  B More than one seasonal maximum
  C A Broad seasonal maximum
3 Phenomenon is aseasonal
  A Occurs more or less constantly around the year
  B Unpredictable irregular variation in occurrence, unrelated to seasons
  C Clear maxima, at predictable subannual or supra-annual intervals, unrelated to seasons

(a) A scheme illustrating contrasts in seasonally hostile periods for selected climatic zones as a consequence of low temperatures (shaded) or low precipitation (stippled). Excessive precipitation (pecked lines) may also select against insect stadia in equatorial habitats. In arid climates, active phases are unpredictable and depend on irregular episodes of precipitation. The severity of temperate environments increases with latitude and towards continental interiors. Note: seasonal arrangements of active and dormant phases in insects are governed by factors other than seasonal climate, such as moisture balance and hostplant phenology (e.g., crucifer pod-feeding butterflies such as temperate *Anthocharis cardamines* (Pieridae) are active only in spring and early summer and dormant from June to April). (b) A phenomenon is regarded as seasonal if it, or its maximum expression, predictably occurs at much the same time of the year, each year that it does occur.

seasonality (voltinism; timing and duration of stages); for instance, larvae are specialized for feeding, whereas adults are adapted for reproduction and dispersal. The seasonal characteristics of butterfly life-cycles are expected to reflect these specializations (Table 2.1). Both the data and the techniques insist on a cautionary note. Most publications are from the temperate zone; furthermore, where observations are largely based on adult records as in the tropics, fluctuations in numbers can be interpreted in different ways: as (i) stage population fluctuations; (ii) shifts in behaviour and habitat attendance; and (iii) phenological patterns (Owen, 1971). Extra-tropical patterns are mainly type 1A or 1B and rarely 1C, whereas in tropical environments virtually all types of seasonality or aseasonality can be found, except usually type 3C (Wolda, 1988; Table 2.1). The distinction reflects the length of the growing season for hostplants and reproductive season for adults, which decline systematically with increasing latitude and altitude in mesic environments. Voltinism declines, seasonality increases, and generally the adult phase starts later and is of shorter duration with increasing latitude and altitude, particularly for species overwintering in the same stage (e.g., *Aricia agestis* and *A. artaxerxes*; Dennis 1992) – even for relatively small regions such as the British Isles (Pollard *et al.,* 1986; Fig. 2.2). This pattern is repeated within the tropics from non-seasonal equatorial habitats to areas with a pronounced dry season. Biennial species (e.g., *Oeneis* spp., Douwes, 1980) and those with supra-annual life-cycles are common at arctic latitudes but are also found in climates with winters and prolonged dry seasons; adult emergence may be seasonal (e.g., *Euphydryas chalcedona*; Scott, 1973) or aseasonal, as in the case of *Papilio rudkini*, whose flights are correlated with rain storms as much as six years apart (Emmel & Emmel, 1973). The frequency of univoltine butterflies increases with altitude and latitude (Shapiro, 1975a), whereas species whose adults are active around the year but with a seasonal maximum seem to be confined to lower latitudes (Wolda, 1988). Unlike numerous Heterocera (e.g., *Operophtera brumata*, Holliday, 1985), cool and cold temperate butterflies cannot breed in winter. Nevertheless, within temperate environments variation in life-history cycles at one location can be marked (Shapiro, 1975a; Scott & Epstein, 1986; Fig. 2.2) even for species using the same hostplant (Wickman *et al.,* 1990); voltinism patterns can also be reversed with latitude as in *Danaus plexippus* which has three more broods at Wisconsin than at Florida, 16° further north (Malcolm *et al.*, 1987). Upper thermal limits as well as low ones occur, and in climates with summer drought (e.g., the Mediterranean) the growing season can be greatly curtailed. Multivoltinism is probably common in tropical climates (Owen, 1971), but in temperate latitudes declining quality of hostplants seasonally selects against multivoltines (Pullin, 1986a, b; 1987a, b) which often require to switch hosts between broods (e.g., *Celastrina argiolus*; Shapiro, 1976). Apparent multivoltinism at high altitude may be the result

(a)

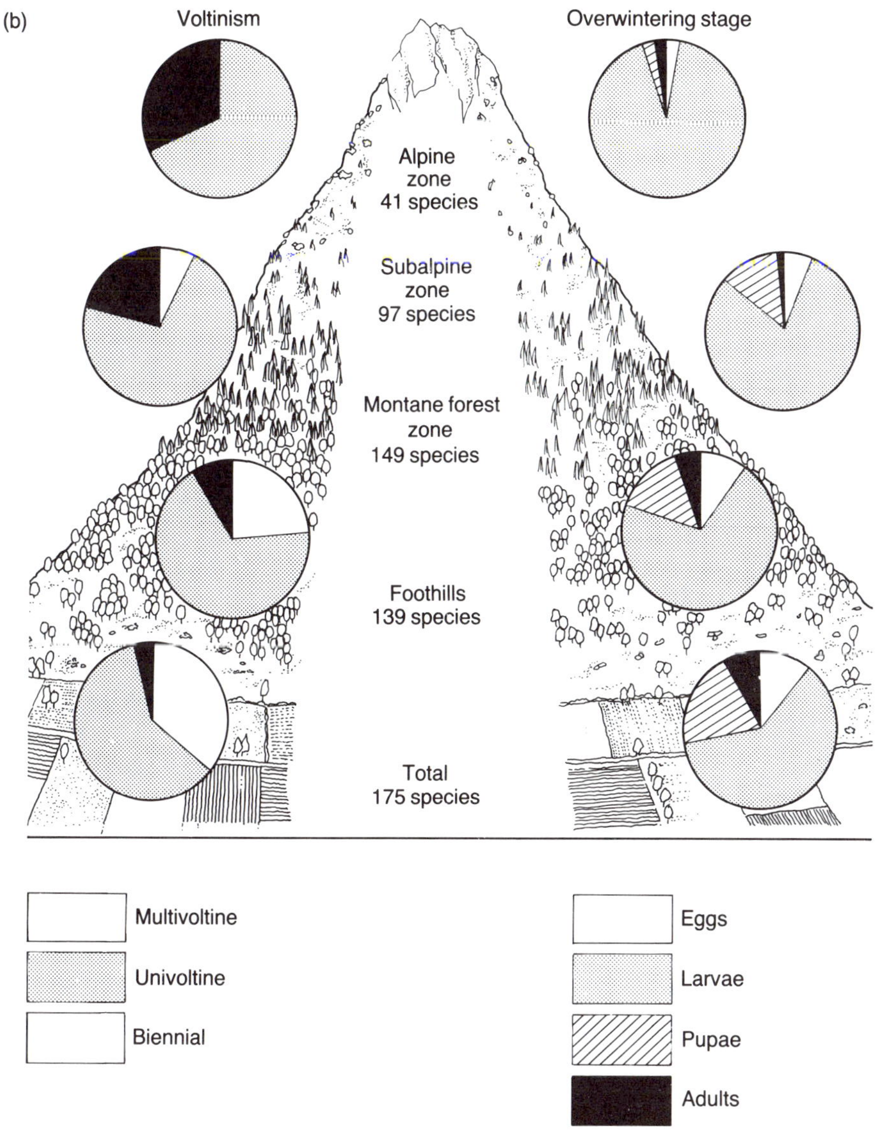

2.2 Spatial patterns in voltinism and overwintering stages with latitude and altitude. (a) Resident butterflies in Britain (100 km grid divisions). Bivoltine species in northern Britain are sparsely distributed and are generally maintained by migrations (e.g., *Pieris brassicae*, *P. rapae*, *P. napi*). Typical migrants to southern Britain (i.e., *Vanessa atalanta*, *Cynthia cardui*) overwinter as adults and do so more successfully in southern Britain than in northern Britain; Baker, 1978). (b) Altitudinal zonation of butterfly voltinism and overwintering stages in Switzerland. (From Geiger, 1987; by courtesy of Ligue Suisse pour la Protection de la Nature, Basle.)

occasionally of 'false broods' of migrating ephemeral adults from lower levels as at the Donner Pass in the Sierra Nevada, USA (Shapiro, 1977).

The general timing of temperate univoltine adult flight periods is not random within the span of the growing season (Shapiro, 1975a); it depends largely on the stage in which species overwinter (viz., eclosion order: adults > pupae > larvae > eggs) and larval growth rates as affected by host phenology and quality (viz., eclosion order: inflorescence feeders > herb leaf feeders > shrub–tree feeders > grass–sedge feeders) (Scott & Epstein, 1986). Moreover, multivoltine butterflies less commonly overwinter as eggs or young larvae. Although emergence may be staggered in relation to latitudinal differences in temperatures and thus to development time, as in *Maniola jurtina* and *Pyronia tithonus* (Brakefield, 1987a), it can be synchronized or the pattern reversed (e.g., *Ochlodes venata*, Heath *et al.*, 1984), particularly when different hostplants are used regionally. The precise timing of eclosion dates can implicate trade-offs between direct weather events and indirect influences on resources sought by adults (Shapiro, 1979). Although in the tropics more particulate seasonal cycles exist in the drier savanna than in the rain forest, seasonality in precipitation results in emergence patterns associated with the transitional wet–dry and dry–wet periods (Emmel & Leck, 1969; Owen, 1971) as well as the main seasons. The disadvantages of the rainy season are understated; heavy cloud and persistent downpours affect flight, nectar sources, dispersal, courtship and egglaying, and may induce 'pluviation' (e.g., *Graphium policenes*; Owen, 1971). The duration of flight periods is greatest for the tropics especially for species with virtually continuous aseasonal life-cycles (e.g., *Charaxes spp*., Owen & Chanter, 1972; *Heliconius ethilla*, Ehrlich & Gilbert, 1973), but the prospect of seasonality remains even in the absence of diapause, as in flesh flies (Denlinger *et al.*, 1991). At temperate latitudes, flight period can be influenced by multivoltinism, especially the occurrence of partial second broods (e.g., *Papilio glaucus*; Slansky, 1974; *Polygonia c-album*, Nylin, 1989), aestivation (e.g., *Maniola jurtina*, Scali, 1971) and polymodality (Waldbauer, 1978; Hagen & Lederhouse, 1985; Shreeve, 1989). The generally U-shaped flight duration curve with season for species at one temperate locality has been suggested to relate to diversity in development rates associated with microhabitats for low sun angles (Scott & Epstein, 1986); but it may also result from trade-offs in timing between extreme weather events and synchronization with resources for which there is greater variation in the spring and autumn. The direct influence of insolation and temperatures on growth rate and development have been implicated in *Aglais urticae* polymodality (Fig. 2.3; Dennis, 1985), but neither this nor variation in microhabitat use, thus microclimate, seem to explain the extended flight period of *Maniola jurtina* on British chalk downland (Brakefield, 1987a; Shreeve, 1989).

The onset and termination of activity periods in insects is usually

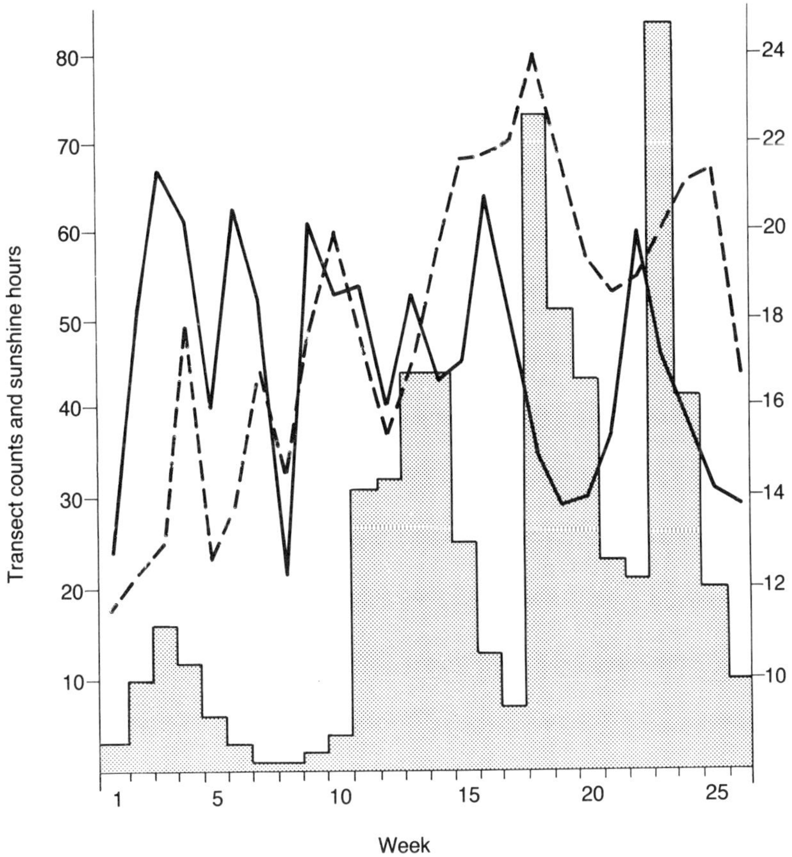

2.3 Voltinism and polymodality within broods for *Aglais urticae* (Nymphalidae) near Swanage in Dorset, UK during 1982. Numbers of butterflies (histograms) are mean weekly counts from fixed transects; temperatures (pecked lines) are mean weekly maxima, whereas sunshine hours (continuous lines) are cumulative weekly totals. Two summer broods occurred in 1982. The second brood, during weeks 17 to 25, is divided into two peaks. The first peak has probably been produced by pupae (from early egg batches) which have developed rapidly in the high temperatures of weeks 15 to 18. The second peak may then represent the arrested development of individuals (from later egg batches) first as larvae, whose thermoregulation may have been affected by low sunshine levels in weeks 18 to 20, and then as pupae affected by lower temperatures in weeks 20 to 22. Polymodality in the second brood was evident as far to the north as Cheshire during 1982. (After Dennis, 1985; by courtesy of the British Entomological and Natural History Society.)

determined by endogenous rhythms (biological 'clocks' and 'counters') that are adjusted by periodic environmental regulators (zeitgebers) (Saunders, 1981, 1982); the neuro-endocrine mechanisms that initiate, maintain and terminate diapause are suggested to be similar in all geographical regions (Denlinger, 1986). In contrast, the token stimuli (e.g., light, temperature, moisture etc.) are not. Most insects in temperate environments use photoperiod progression pattern as a cue for hibernal and aestival diapause (Beck, 1980; Tauber *et al.*, 1986); seasonal changes in daylength and light intensity are large and highly correlated with seasonal changes in temperatures, precipitation, food supply, enemies and other factors affecting development and survival. Termination of seasonal activity by most species is conservative; implying that fitness is adjusted to the variance as well as to means in meso-scale climatology (Levins, 1968; Shapiro, 1975a). In the arctic, spectral composition may replace light intensity as the main zeitgeber (Nordtug & Melo, 1988). As equatorial seasonal patterns of activity are greatly more varied, in many insects photoperiod is probably replaced by a menu of cues tailored to foretell the advent of specific seasons (Denlinger, 1986) particularly temperature (Cantello, 1974; Jacquemard, 1976; Sims & Shapiro, 1983) and, at least involved in terminating diapause, precipitation (see Denlinger, 1986 for references). *Lepidochrysops* (Lycaenidae) emergence from pupa occurs only after the grass has been burnt; the hostplant *Becium* (Labiatae) remains dormant unless there is a fire (Carcasson, 1964). However, photoperiod is still a very important zeitgeber in some tropical insects (Tanaka *et al.*, 1987; Pullin & Knight, 1992). Although thermoperiod is not the major token stimulus among temperate and arctic insects (Tauber *et al.*, 1986), temperature influences growth rates and development times (Wolda, 1988) and therefore exerts a strong effect on the proportion of individuals completing their photosensitive stages before or after the occurrence of diapause-inducing photoperiods (e.g., *Aglais urticae*; Danilevskii, 1965; Beck, 1980; Saunders, 1982; Fig. 2.4), even in species (e.g., *Pararge aegeria*) in which growth rate is finely tuned to photoperiod (Nylin *et al.*, 1989). Partial broods, implicating trade-offs between current and future reproduction, are produced as in Californian *Battus philenor*, whose *Aristolochia* hostplant depends critically on seasonally variable precipitation (Shapiro, 1975a). Seasonality in activity patterns has also been successfully modelled on temperatures using minimum development threshold temperatures T_0min and day-degrees (Malcolm *et al.*, 1987; Higley *et al.*, 1986).

Insects have evolved numerous ways to exploit environmental stimuli as cues for development, morphology and behaviour (Danilevskii, 1965; Lees, 1968; Beck, 1980; Saunders, 1982). Diapause is a multifaceted phenomenon and different aspects of it (i.e., initiation, intensification, maintenance, termination, post-diapause development) as well as accompanying adaptations (e.g., cold and heat hardiness; seasonal polyphenisms; dormancy) may have the same or different triggers. Photoperiod and temperature (or other) reactions of butterflies can be both diverse and dynamic, and temperature may interact with photoperiod to enhance or arrest diapause effects (Tauber *et al.*, 1986; Fig. 2.4). Although the significance of findings and the latent potential revealed by experimental conditions applying only stationary and nonsense photoperiods (e.g., continuous light) is still unclear (cf., Beck's, 1980 critical photoperiod response types *versus* Tauber *et al.*, 1986 types based on changing light regimes), reactions to long, short or changing photophases have logical adaptive bases (Pullin, 1986a; Nylin *et al.*, 1989; Fig. 2.4). Regulation of adult reproductive diapause and seasonal polyphenism in Swedish *Polygonia c-album* is controlled by changing daylengths (Nylin, 1989). Multiple responses to critical photoperiods may also be adaptive. For example, Beck's type III response photoperiod, which shows two well-defined critical daylengths, is argued to have a selective advantage in climates with relatively benign winters; it allows continuous development for offspring which result from autumnal oviposition, as in Californian *Pieris rapae* and *Papilio zelicaon*, rather than carry-over pupal diapause with potentially higher mortality (Shapiro, 1984a). Butterflies vary enormously in their phenotypic plasticity and sensitivity to environmental cues. Some species are relatively more refractory ('obligative') to different stimuli (e.g., *Battus philenor*, Sims & Shapiro, 1983; *Lasiommata megera*, Wickman *et al.*, 1990) than others ('facultative': *Coenonympha pamphilus*, Lees, 1962b; Wickman *et al.*, 1990). Many species show geographical differences in environmental regimes which induce or inhibit diapause (Fig. 2.5), the basis for which is intra-population genetic variance in sensitivity to token stimuli (see Pullin, 1986a). The sensitivity and developmental and phenotypic plasticity of diapause and polyphenism *sensu* Tauber *et al.* (1986) are probably determined by polygene modifiers for the threshold of expression of conservative physiological mechanisms under a major gene (Masaki, 1978; Roff, 1980; Shapiro, 1980a, b, c; Tauber *et al.*, 1986). Polyphenism as opposed to polymorphism allows rapid adjustment to predictable environmental cycles. For example, two levels of response, controlling diapause or continuous development and variable rates of growth and development, enable *Pararge aegeria* to follow one of four major developmental pathways, to aestivate as well as hibernate and to overwinter as a pupa or third instar larva (Nylin *et al.*, 1989). Such fine-tuning permits rapid adjustment either to changing environments or during colonization, as in the case of *Pieris rapae* in North America (Shapiro, 1980a).

Diapause-dependency increases and variability in diapause response decreases with increasing latitude as abiotic adversity intensifies. A similar pattern occurs with altitude, especially at cool temperate latitudes (Lees, 1970), but may be reversed in dry climates where heat stress is reduced and moisture increases with elevation (e.g., *Maniola jurtina* in Tuscany, Scali, 1971). Although more tropical species lack diapause (Shapiro, 1978; Owen & Chanter, 1972; Scott, 1979), it occurs at all latitudes and potentially

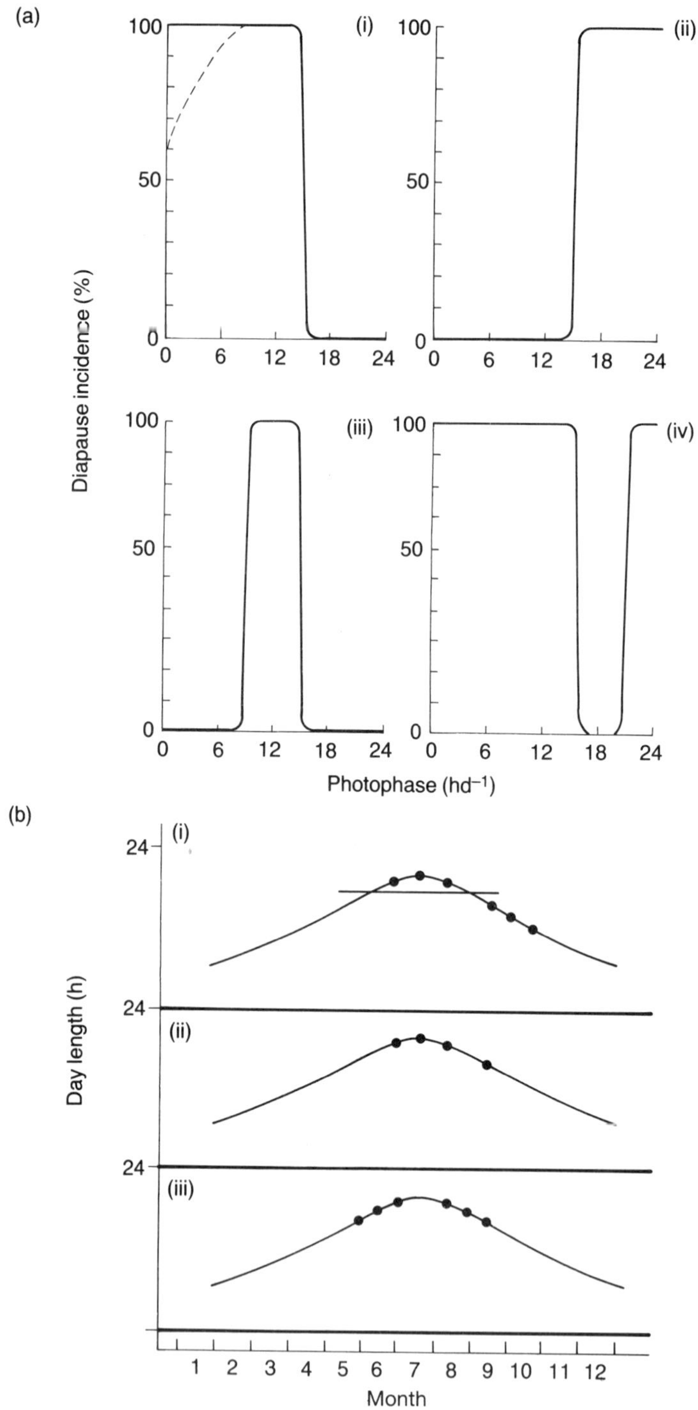
(a)
(i)
(ii)
(iii)
(iv)
100
50
0
0 6 12 18 24
Diapause incidence (%)
Photophase (hd⁻¹)
(b)
(i)
(ii)
(iii)
24
Day length (h)
1 2 3 4 5 6 7 8 9 10 11 12
Month

2.4 Influence of photoperiod on diapause-induction in insects. (a) Diapause-induction curves for fixed photoperiods. (i)–(iv), types I to IV: continuous growth and development tends to be induced in type I by relatively long daylengths, in type II by relatively short daylengths, in type III by very short or very long daylengths and in type IV by a very restricted range of relatively long daylengths. Types III and IV have two well-defined critical daylengths. In type I, low T_as tend to cause the response curve to be shifted to the right and high T_as to the left. The effect of T_a on type II response is usually opposite to its effect in type I. In type III, low T_as tend to broaden the range of diapause-inducing photophases, and high T_as tend to reduce this range. The effect of T_as on type IV response patterns has not been determined. Whereas type I induction curves are typical of a great many species (e.g., *Aglais urticae*), type II (e.g., possibly some Mediterranean species), type III (e.g., *Pieris brassicae*) and type IV are less widespread. (From Beck, 1980 by courtesy of Academic Press). (b) Hypothetical circumstances which favour responses to changing as opposed to fixed photoperiods. With the exception of the summer solstice, every absolute value of daylength will be repeated, before and after summer solstice. The curves show changes in daylength over the year. The positions of the curves marked by dots indicate those photoperiods experienced by a developmental stage which is sensitive to photoperiod. (i) case 1: the sensitive stage of the first generation occurs at summer solstice but since the second generation is complete, direct development may be induced by any critical daylength which is passed between generations (i.e., horizontal line); (ii) case 2: the second generation is partial and only the earliest members of the first generation should ideally develop directly and contribute to it; (iii) case 3: the sensitive stages of the first and second generations occur at similar daylengths. In the latter two cases, only a system for photoperiodic regulation which is sensitive to the direction of daylength will be able to distinguish photoperiods before and after summer solstice (e.g., voltinism and polyphenic morphs in *Polygonia c-album*). (From Nylin, 1989, by courtesy of Blackwell Scientific Publications Ltd.)

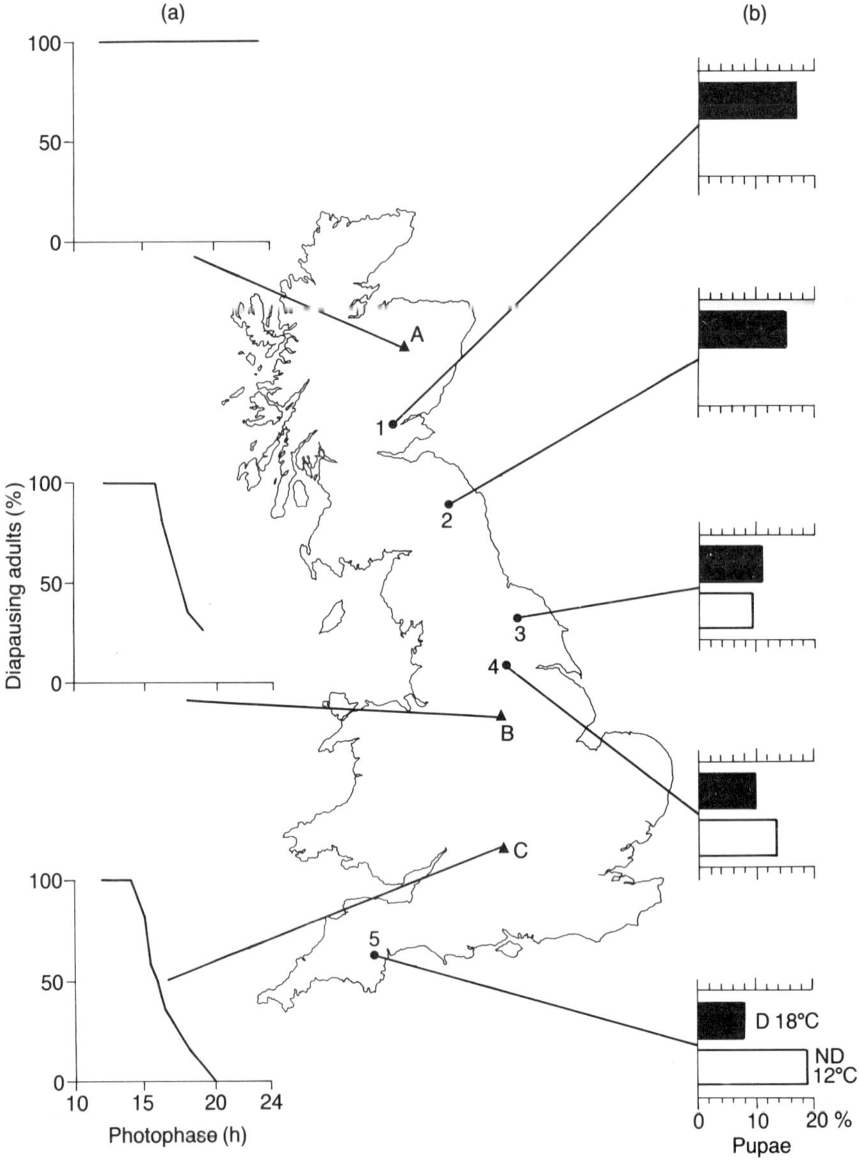

2.5 The influence of photoperiod and temperatures on diapause-induction of two British butterflies. (a) Photophase induction of diapause in *Aglais urticae* for a constant temperature of 20°–24°C: A, Braemar; B, Chesterfield; C, Oxford. (Adapted from Pullin, 1986c; courtesy of A.S. Pullin.) (b) Number of diapause pupae at 18°C and non-diapause pupae at 12°C produced among progeny in *Pieris napi* for an 18 h photophase: 1, Dollar; 2, Bellingham; 3, Strensall; 4, Menston; 5, Teignmouth. (Adapted from Lees & Archer, 1980.)

during and across all seasons (Denlinger, 1986). Compared to hibernal diapausing insects, aestivating and tropical diapausing species need to reduce metabolic requirements without the aid of low T_a; moreover, high temperatures promote water loss and permit continued high predation by invertebrates and parasitoids. Many adaptations are similar for all diapausing regimes: (i) fixed life-history stage dependency; (ii) reduced metabolic syndrome culminating in dormancy; (iii) biochemical and physiological adjustments to climatic extremes; (iv) morphological and colour modifications; (v) changes in behaviour and microhabitat, and (vi) migration.

Whereas non-diapausing species may be found developing and 'overwintering' in any life-history stage (e.g., *Ascia monuste*, Scott, 1979), there is distinctive taxonomic patterning for diapause (Table 2.2). Very few species can overwinter (e.g., *Pararge aegeria*, Shreeve, 1986b) and aestivate (e.g., *Papilio machaon*, Wiltshire, 1958) in more than one stage. The frequency of diapausing stages and of mobile and immobile phases among insects is similar in the tropics (7 per cent eggs; 32 per cent larvae; 30 per cent pupae; 31 per cent adults; N = 73) and temperate zone (14 per cent eggs; 32 per cent larvae; 21 per cent pupae; 30 per cent adults; N = 317) (Saunders, 1982), but among butterflies bias to adults and against ova increases for aestivation and tropical diapause compared to hibernation (e.g., *Hipparchia parisatis*, Oman; Larsen & Larsen, 1980; Masaki, 1980). High temperatures permit adult activity for refuelling and predator-escape in warmer climates, whereas eggs, with a high surface area : volume ratio, may desiccate without the means to replenish water losses. To some extent butterfly diapause stage may be adapted to hostplant phenology and distributions (Scott, 1979). Diapausing pupae and adults are not so dependent on the proximity of hostplant resources once diapause is terminated. Butterflies which overwinter as eggs typically have perennial hostplants and lay on woody hostplant parts or on non-hostplant substrates (Wiklund, 1984). Proportionately more species overwinter as larvae at high latitudes; larval overwintering is obligate in species with life-cycles exceeding one year, which need to compensate for the limited growing season (Danks, 1978; Shreeve, T.G., pers. comm.).

Apart from avoidance (e.g., microhabitat use, migration), two mechanisms have evolved in insects to withstand cold (i) overwintering cold hardiness, and (ii) rapid cold hardening (Table 2.3). The former divides species into those which are freeze-intolerant,* which die if they fall below the supercooling point (T_c) and freeze, and those which are freeze-tolerant, able to survive extensive extra-cellular ice formation within body organs and tissues (Lee, 1989; Block, 1990; Storey, 1990; Duman *et al.*, 1991; Lee & Denlinger, 1991), though some insects have the capacity for engaging both (Kukal &

* Freeze-intolerant insects die if they freeze but do not necessarily survive if they do not freeze, so survival is not always dependent on the T_c (Pullin, A.S., pers. comm.).

Table 2.2 Hibernal diapause stages (number of species) of North American and European butterflies (Data from Scott, 1979; Chinery, 1989)

	North America				*Europe*					
	Egg	*Larva*	*Pupa*	*Adult*	*Egg*	*Larva in egg*	*Larva*	*Pupa*	*Adult*	*In > 1 stage*
Papilionidae	3	2^-	18^+	0	0	3	3^-	7^+	0	3
Pieridae	1^-	10^-	13^+	1	1^-	0	14^-	22^+	3	5
Lycaenidae	30^+	18^-	31^+	0^-	12^-	3	44^-	10	1^-	7
Nymphalidae	0^-	79^+	1^-	14^+	1^-	2	116^+	4	11^+	5
Hesperiidae	3^-	59^+	7^-	0^-	2	1	19^+	1^-	0	1

The frequencies for overwintering stages difer significantly for the higher taxa; for larvae and pupae only: North America, $\chi^2_{(4)} = 112.1$, $p < 0.001$; Europe, $\chi^2_{(3)} = 75.8$, $p < 0.001$. For the European fauna, the figures include the multiple classification of species believed to overwinter in more than one stage; data are unavailable for 94 European species. Direction in bias for a null hypothesis of homogeneity among taxa: $^+$, more than expected; $^-$, less than expected. Bias for overwintering stage is more evident for lower taxa (e.g., egg: Theclinae, Lycaenidae; larva: Satyrinae, Nymphalidae; pupa: Papilioninae, Papilionidae; adult: Nymphalinae, Nymphalidae).

Table 2.3 Summary of characteristics associated with cold-hardening for winter and the rapid cold-hardening response that protects against injury due to cold shock (From Lee, 1989; © 1989 The American Institute of Biological Sciences; by permission)

Characteristics	*Winter cold-hardening*	*Rapid cold-hardening*
Differences		
Cold tolerance	Long-term freeze tolerance and/or supercooling	Short-term prevention of non-freezing injury
Stage	Only in overwintering stage	Present in more than one developmental stage (larva, pupa, adult)
Activity	Inactive (diapause, quiescence)	Active feeding and reproduction (sometimes diapause)
Timing	Seasonal	All year
Rate of cold-hardening	Slow (weeks-months)	Rapid (minutes–hours)
Similarities		
Cryoprotectant	Present	Present
Induction trigger	Low temperature	Low temperature

Duman, 1989; Fig. 2.6). In freeze-intolerant species, T_c often lies within the range of – 15° to – 35°C (Somme, 1982; Block, 1982; Bale, 1987; Jackson *et al.*, 1990); some insects in Alaska supercool to – 60°C (Miller, 1982). T_c min for the few diapause butterflies tested to date is not so low (e.g., *Pieris brassicae* pupae: – 23 to – 25°C; Pullin *et al.*, 1991; *Inachis io* and *Aglais urticae* adults: – 20° to – 24°C; Pullin & Bale 1989a, c), and arctic species must have much lower values. Species which overwinter high above the soil surface can withstand very low temperatures of – 32°C (e.g., Finnish *Papilio machaon*), but the overwintering stages of many high latitude butterflies may be insulated from the most extreme conditions by snow cover (Ekholm, 1975). However, non-diapause butterflies may have little tolerance of cold conditions; prolonged chilling at 4°C is fatal for summer *P. brassicae* (Gardiner, 1980). Freeze-tolerance, not identified in butterflies as yet, refers to the specific temperature range in which tissue freezing is tolerated, which can be as low as – 70°C (Miller, 1982; Storey & Storey, 1988); cooling below it may result in death from excessive cellular dehydration (Mazur, 1984). Both strategies use cryoprotectants or polyols (e.g., glycerol, sorbitol, trehalose etc.); some species also produce thermal

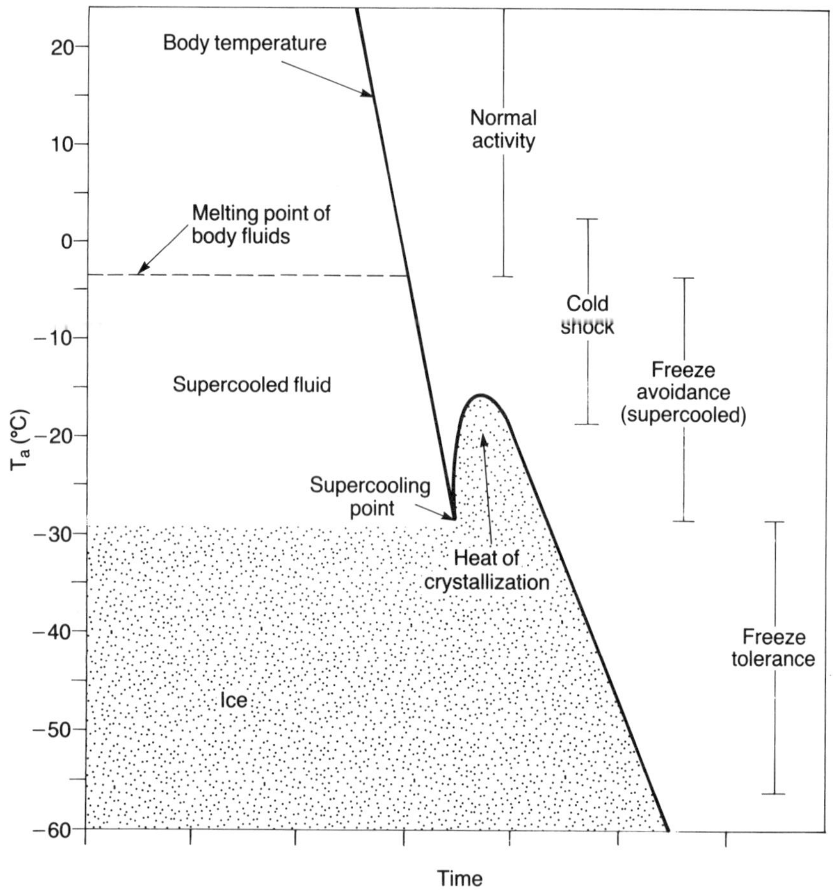

2.6 Generalized response of an insect cooled to subzero temperatures. Insect body temperatures (heavy line) in relation to the melting point, the supercooling point, and the nucleation of ice in body fluids. Although the bars on the right convey general ranges of insect response to low temperatures, the top of the bar for the range of freeze tolerance and the bottom of the bar for freeze avoidance correspond to the supercooling point value illustrated in the centre of the figure. However, a number of freeze-tolerant insects have supercooling points in the range of – 8° to – 10°C, whereas some freeze-susceptible species supercool extensively, to – 60°C or below. (After Lee, 1989; © American Institute of Biological Sciences: by permission.)

hysteresis proteins (Duman & Horwath, 1983). In cold winter hardening, the acclimation process is anticipatory, but in some insects short-term chilling (i.e., at 0°C for 20 min) before cold shocks (i.e., – 10°C) greatly enhances survival (Chen *et al.*, 1987; Lee *et al.*, 1987) and provides protection from low temperatures above T_c for any stage when insects are active. This process needs to be studied in butterfly species and the costs accounted (see Shapiro, 1982a).

Ability to attain $T_c$min is the only protection that most hibernating species possess; it can be influenced by conditions during acclimation and polyol accumulation (Zachariassan, 1985; Pullin & Bale, 1989c), the duration of diapause (Salt, 1966a; Somme, 1982; Pullin & Bale, 1989a, c), the presence of nucleating agents (i.e., from feeding, meconial fluids; Salt, 1966a, 1968; Pullin & Bale, 1989a); moisture content (Salt, 1956), inoculative freezing from surface water (Salt, 1956; Pullin & Bale, 1989a) and cooling rate (Salt, 1966b). The lowest temperature regimes may not produce the lowest T_c (Pullin & Bale, 1989a). Too much reliance can be placed on T_c for winter survival; cold can lead to injury and death in the absence of freezing (Bale, 1987) for a variety of causes (Prosser, 1973) despite mechanisms for cold-hardening. The latter, together with potential interactions among triggers and the overwintering acclimation process, which is complex in nymphalid butterflies, may account for much variation in cold responses (Baust, 1982; Pullin & Bale, 1989a, b). Once supercooling has been attained, mortality will depend on $T_c$min and winter cold extremes and duration, body weight, and the microclimate of habitats (Bale, 1987; Tenow & Nilssen, 1990).

Diapause heat tolerance is unlikely to be so critical; tropical dry and wet seasons can be equally hot and diapause may relate more to 'die-back' and resource limitations than inability to be active. The use of cool microhabitats may generally provide adequate refuge from heat but specialized adaptations are often required to reduce water loss and depletion of fat reserves for both active and inactive phases of the life-cycle (e.g., special structures such as oral and anal 'plugs' found in diapausing *Euphydryas* spp.; see Sbordoni & Forestiero, 1984: 238). Nevertheless, butterflies have upper limits of tolerance in both active and inactive stages. Many species cease normal activities when temperatures (T_{sc}) exceed 30°C and roost in cool microhabitats (see section 1.2B). In *Pieris brassicae*, eggs laid are infertile when $T_a > 33$°C, a temperature which is also lethal for larvae (Gardiner, 1988).

Response to divergent seasonal conditions involves a co-ordinated system of adaptations including morphology, behaviour and life-history traits. The system is highly flexible as indicated by: (i) the extent of developmental plasticity (polyphenism) under endocrine direction as in diapause (Shapiro, 1976; Endo, 1972; Endo & Funatsu, 1985; Endo & Kamata, 1985); (ii) alternatives in predictive and interdictive triggers (Shapiro, 1976, 1982b; Tauber *et al.*, 1986) which are varied and usually seasonal (e.g., light,

Shapiro, 1976; T_a, McLeod, 1968, 1984; moisture, Mell, 1931; Owen, 1971), but which can also be aseasonal (e.g., background colour, texture and chemical cues in pupae, West *et al.*, 1972; Hazel & West, 1979; West & Hazel, 1985) yet indirectly influenced by seasonal conditions; (iii) the degree of interaction for token stimuli (Jones *et al.*, 1985; Rienks, 1985); (iv) varying interdependence or independence of the various adaptations (e.g., morphology, diapause, migration etc.) and their control by environmental cues which can be tightly coupled (e.g., morphology with diapause in *Polygonia c-album*, Nylin, 1989) or relatively uncoupled (e.g., *Nathalis iole*, Douglas & Grula, 1978); (v) the variety of adaptations and degree of manifestation in different life-history stages (see below); and (vi) the varying degree of genetic assimilation and canalization of phenotypes which allows varied responses in transitional or extreme conditions (e.g., *Melanitis leda*, Brakefield, 1987b). Adverse seasons can result in dormancy or reductions in activity, development, growth and maturation. Increased weight associated with increased lipids and fat bodies is typical of diapause stages (Scali, 1971; Gilbert & Singer, 1975; Denlinger, 1986; Pullin & Bale, 1989a). Seasonal polymorphism in morphology occurs for a wide range of attributes in eggs (e.g., egg size in *Parnara guttata*, Nakasuji & Kimura, 1984), larvae (e.g., colour in *Plebejus icarioides*; integumen thickness in *Euphydryas editha*; instar number in *Asterocampa* spp., Scott, 1979), pupae (e.g., colour in *Papilio polyxenes*; Hazel & West, 1983) and adults (e.g., wing size, shape, pattern and colour in *Precis octavia* and *Melanitis leda*; Brakefield & Larsen, 1984; Brakefield, 1987b). Some degree of seasonal polyphenism in adult morphology is probably characteristic of most multivoltine species in seasonal climates and is even detectable in dull, dark *Erynnis* spp. (Burns, 1964; see section 3.4 for significance).

Behaviour changes often accompany seasonal polymorphism for morphology, especially where seasonal contrasts are marked. During adverse seasons behaviour is geared to escape *in situ* (e.g., adoption of shade in *Maniola jurtina*, Scali, 1971; anachoresis in *Melanitis leda* and *Precis octavia*, Brakefield & Larsen, 1984; adult aggregations in humid forest in *Smyrna karwinskii*, *Euploea* spp. and *Eurema daira*, Muyshondt & Muyshondt, 1974, Monteith, 1982, Janzen, 1983; changes in resting posture and background in *Junonia almana*, Wynther-Blyth, 1957) or ex-habitat (e.g., migration of *Gonepteryx rhamni* between hibernating and oviposition sites in Lebanon, Larsen, 1974). Some movement is associated with seasonal environments though this may often be small (Tauber *et al.*, 1986). Movements can be coupled to diapause or not (e.g., *Danaus plexippus*, Malcolm *et al.*, 1987; *Parnara guttata*, Nakasuji & Kimura, 1984) and in many cases apparently aseasonal 'pied-piper' one-way flights occur altitudinally (Shapiro, 1973) and latitudinally (Dingle, 1982; Baker, 1984; see section 4.3C). Although Tauber *et al.* (1986) argue against interpreting adult seasonal polyphenism for morphology, dormancy and migration as alternative adaptations to adversity, they are often only partly coupled or uncoupled in species, in some regions dormancy and seasonal polyphenism *sensu lat.* affect different stages in different seasons and migrant species are often not polyphenic (Shapiro, 1976). What is known of geographical disparity in seasonal adaptations suggests that as seasonal adversity increases in magnitude and duration (drought) in continuously warm environments, dormancy and then migration will dominate responses (see Larsen, 1984a: 91, b). In climates with cold seasons, dormancy may be paramount.

2.2 **Atmospheric influences on butterfly populations**

The ability of populations to persist relates to their mean size (viz., available resources), isolation (viz., movements) and fluctuations in size (Andrewartha & Birch, 1954; MacArthur & Wilson, 1967). Agents that cause fluctuations are particularly important inasmuch as they may also influence movements and thus isolation as well as affecting the resource base, thus long-term and short-term population trends. Climatic agents may influence population fluctuations directly or indirectly, either via other agents (e.g., predators, parasites) or resources (e.g., quality of hostplants). Attempts to understand fluctuations have employed two basic methods: (i) by association using synoptic recording schemes and (ii) by causation applying autecological survey, although the two can be and have been dovetailed into occasional research programmes (Pollard, 1979; Warren *et al.*, 1986).

A *Weather and population fluctuations: trends and associations*

The butterfly monitoring scheme (BMS, UK) is based on transect counts for 26 weeks in each year from 1 April to 1 October at some 80 sites through Britain from 1976 (see Pollard, 1977; Hall, 1981 for method). The scheme allows investigation of national and regional trends and changes at individual sites where transects are divided into sections based on habitat (Pollard, 1982a). Relative abundance of species at sites is given as an index (I, summation of mean weekly counts) and individual site index values are collated to give 'regional' and 'national' trends (Pollard & Lakhani, 1985). Transect counts and index values have been shown to correspond closely with changes in population size whereas variation and direct weather effects, above the criteria set for transects, do not invalidate counts as measures of population change (Pollard *et al.*, 1986; Pollard, 1984; 1988). The BMS data have revealed several important trends: (i) intra- and inter-covariation in population fluctuations regionally and nationally in which groups of species behave similarly ($r_P = 0.41$ for 31 species), contrasting with the asynchronous and disparate population changes in Heterocera where high and low population zones change erratically (Taylor & Taylor, 1977); (ii) interspecies differences in the magnitude of annual fluctuations, migrant (e.g., *Vanessa atalanta, Cynthia cardui*) and multivoltine species (e.g., *Celastrina argiolus*) being substantially more variable than relatively sedentary and univoltine species (e.g., *Melanargia galathea, Maniola jurtina* and *Anthocharis cardamines*) (Pollard, 1984); (iii) consistency in size of counts between sites relating to transect characteristics and available resources, regardless of fluctuations and exceptions which are accounted in site habitat changes and management (Pollard *et al.*, 1986); (iv) larger fluctuations in counts for northern than southern sites (e.g., *Maniola jurtina*, Pollard *et al.*, 1986); (v) evidence, albeit indirect, for the correspondence of colonization events with population increases in some species (e.g., *Polygonia c-album* and *Celastrina argiolus*; Pollard, 1984; Pollard & Yates, 1992) but not all species (e.g., *Pyronia tithonus*; Pollard, 1991a) and (vi) correlations between counts of successive broods (e.g., *Pieris brassicae*, within year $r_P = 0.83$; between years $r_P = 0.64$).

Covariation in population fluctuations between sites and species implicate weather as a cause. Pollard (1988) has applied a partial regression model to 31 species for 11 years' data:

$$\log Y = a + b_0 \log X + b_1 w \qquad \text{(Pollard \& Lakhani, 1985)}$$

where Y, the current index for a species, is assumed to depend on the previous index X and a weather variable w; b_1 is the partial regression coefficient. Temperature and precipitation have been examined (single monthly and three-monthly pooled data) in a series of equations for months in the current and previous years. Except for single monthly rainfall figures, more significant associations between weather and butterfly abundance emerged than expected by chance ($p < 0.05$); also more significant associations occurred for temperature than rainfall, in current rather than previous years for temperature, but previous rather than current years for rainfall (Table 2.4). Of paramount significance, many species (13 of 30 reported) showed increased index values in warm summers. The flight periods were predominantly in July and August, while the majority of significant results, using simple monthly means, were in June, suggesting that the largest effect of temperature may be on late larval and pupal development (Pollard, 1979). Some other species showed either a positive association (e.g., spring-flying *Pyrgus malvae, Erynnis tages* and *Gonepteryx rhamni*) or a negative association (e.g., *Aphantopus hyperantus, Pararge aegeria*) with temperatures in previous years. Positive associations evoke potential effects on egglaying or early larval survival for univoltine vernal species but possibly on nectar feeding for multivoltine adult hibernators. Negative associations imply an interaction between temperatures and rainfall, inasmuch as warm summers tend to be dry ones and the two species so effected occur in moist, sheltered habitats. Significant associations with rainfall were exclusively negative (high precipitation with reduced I) in the current year and generally negative in the latter part of the

Table 2.4 Significant pcsitive and negative associations between butterfly index values from the Butterfly Monitoring Scheme and (a) temperature and (b) rainfall, using multiple regression

	Previous year												*Current year*							
(a)	*Jan.*	*Feb.*	*Mar.*	*Apr.*	*May*	*Jun.*	*Jul.*	*Aug.*	*Sep.*	*Oct.*	*Nov.*	*Dec.*	*Jan.*	*Feb.*	*Mar.*	*Apr.*	*May*	*Jun.*	*Jul.*	*Aug.*
(i) *Thymelicus sylvestris*																		+	+	
Erynnis tages						+	+													
Pyrgus malvae							+													
Anthocharis cardamines				–							+									
Lysandra coridon									+					–	–					
Ladoga camilla														–						
Argynnis paphia													–							
Melanargia galathea																		+		
Hipparchia semele		+	+						–										+	
Pyronia tithonus																		+	+	
Maniola jurtina																		+	+	
Aphantopus hyperantus					–	–	–													
(ii) *Pieris brassicae*													–	–	–					
Pieris rapae																–				
Lycaena phlaeas																		+	+	
Polyommatus icarus																		+	+	
Aricia agestis																		+	+	
Lasiommata megera									–									+	+	
(iii) *Gonepteryx rhamni*							+	+												
(iv) *Vanessa atalanta*																	+	+		
Polygonia c-album						–											+	+		
Pararge aegeria						–	–											+		
Coenonympha pamphilus																		+	+	

(b)		Previous year												Current year							
		Jan.	Feb.	Mar.	Apr.	May	Jun.	Jul.	Aug.	Sep.	Oct.	Nov.	Dec.	Jan.	Feb.	Mar.	Apr.	May	Jun.	Jul.	Aug.
(i)	*Thymelicus sylvestris*											–								–	
	Ochlodes venata					+	+										–				
	Erynnis tages							–													
	Pyrgus malvae						–	–													
	Anthocharis cardamines								–												
	Callophrys rubi				+										–	–					
	Lysandra coridon												+								
	Ladoga camilla															–					
	Argynnis aglaja						–	–								–		–			
	Argynnis paphia										–										
	Melanargia galathea				+											–					
	Hipparchia semele																			–	
	Pyronia tithonus																			–	
	Maniola jurtina											–								–	
	Aphantopus hyperantus		+	+		+															
(ii)	*Pieris napi*			+																	
	Polyommatus icarus																			–	
	Celastrina argiolus							–													
	Lasiommata megera			+			+	–													
(iii)	*Gonepteryx rhamni*									+											
	Inachis io														–						
(iv)	*Aglais urticae*															–					
	Polygonia c-album			+											–						
	Pararge aegeria			+		+	+														
	Coenonympha pamphilus					+															

(i) species with one flight period each year; (ii) species with two flight periods representing different generations; only the second generation is used in this analysis; (iii) species with two flight periods, but only one generation; adult overwintering; (iv) species with two or more flight periods, but only a single annual index is given because flight periods are not readily separated. Significance level, $p < 0.05$. Temperature and rainfall data are the pooled means of three successive months. The first and last month's (January and August) data were thus used only in formulating means for adjoining months (From Pollard, 1988 by courtesy of the *Journal of Applied Ecology* and the British Ecological Society previous year. But, in the early part of the previous year, nearly all associations were positive, especially those for *A. hyperantus* and *P. aegeria*. The effect of rain may be a balance between indirect beneficial effects on plant growth and direct harmful effects on butterflies (Pollard, 1988). The drought of 1976 caused sharp declines in I for several species (Pollard *et al.*, 1986). The BMS data confirm two of Beirne's (1955) findings from a qualitative assessment of relationships between weather and abundance of British Lepidoptera 1864 to 1952: the beneficial influence of warm dry summers and the detrimental impact of drought. However, there is no evidence for the beneficial effects of 'severe cold in winter'. In Finland, Ekholm (1975) demonstrated that the annual number of 'frequent' species increased in dry warm summers preceded by moist summers. Pollard stresses the need for caution in the interpretation of results based only on 11 years' data; both chance occurrences and some extreme events may influence associations and more complex non-linear relations may occur. Ideally, this invaluable research programme needs to be extended over the continent and further into non-reserve sites, so that spatial differentiation in population fluctuations can be examined in relation to regional climates and human agencies. Generalizations based on findings in small regions (UK) cannot be applied over wider areas; indeed reversals may be expected, as in central Asia where extreme low temperatures ($< -25^{\circ}C$) could have adverse effects especially when coupled with warmer than average summers (see Ekholm, 1975).

B *Life tables, key factors and population dynamics*

Insufficient space precludes a detailed review of population dynamics but a brief assessment is necessary to determine the factors occupying prominent roles in competing theoretical structures and their influence by climate and weather. A clear distinction exists between the concept of 'regulation', in which populations are maintained in dynamic equilibrium density by density-dependent processes (Varley *et al.*, 1973; May, 1981) and that of 'limit' (LBC) in which populations fluctuate between a 'ceiling' determined by the carrying capacity of the habitat and a 'floor' below which extinction is certain, in response to a complex of interacting factors, including density-independent as well as density-dependent processes (Milne, 1957; Table 2.5). Semantic issues bedevil the concept of population regulation. Herein, there is insufficient space to explore these problems, but the reader should be aware of variants in the definition of 'regulation' (e.g., populations maintained below their carrying capacity by density-dependent processes; Hassell, 1985). In a detailed review of Lepidopteral ecology, Dempster (1983) rejected 'regulation around an equilibrium' (RAE) as a viable mechanism. Of 24 studies, eight revealed no evidence of density-dependence, and in a further 13 the only density-dependence demonstrated was due to intraspecific competition for resources, a process he argued incapable of

Table 2.5 Features distinguishing models of population control (From Dempster, 1983; Courtney, 1986)

	Regulation around an equilibrium (RAE)	*Limitation set by a ceiling (LBC)*	*Delayed density-dependence in a seasonal environment (DSE)*
Population	Determined by equilibrium level at or below the carrying capacity	Determined by ceiling set by carrying capacity	Determined by oscillating resource lower than ceiling
Density dependence	Acts in all populations, and at all densities	Acts in some populations only, and then over a limited range of densities	Acts over wider range of densities than LBC
Density-dependent factors	Intraspecific competition and natural enemies	Intraspecific competition only	Intermediate between alternatives
Resource limitation	Infrequent	Frequent	Occasional only
Local extinctions	Rare	Relatively frequent	Frequent

regulating populations at low density. The ability of enemies to function in a density-dependent manner in the three remaining studies was questioned.

A striking feature of a wider review of 58 insect species, including seven butterflies, was the absence of any density-dependence factors in half of the studies (Stiling, 1988). Any given type of mortality or population process tended to perturb densities away from a mean level rather than to be density-dependent. In cases where density-dependence was detected, it was more likely to be due to a source of mortality acting from the trophic level below (i.e., the plant) than from that above (i.e., predators, parasites). This finding supports the LBC model rather than that of RAE.

Dempster's review did not consider delayed density-dependence which Turchin (1990) revealed as prominent in eight of 14 long-term single site studies of forest insects, nor patchy density-dependence (Jones *et al.*, 1987); both are difficult to detect using life table analysis. Nevertheless, evidence of density-dependence in populations is not strictly proof of population RAE, which would require density-dependent responses at all densities beyond some equilibrium value (Dempster, 1983). Courtney (1986) had previously referred to studies on Pieridae which had repeatedly demonstrated delayed density-dependent relationships of diseases or parasitoids and their hosts, but pointed out that such lags together with periodic disturbances in environmental conditions (e.g., winter) militate against the regulation of the host population around an equilibrium density. Courtney's more realistic 'delayed density-dependence in a seasonal environment' (DSE) is expected to display more features in common with 'limit' than 'regulation', with the exception that the carrying capacity of habitats will be relatively unimportant in seasonal systems.

The likelihood of RAE occurring, *sensu* Varley *et al.* (1973), would seem to depend on environmental stability (see section 1.1). Regulation may often be counteracted by unpredictable weather systems and seasonal adversity in temperate environments in which nearly all studies have been undertaken, compared to tropical systems which may also be more stable as a result of greater diversity of competitors and enemies (Ricklefs, 1979; Gilbert, 1984; Gee & Giller, 1987). However, Pollard (1991b) suggests that the close synchrony of fluctuations in local and regional butterfly populations in Britain indicates the presence of density-dependence. He argues that, although some synchrony between annual fluctuations due to weather is expected, populations without any density-dependence would follow 'random walks', the trends dependent on the growth rates of individual populations; as underlying growth rates are unlikely to be similar in the range of different habitats occupied by individual butterfly species in the same region, fluctuations may well be moderated by density-dependent factors. Assuming this view to be correct, it does not, of course, provide necessary proof of population regulation.

Locational bias and paucity both of extensive surveys to determine any spatial variation of key factors (Morton, 1985) and of long-term life tables on butterflies, essential for testing the RAE model and for determining realistic bounds of population fluctuations (Dempster & Pollard, 1986; Pimm & Redfearn, 1988; Hassell *et al.*, 1989), make further comparison of population dynamics models unrewarding (see Stewart-Oaten & Murdock, 1990). Nevertheless, some cautionary points emanate from a consideration of the effects of climate on the main components of the models. Firstly, weather and climate are usually envisaged as influencing populations stochastically. A prominent key factor among butterfly and moth population studies, egg shortfall, is believed to be the result, primarily, of a density-independent influence of weather limiting time for nectar feeding and oviposition rather than density-dependent influences on egg production or movement, but measurements of egg shortfall and its components are often unsatisfactory. Only in *Thecla betulae* has egg shortfall been shown to have a density-dependent basis, probably related to dispersal (Thomas, 1974).

Secondly, agents involved in density-dependent relationships can be substantially influenced by weather and climatic variation; directly as in the impact on populations of predators or parasitoids (Courtney, 1986) or on larval development time and exposure to enemies (e.g., *Ladoga camilla*, Pollard, 1979a); indirectly, by affecting resource quality (i.e., hostplants, mate-location sites, roosts, nectar etc.), thus the carrying capacity of habitats and intraspecific competition, either on a short-term seasonal basis (e.g., synchronization of crucifer flower buds with *Anthocharis cardamines* and *Euchloe simplonia* oviposition; Dennis, 1983a, unpub. data) or in the long term (e.g., through eco-climatic constraints on symbionts for *Maculinea arion*; Thomas, 1989; see section 2.3). Thus, it is possible that many density-dependent relationships may be affected by the stochastics of atmospheric processes. Analyses neither for density-dependence nor for delayed density-dependence take into account stochastic events, the latent effects of which may be incorporated in estimates of multiple correlations and regression parameters for key factors or autocorrelates (Turchin, 1990). It is nevertheless possible for weather and climate to function in a density-dependent manner in situations when adverse seasonal weather selects against denser populations disproportionately, owing to prior occupancy of suboptimal niches, which may be suitable and become colonized in favourable conditions (e.g., *Lasiommata megera*; Dennis, pers. obs.).

Thirdly, weather and climate have an important role in causing over-exploitation of resources and extinctions, particularly in environments which are climatically marginal (Ehrlich *et al.*, 1972). The two can be closely associated since seasonal conditions may curtail carrying capacity and this in turn may severely limit effective population size (e.g., drought on the host-plant *Collinsia tinctoria* for *Euphydryas editha*; Singer & Ehrlich, 1979). Such weather-induced reductions in resources can result in density-dependent mortality where localized high larval density results in over-exploitation and intraspecific competition (e.g., *E. editha*; White, 1974), but mortality may also simply result from starvation, the failure of larvae to find alternative sources and thus be density-independent (Singer, 1972). Of course, not all intraspecific competition is due to limits placed by carrying capacity (e.g., oviposition edge-effects and cannibalism in *Anthocharis cardamines*; Courtney, 1980), nor by any means are all changes in carrying capacity due to weather and climate (e.g., *Mellicta athalia*; Warren *et al.*, 1984). An important issue is that many of the examples of resource over-exploitation and of population extinction in butterflies, qualifiers for the LBC model, occur in climatically marginal areas for the species studied (Dennis, 1977; Pollard, 1979; Singer & Ehrlich, 1979). In Britain, although population extinction events are poorly recorded, many may have an explanation which involves climatic constraints (Dennis, 1977; Pollard, 1979; Thomas *et al.*, 1986).

Finally, the contribution of dispersal to butterfly dynamics has been generally ignored, and may well account for population fluctuations of many species (e.g., *Anthocharis cardamines*, Dempster, 1989). Population extinctions and colonizations due to influences on inter-habitat movements may also be more frequent than is appreciated. Techniques of mark – release – recapture are notoriously inadequate for determining successful long-distance dispersal. Local populations may well be embedded in more broad-scale 'metapopulations' – local units of which are dynamic entities subject to frequent extinction and re-establishment (Harrison *et al.*, 1988; Hanski & Gilpin, 1991). Seasonal climate and short-term weather conditions may greatly affect dispersal; directly by determining flight characteristics and indirectly through resources. Conditions may be relatively favourable, leading to the build-up of large populations from which proportionately more adults successfully disperse over larger distances (Dennis, 1982; Dempster, 1989) or adverse, when adults are forced to seek out nectar sources or oviposition sites during drought (Horton, 1977). The effects can be either density-dependent or density-independent depending on changes in the resource base, including, unfortunately, how it is measured. Moreover, the quality and distribution of environmental patchiness, associated with metapopulations, within which stabilizing influences may be operating, can also be influenced by climatic conditions (but see Kareiva, 1987).

2.3 **Climate, species diversity and distributions**

Gradients of species diversity, herein treated simply as the number of species per unit area (see Routledge, 1983; Brown, 1988), implicate nonrandom covariation of species' distributions for limiting factors. The marked space–time scale dependence in the relative influence of such limiting factors (Gee & Giller, 1987; Auerbach & Shmida, 1987) generates a structure for this section, which is applied to case material from the British mainland, following a brief review of global issues.

A *Climate and species diversity gradients*
Apart from scale, explanations for diversity gradients must account for variation in the type of gradients and the location of peak abundances, reversals in gradients and the wide range of organisms affected (Pianka, 1966, 1974; Colinvaux, 1973; Ricklefs, 1979; Brown, 1988). Butterfly macrodiversity gradients occur with latitude (e.g., Papilionidae, Scriber, 1973; 1984; Collins & Morris, 1985; Fig. 2.7), elevation (Geiger, 1987; Fig. 2.2); into continental interiors (Larsen, 1984a, b; Dennis *et al.*, 1991), and over peninsulas and islands (e.g., Antilles, Caribbean, Munroe in Gilbert, L.E. 1984; Scott, 1972; Baja California, Emmel & Mattoon, 1990). The most distinctive gradient is that from equator to poles; although peak abundances are usually associated with tropical latitudes (Lamas, 1983; Brown, 1984; Emmel, 1989), some lower taxa have their maximum diversities within the temperate or arctic zones (e.g., subfamily Parnessiinae, Tribe Erebiini and Oeneini) and, significantly, clear reversals in species diversity with latitude occur for different polyphyletic faunal groups (Dennis *et al.*, 1991). Butterfly diversity often declines into continental interiors (e.g., Arabia, Larsen, 1984b), but multivariate analysis of the butterfly fauna of large regions has revealed a variety of diversity gradients for faunal subsets, even reversals of this pattern (Dennis *et al.*, 1991). Variance in elevation is a strong positive predictor of species diversity as in alpine Europe (Dennis *et al.*, *ibid.*) whereas butterfly diversity generally declines with elevation, though this is influenced substantially by the negative relationship between latitude and the elevation at which peak richness occurs, climatic zonation, maximum elevation and sampling strategies in relation to seasonality (Shields, 1967; Wolda, 1987; McCoy 1990). Peak diversity occurs at lowest altitudes in northern Britain but at mid altitudes in the Alps (see Fig. 2.2b), in Lebanon, on Tenerife and in Costa Rica (Dennis, 1992; Leestmans, 1975; Larsen, 1974; Geiger, 1987; DeVries, 1987). Gains and losses can be cumulative, inasmuch as the loss of aposematic models with altitude in the tropics (e.g., Ithomiinae) can influence the diversity of mimics (e.g., Dismorphiinae) (DeVries, 1987).

The variety of diversity gradients is matched by the multiplicity of hypotheses for the latitudinal decline in species, emphasizing such factors as: geological history (Wallace, 1876), physical heterogeneity (Legg, 1978; Cox & Moore, 1985), productivity (Connell & Orias, 1964), stress, harshness, favourableness and environmental variability or stability (Slobodkin & Sanders, 1969; Stevens, 1989; Runkle, 1989; Adams, 1989); energy (Hutchinson, 1959; Brown, 1981; Wright, 1983; Turner *et al.*, 1988); biotic structure and architecture (Terborgh, 1985; Lawton, 1986; Morse *et al.*, 1985) and community organization and dynamics (Dobzhansky, 1950; MacArthur,

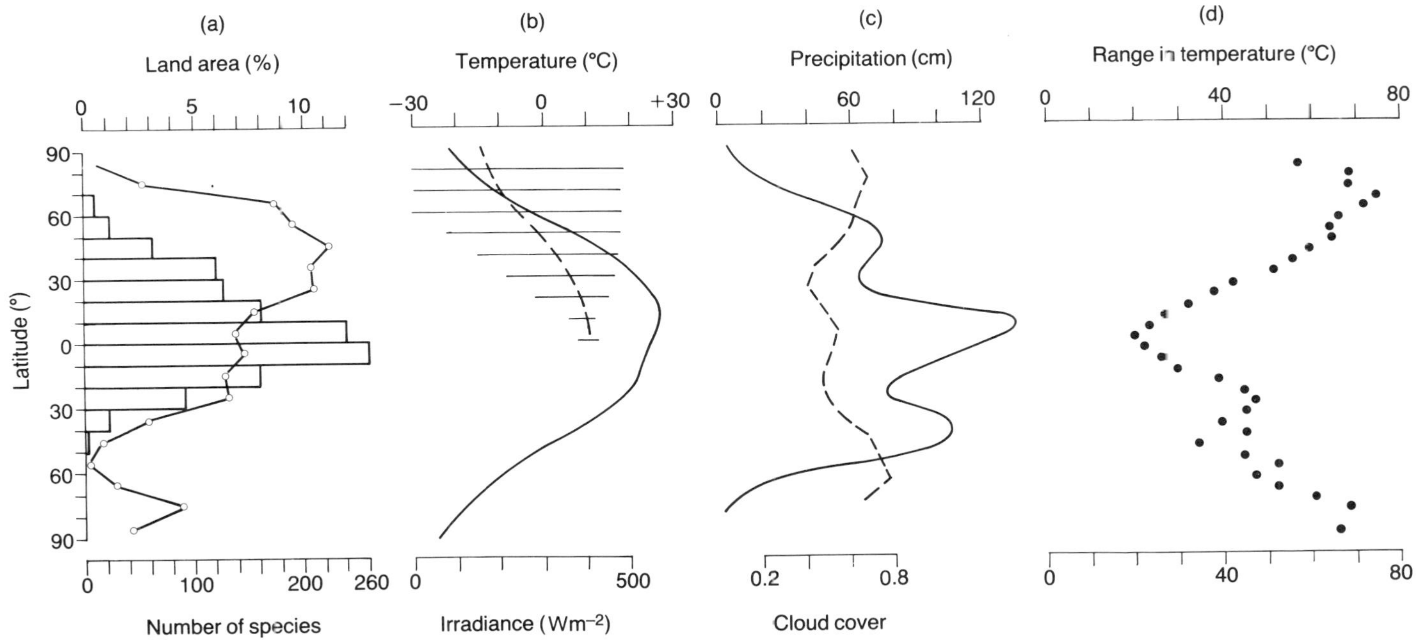

2.7 Climatic correlates of latitudinal gradients in species richness among the Papilionidae. (a) Papilionidae; numbers of species for latitudinal zones (histogram) (adapted from Collins & Morris, 1985) and land area as a percent of total global landmass (line graph). (b) Mean annual temperature (solid line) and mean annual irradiance and range in mean monthly irradiance (pecked line and vertical bars). (Redrawn from Lamb, 1977.) (c) Mean annual precipitation (solid line) and mean annual cloud cover (pecked line). (Redrawn from Tallis, 1991a.) (d) Mean absolute range of temperature. (Adapted from Stevens, 1989.)

1962; Huston, 1979). A similar set has been expounded for species diversity with elevation (Janzen, 1973; Janzen *et al.*, 1976; Lawton *et al.*, 1987; McCoy, 1990). Reviews of these ideas have had limited success for several reasons (i.e., Pianka, 1974; Colinvaux, 1973; Ricklefs, 1979; Giller, 1984; Cox & Moore, 1985; Begon *et al.*, 1986; but see Brown, 1988): rarely are they mutually exclusive; many lack clear definition of factors or mechanisms; few are unequivocally supported or rejected by logic or data; many solutions are either proximal or distal accounts, relate to limited scales or to a single, ecological or evolutionary, level. More obviously, phytophage diversification may be promoted by plant diversity and structural complexity (viz., biomass; fractal dimensions; hostplant chemistry; Pianka, 1966; Connell, 1978; Haber in Gilbert, 1984; Mitter *et al.*, 1988), but trophic dependence simply transfers the question of ultimate controls to a different stage.

Explanations of species diversity benefit from a division of primary factors (abiotic gradients) and secondary factors (biotic interactions) and of ecological processes (net species maintenance per unit area) and evolutionary issues (net species production per unit area). Net species maintenance depends on the balance between population extinction and immigration. The probability of population extinction is inversely related to population size; small populations are vulnerable to fluctuations. Population and distributional status tend to be closely bound together (Hanski, 1982); empirical evidence exists for significant positive correlations between local population abundance, local population variability and regional distribution (Brown, 1984; Glazier, 1986; Gaston & Lawton, 1988). Thus, the likelihood is that small populations whose carrying capacity is minimized by some limiting factor(s) are part of a marginal metapopulation – isolated, sparse, less fluid, thus emitting and receiving fewer propagules, and perhaps also experiencing relatively larger unpredictable perturbations (Brakefield, 1984; Pollard *et al.*, 1986). All diversity gradients correlate largely with variation in the contemporary physical environment; as severity of the limiting factors decreases, so the impact of fluctuations and unpredictability decreases. Limiting factors may include supply of energy, nutrients or a substrate facility but not energy alone (Wright, 1983) except inasmuch, axiomatically, that the entire earth – atmosphere system is insolation– dependent. Inverse species diversity – latitude gradients (e.g., arctic butterfly faunal unit) have only an indirect link with energy supply (i.e., vegetation succession and habitat exclusion) as does decline in species diversity with aridity (i.e., water balance). Significantly, climate may fulfil several co-varying, potentially synergistic roles to cause marginal conditions and limit butterfly diversity along geographical gradients via population dynamics, as: (i) an energy source, (ii) conditions mediator, (iii) hazard perpetrator and (iv) a nutrient conveyor.

Any complete explanation for global differences in species diversity must invoke 'species dynamics' – rates of species origination and extinction, and therefore the history of refuges and other landscape disturbances (Brown, 1988; section 4.3B). However, unlike population dynamics, where extinction and immigration may be similarly affected by limiting conditions, any process which enhances extinction presumably also provides the circumstances that encourage speciation (Myers & Giller, 1988; Dennis *et al.*, 1991). The two are intimately linked inasmuch as the probabilities of extinction and evolutionary integrity are decreasing functions of distributional dimensions, *ceteris paribus* (Lewin, 1989). Although low-frequency, high-magnitude environmental perturbations (e.g., glacial cycles) affect the entire biosphere there is some evidence that this influences origination and extinction rates and thus net production rates differently between tropics and poles (Grubb, 1987; Sauer, 1988; Adams & Woodward, 1989). Regional variability (i.e., variance in relief) may substantially influence this balance as suggested for South America (Haffer, 1969; Legg, 1978) and European alpine butterflies (section 4.4). None of this denies differential evolutionary rates between tropical and temperate fauna or secondary net species maintenance involving hotly debated community mechanisms and structural derivatives (Huston, 1979; Terborgh, 1985; Brown, 1987; references in Stork, 1988; Gee & Giller, 1987; Smith, 1989), including postulated mutual reinforcement of plant, phytophage and parasitoid diversity by 'coevolutionary' feedbacks (Ehrlich & Raven, 1964; Zwolfer, 1978; Strong *et al.*, 1984; but see Jermy, 1984). Rapoport's rule (viz., equatorial species have narrower latitudinal ranges) and the increasing frequency of hostplant specialists at lower latitudes also suggests narrower adaptive responses to relatively stable tropical climates that may influence diversity (Pianka, 1974; Scriber, 1984; Stevens, 1989).

B *Species richness along a latitudinal gradient: correlations and causation*

Scale, geographical position and history largely establish the explanatory framework for regional species diversity gradients. Relevant points that emerge for the small latitudinal expanse of the British mainland (8° range) are: (i) species maintenance but not species production is applicable; (ii) sufficient time has elapsed since initial Holocene colonization for equilibrium to occur between species' distributions and environmental conditions (Dennis, 1977); (iii) butterfly diversity is evidently not limited by hostplant distribution *per se* (Fig. 5.5); and (iv) population studies indicate that butterfly diversity is not primarily accounted for in community interactions among phytophages and their enemies (section 2.2B). This does not mean that historical influences have no effect on overall diversity levels (Dennis, 1977), that species are now entirely in equilibrium with resources, nor does it imply that other hostplant effects are negligible.

Interesting vantages are provided by two independent surveys which differ in sampling frames, numbers of predictors and manipulation of data

(Dennis & Williams, 1986; Turner, 1986; Turner *et al.*, 1987; Table 2.6). Correlations between diversity and virtually all predictors are substantial and significant. Latitude is the best overall predictor; summer temperatures and sunshine jointly account for as much variance as latitude (Fig. 2.8). Modest to high correlations occur for other climatic variables, including January temperatures. This contrasts with Turner's data and may result from a bias in his sampling frame to low elevation and urban sites as well as differences in the climatic data for the two periods, but is more likely due to the clustering of meteorological stations and the inclusion of small offshore islands which have a western distribution and are deficient in species (Dennis, 1977). It is noteworthy that correlations between numbers of species and January mean sea level temperatures remain positive for the current data set. Correlations involving land variables (e.g., elevation, calcareous bedrock, calcareous soils, acid, peaty and immmature soils) make good biological sense; for others, reversals in sign (e.g., per cent rough grazing; per cent urban cover; per cent crop cover) from those expected suggest interaction effects of a third variable(s) (see below). Substantial correlations also exist between species richness and longitude and between geographical location, elevation and most predictors.

Extensive and high multicollinearity between predictors presents a serious obstacle to partitioning variance meaningfully; a multicausal nexus may be inferred for butterfly diversity on mainland Britain involving numerous indirect as well as direct pathways associated with environmental gradients, orientated primarily south-east to north-west, via population dynamics processes discussed earlier (section 2.3A). The most obvious link is a direct one between summer energy supply and adult activity; lower R and T_a northwards adversely influences the capacity of adults to reproduce, disperse and escape from predators; moreover, it affects the development and growth of early stadia, thus their susceptibility to parasitoids and disease (section 1.3). Conditions are also influenced by declining sun and heat. Predators and parasitoids may be no less active in cooler temperatures (section 1.3) though both symbionts (e.g., ants for lycaenids) and hostplant quality may be adversely affected. The shorter 'growing' season northwards significantly limits broods (Fig. 2.5), curtails r, the intrinsic rate of increase, and restricts the distribution of species with relatively inflexible voltinism profiles (e.g., *Lasiommata megera*), especially those flying late in summer (e.g., *Lysandra bellargus*; *Thymelicus acteon*; Pollard *et al.*, 1986). An increase in diversity with increased winter warmth may simply reflect the longer development period in summer and thus increased r, but the probability of rapid falls in temperature and lower critical limits being exceeded increases with altitude and thus with lower mean winter temperatures. Although energy use (i.e., lipids) increases with T_a in the diapausing winter stages, this may be offset by an inverse correlation between winter warmth and the duration of winter and shorter periods of dormancy. A corresponding decline in winter and summer temperatures with latitude is more common in continental interiors and east coast margins; T_c for species may then become critical (section 2.1B).

Spring to autumn conditions northwards may also be more 'hazardous' for butterflies owing to more frequent, unpredictable and sudden weather changes – depressions, intense downpours, high winds (Fig. 2.8e), unseasonal snowstorms and frost. These may be deleterious to nectar sources and hostplants, but the most significant impact may result from the limited available response time and the failure of adults to adopt appropriate cover. Even if a negative correlation did exist between winter warmth and diversity (i.e., lower diversity in regions with warmer winters), this would not preclude a hazard effect; regions with warmer winters also have a higher probability of 'killing' frosts following mild spells and experience higher rainfall, dampness and storms conveying several inferences for butterfly survival (Beirne, 1955). The paramountcy of climate is suggested from distributional changes in response to climatic change: the dominance of summer warmth over geometrical factors (i.e., area, isolation) in accounting for inter-island variance in species richness; also, from the way in which climate overwhelms the expected influence of land variables, such as crop cover, rough pasture and urban cover, whose signs are reversed from those expected in correlations with species richness (Table 2.6) (Dennis, 1977; Dennis & Williams, 1986). Climatic contrasts are exaggerated due to the fortuitous correspondence of calcareous substrate and increased elevation with the south and the north-west respectively (Table 2.6; see Fig. 5.3 and section 2.3C), evoking a direct impact on climate and an indirect one through anthropogenic agencies. Not only have northern landscapes been more seriously degraded (Simmons, 1990), but the same land-use can have reversed effects on northern and southern species as in the case of grazing; this has adversely influenced microclimates via decreased soil infiltration capacity in northern uplands, but has enhanced maximum ground temperatures essential for the perpetuation of several species in southern Britain (e.g., *Maculinea arion*; *Lysandra bellargus*; *Hesperia comma*: see section 4.3B; Fig. 4.4).

These observations lend support to the influence of a wider variety of interdependent factors on species richness in Britain than the energy hypothesis (Turner, 1986), although summer heat and sunshine are paramount. Caution is advised in interpreting the results since there is geographical bias in recording and such surveys promote correlation not causation. Autecological research highlights post-1945 habitat transformations from changes in land management as the most significant factor in range contraction and distribution changes in British butterflies (Heath *et al.*, 1984; Thomas, 1984). It is very unlikely that 'the total number of species is limited by energy supply' (Turner *et al.*, 1987: 203), if only because there are much greater constraints to recolonization either from the continent

Table 2.6 Correlations and regression parameters between butterfly species and selected environmental variables on mainland Britain

(a)

	Species	Grid east	Grid north	Elevation	January temperature	July temperature	July sunshine	Frost-free days	Precipitation	Windspeed	Annual temperature range	Steep slopes	Calcareous rocks	Calcareous soils	Immature soils	Rough grazing land	Urban cover	Arable cover
Species		0.48	−0.89	−0.61	0.59	0.85	0.84	0.58	−0.49	−0.16	0.49	−0.49	0.54	0.60	−0.77	−0.70	0.72	0.53
Grid east	0.47		−0.47	−0.64	0.09	0.63	0.54	0.17	−0.83	−0.21	0.83	−0.48	0.55	0.59	−0.46	−0.70	0.58	0.74
Grid north	−0.91	−0.43		0.55	−0.67	−0.94	−0.83	−0.59	0.43	0.15	−0.49	0.58	−0.49	−0.49	0.67	0.68	−0.74	−0.48
Elevation	−0.61	−0.61	0.51		−0.33	−0.59	−0.74	−0.58	0.79	0.08	−0.47	0.56	−0.47	−0.53	0.72	0.86	−0.67	−0.81
January temperature	0.71	0.37	−0.74	−0.80		0.48	0.60	0.80	−0.03	−0.32	0.22	−0.41	0.13	0.10	−0.42	−0.32	0.47	0.11
July temperature	0.79	0.76	−0.78	−0.84	0.82		0.79	0.48	−0.59	−0.33	0.72	−0.60	0.56	0.55	−0.68	−0.78	0.80	0.60
July sunshine	0.84	0.51	−0.85	−0.75	0.87	0.85		0.74	−0.59	−0.12	0.41	−0.60	0.47	0.49	−0.74	−0.72	0.70	0.65
Frost-free days	0.59	0.19	−0.59	−0.63	0.81	0.60	0.75		−0.31	−0.47	0.04	−0.45	0.23	0.20	−0.53	−0.48	0.54	0.39
Precipitation	−0.50	−0.85	0.39	0.81	−0.57	−0.82	−0.59	−0.33		0.17	−0.70	0.64	−0.49	−0.57	0.62	0.84	−0.60	−0.91
Windspeed	−0.61	−0.68	0.54	0.64	−0.51	−0.79	−0.54	−0.23	0.79		−0.55	0.04	−0.09	−0.20	0.11	0.17	−0.11	−0.09
Annual temperature range	0.49	0.51	−0.45	−0.43	0.20	0.70	0.41	0.02	−0.70	−0.37		−0.36	0.54	0.58	−0.47	−0.67	0.52	0.62
Steep slopes	−0.51	−0.50	0.56	0.61	−0.68	−0.69	−0.60	−0.42	0.62	0.63	−0.34		−0.37	−0.37	0.57	0.65	−0.61	−0.61
Calcareous rocks	0.54	0.61	−0.48	−0.48	0.36	0.59	0.48	0.24	−0.51	−0.45	0.56	−0.35		0.64	−0.46	−0.56	0.57	0.47
Calcareous soils	0.60	0.61	−0.49	−0.53	0.42	0.64	0.50	0.23	−0.57	−0.54	0.59	−0.32	0.69		−0.53	−0.62	0.47	0.59
Immature soils	−0.78	−0.51	0.68	0.75	−0.71	−0.77	−0.78	−0.53	0.67	0.67	−0.48	0.64	−0.49	−0.53		0.76	−0.60	−0.69
Rough grazing land	−0.70	−0.72	0.64	0.84	−0.69	−0.89	−0.71	−0.45	0.85	0.82	0.68	0.68	−0.57	−0.61	0.78		−0.78	−0.85
Urban cover	0.71	0.60	−0.71	−0.66	0.65	0.78	0.68	0.50	−0.60	−0.62	0.52	−0.61	0.54	0.46	−0.62	−0.77		0.54
Arable cover	0.53	0.75	−0.46	−0.81	0.63	0.81	0.66	0.40	−0.91	−0.74	0.63	−0.61	0.51	0.60	−0.74	−0.85	0.54	

(b)

(i) Sampling frame: 50 km grid intersections; stepwise entry

Variables	b	SE_b	Beta	T	p	SE_y	AjdR^2	DF	F ratio
1 10 × 10 km squares									
July sunshine	6.24	0.74	0.44	8.46	< 0.0001	7.95	0.60	1/356	546
July temperature	4.73	0.57	0.43	8.24	< 0.0001	7.29	0.67	2/355	358
a = − 86.4									
2 20 × 20 km squares									
July sunshine	6.11	1.07	0.47	5.73	< 0.0001	6.51	0.69	1/91	209
July temperature	4.82	0.84	0.47	5.71	< 0.0001	5.61	0.77	2/90	157
a = − 87.3									

(ii) Sampling frame: climatic stations; sequential elimination of variables

Variables	b	DF	F ratio	p	R^2	DF	F ratio	p
1 10 × 10 km squares								
Summer sunshine	0.02	1/141	6.7	0.011	0.55	2/141	87	0 (10–25)
Summer temperature	3.31	1/141	52	0 (10–9)				
a = − 182.2								
2 30 × 30 km squares								
Summer sunshine	0.03	1/141	23	0 (10–6)	0.79	2/141	259	0 (10–48)
Summer temperature	3.69	1/141	147	0 (10–6)				
a = − 198.3								

Winter sunshine and winter temperature were rejected variables. For statistical terms, see Glossary of symbols.

(a) Correlations for tetrads (mean values) based on a systematic sampling frame of 50 km grid intersections; N = 93. Upper matrix: Pearson product–moment correlations for transformed variates (normalized) with climatic variables uncorrected for altitude. Lower matrix: Spearman correlations for raw variates corrected for altitude on mean value of the highest contour per spatial unit. For both matrices, $r_p < \pm 0.17$, $p > 0.05$; $r_p < \pm 0.23$, $p > 0.01$. Variables: species, mean number of species per 10 km unit; elevation, land > 245 m (per cent); January and July temperatures (°C), correction for altitude 0.6°C per 100 m; frost-free days, mean number of days with screen temperatures < 0°C; precipitation, annual precipitation (mm); windspeed, hourly windspeed (ms^{-1}) exceeded 50 per cent of time, correction for altitude × 7 per 100 m; annual temperature range, difference between January and July means (°C); steep slopes, land > 11° (per cent); calcareous rocks, cover by limestone and chalk (per cent); calcareous soils, cover by calcareous brown earths and rendzinas (per cent); immature soils, covers by peats, peaty podsols and skeletal soils (rankers) (per cent); rough grazing land, unimproved land (per cent); urban cover, built-over land (per cent); arable cover, land under crops (per cent). Correlations for individual 10 km squares are lower but differ less than $r_p = 0.1$ from the correlations tabulated. July maximum temperatures correlate highly with July mean temperatures (r_p, 0.90; r_s, 0.97) and January minimum temperatures with January mean temperatures (r_p, 0.96; r_s, 0.97). (From Dennis & Williams, 1986; unpublished data.) (b) Summary multiple regression parameters of butterfly species on sunshine and temperatures. (i) From Dennis & Williams, 1986; unpublished data. Variables as in (a). Data are

Table 2.6 (cont.)

untransformed and uncorrected for altitude for comparison with analyses in Turner *et al.*, 1987. Transformed data have no effect on the order of entry of variables into equations, and make little difference to the size of R^2 and the overall level of significance of parameters. The utilization of more appropriate regression models (i.e., Poisson log. link model) does not significantly improve upon the predicted estimates for species numbers determined in the orthodox normal model. On introducing grid north into stepwise regression equations with a full array of climatic variables, it is entered first for analyses at both 10 km square and 20 km square resolutions and accounts for most of the variance (10 km squares; 67 per cent; 20 km squares: 80 per cent). Only July sunshine added a significant amount (2–3 per cent) of additional variance. (ii) From Turner, 1986; Turner *et al.*, 1987. Temperatures (°F) and sunshine (h) accumulated over 5 months (May to September). On introducing latitude into equations with summer and winter sunshine and temperatures, climatic variables (summer temperature and winter sunshine) still accounted for significant contributions to R^2. Note: the regression parameters provide a guide only to the relative significance of variables; substantial multicollinearity between the independent variables detracts from the value of regression techniques in modelling such a complex biological system (see Vincent, 1981). The gradient in species richness northwards may be attributed to a number of covarying climatic and environmental factors, at least involving energy supply, conditions and hazards (see text).

(e.g., *Maculinea arion*; *Cyaniris semiargus*; *Aporia crataegi*) or due to habitat loss northwards throughout Britain (e.g., *Leptidea sinapis*) when conditions improve, than there are to population extinctions. Empirical findings of population dynamics and autecology also fail to support the view; the latter demonstrates the importance of resource limitations unrelated to climate in many species (Dennis & Williams, 1986). It seems more likely that diversity reflects upon the general adaptability of individual species (Stehli *et al.*, 1969); some may simply lack sufficient genetic variability; most may have it but experience critical life-history trade-offs in metabolism or phenotype (section 3.3) or in the marginal environments fail to persist long enough to incorporate innovations.

C *Ecoclimatic constraints on butterfly distributions and diversity*

Climate may influence distribution and diversity patterns indirectly over smaller regions than the main gradients (i.e., latitude; altitude), for example: (i) residual patterns at the same resolution, and (ii) higher resolution patterns subordinate to the sampling frame. Intermediate factors involved include substrate (geology & soils), geomorphological history, relief and topography, anthropogenic activities and environmental history (seres and vegetation succession). However, substantial difficulties arise when attempting to partition effects due to chains in causation (ultimate *versus* proximal factors) and interactions among variables (Dennis, 1977; 1992). The main residual patterns for butterfly richness on summer warmth and sunshine in Britain correspond with rich farmland at low elevation (negative: e.g., East Anglia; Lancashire Plain) and areas with greater habitat heterogeneity (positive: e.g., North Downs; Chilterns; North Wales; Arnside). Urban areas are not highlighted though more detailed mapping indicates a decline towards the centre of Sheffield and London (Garland, 1981; Plant, 1987). Not only do urban areas often provide varied habitats for butterflies, but they may be warmer, if not sunnier, and drier depending on their location *vis-à-vis* regional climates (Barry & Chorley, 1982; Ruszczyk, 1987; section 1.1A). The extensive cropping areas simply lack butterfly habitats; higher windspeeds and lower ground temperatures must also be symptomatic of level terrain absented of tree and hedge cover. Surplus areas are typically regions of varied geology, slope angle, aspect, soil cover and land-use (e.g., Carboniferous limestone scars) and promote habitat variability. But topographic heterogeneity has at least two important effects in maintaining high butterfly diversity. Firstly, rock outcrops, with their variety of slope facets and aspects, enhance extremes and variability in microclimates for substrates and the boundary layer in which butterfly stadia function. Significantly warmer and drier (steep equatorward slopes) and cooler and damper (shaded and poleward footslopes and basins) conditions occur in these areas compared to locations of similar latitude but lacking topographic heterogeneity; thus the interdigitation of northern (e.g., *Coenonympha tullia*) and southern (e.g., *Hamearis lucina*) faunal elements near Arnside, Lancashire. Secondly, topographic variability provides an essential resource during climatic change since survival can be achieved by dispersing short distances to adjacent sites. Geology, slope angle and aspect can combine to permit outliers or thermal 'relicts' beyond the main distribution for northern or southern ranges (e.g., *Argynnis adippe* and *Erebia aethiops* respectively at Arnside; Dennis, 1977); a recent example has been described for *Idaea dilutaria* (Geometridae) on the Kullaberg Peninsula (Ryrholm, 1988).

At larger scales (100 m^2–2 km^2), non-climatic factors can be held constant and it becomes possible to test the influence of local climate on niche components. Particularly evident is the joint impact of slope angle and aspect on heat and moisture budgets and these in turn on hostplant distribution and habitat quality. Warm dry south-westerly slopes tend to be at a premium at high latitudes (e.g., *Plebejus argus* and *Hipparchia semele* on the Great Orme's Head, North Wales; Dennis, 1977; 1992), but at lower latitudes shaded and cool, moist northerly slopes may be essential for hostplant survival (e.g., *Aristolochia cretica* for *Zerynthia cretica* in South Crete; Dennis, unpub. data). Complex situations occur when butterfly stadia and their hostplants are affected disparately by heat and moisture budgets of different slope elements. In the Coastal Ranges habitat for *Euphydryas editha* in Central California, the superimposition of unpredictable seasonal rainfall distribution on summer drought places a premium on the

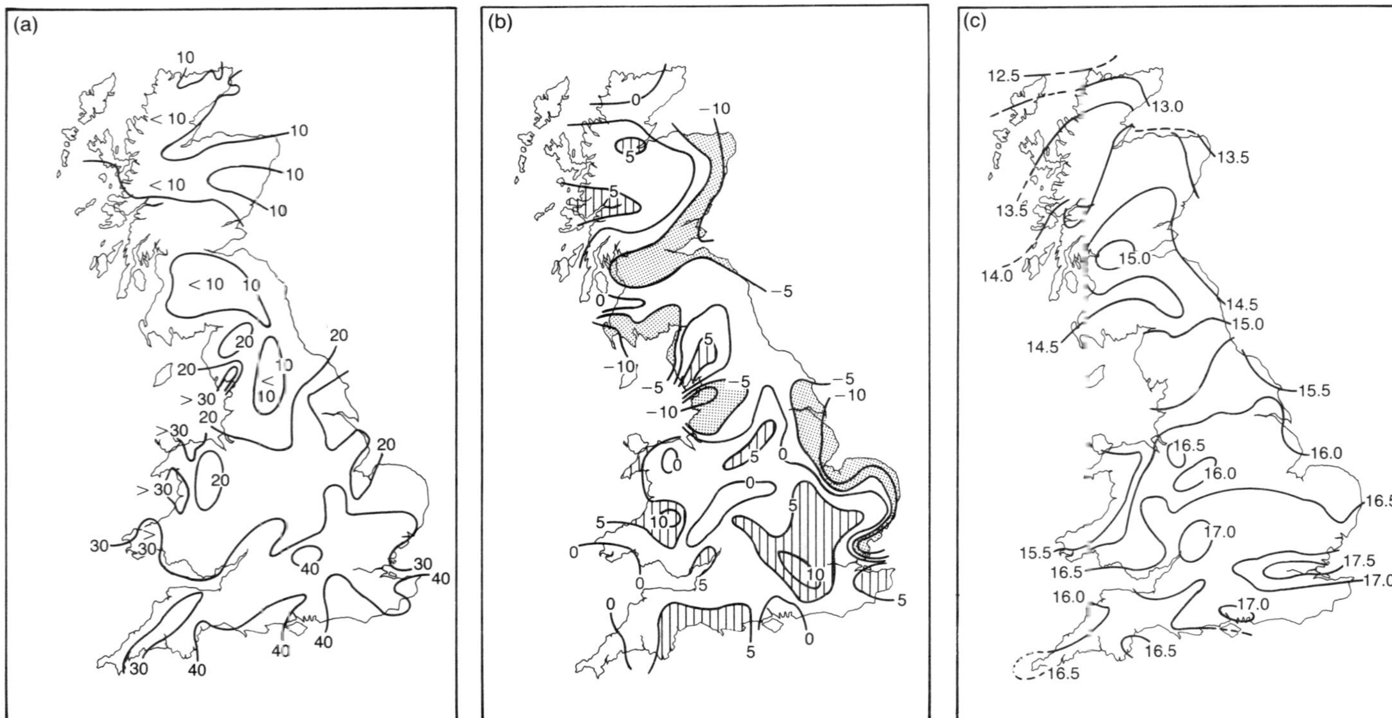
(a)
(b)
(c)

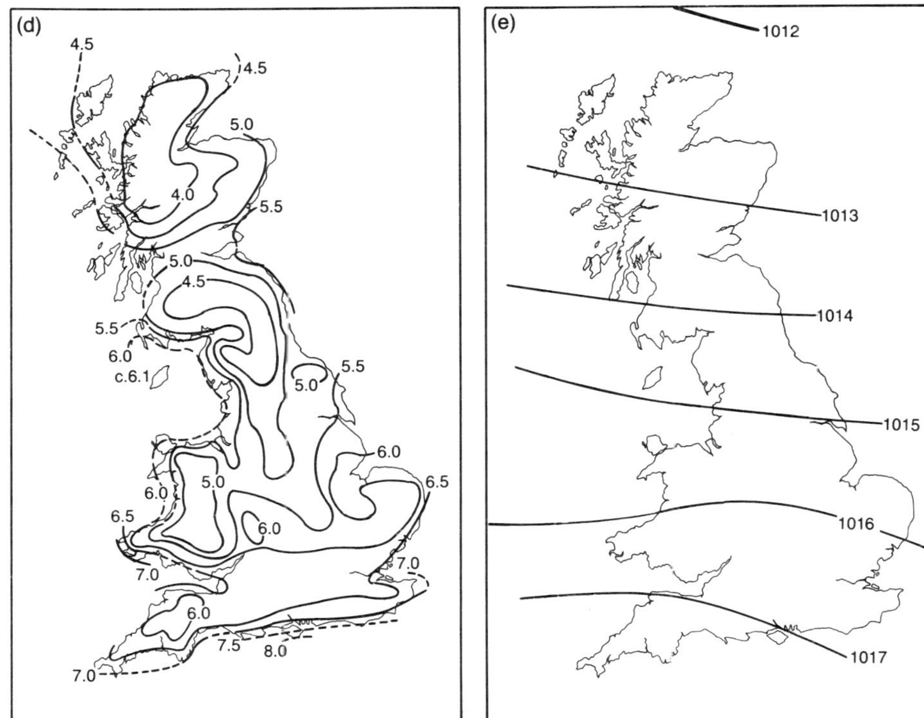

2.8 Generalized trends in butterfly species richness and climatic variables on the British mainland. (a) Species richness (number). (b) Residuals of species richness on mean July sunshine and temperature (number). Isolines on maps (a) and (b) are plotted for mean values of tetrads for 10 km squares around 50 km grid intersections. (c) July sea level mean temperatures (°C); the distribution of upland areas over 400 m in altitude is shown in Fig. 5.3a. (d) July mean sunshine (h d^{-1}). (e) July mean sea level barometric pressure (mb). (Climatic data adapted from Meteorological Office 1973, 1975a, b, by permission of the Controller of HMSO.) July summary climatic data correlate highly with data for the remaining summer months from May to September. The mean barometric pressure pattern suggests an increase in the frequency and severity of depressions northwards, as well as increasing potential air mass instability (i.e., ana versus kata fronts), and thus the higher probability of unsuitable weather conditions for butterflies.

intedigitation of diverse fine-grained slope elements (wavelength c. 20 m) which allow larvae to disperse between microclimates and hostplant resources (Weiss *et al.*, 1988).

The diversity of butterflies within small areas depends largely on the distribution of habitats and the diversity of their resources *sensu lat.* within them, therefore to their representation in different seres and the stochastics of patch dynamics. With reference to hostplants alone, in Britain, the majority of species are restricted to herbs (78 per cent, Table 5.3); although some of these persist under light canopy and permanently on geomorphologically unstable landforms (e.g., scarps, screes), most are adversely affected by shade and exist in habitats with short life-spans (<5a) and require regimes of ubiquitous and frequent disturbances (see Greatorix-Davies *et al.*, 1992 for references). The continued existence of several species now relies on anthropogenic activities and management (Warren; 1984, 1989), but weather and climate can contribute to this process in numerous ways by causing tree fall (storms: Cannell & Coutts, 1988), tree senescence and death (drought), fires (lightning) (Simmons, 1990), mass wasting and landslips, floods and silt accumulation (cryoturbation cycles, high rainfall, sudden snow melts, downpours). Thus, maximum diversity may not correspond with maximum productivity (Abramsky & Rosenzweig, 1984), but the potential significance of patches within different seres and the factors involved varies enormously for regional butterfly faunas (Owen, 1971; Montesinos, 1985; DeVries, 1987).

3
Morphological adaptations to climatic gradients

Butterfly morphologies provide a valuable tool with which to study pattern and process in evolution. Although variation, particularly that of the adult phenotype, is highly diverse and apparently complex, it adheres to relatively simple models of design (Nijhout, 1985). Diversity incorporates several scales of variation: intra-individual (e.g., between wing cells, wings and surfaces), inter-individual (e.g., polymorphisms; sexes), seasonal and geographical variation, as well as taxonomic differentiation within and between habitats. Yet, despite this and the large number of nominal species, over 12,000 (Shields, 1989), there is substantial wing pattern inertia for higher as well as for lower taxa and potential for convergence (e.g., Batesian and Müllerian mimicry). The present chapter focuses on the adult phenotype. An introduction to the structural components of and selective influences on wing patterns is followed by a brief review of global patterns in wing diversity gradients. Taxonomic incongruity for modifications in wing pattern associated with environmental gradients has long evaded explanation; the last two sections investigate potential solutions for geographical and seasonal differentiation in interactions between behavioural postures, habitat use and climatically geared selective agents.

3.1 **Determinants of butterfly wing morphology**

A *Structure and formation of wing colour patterns*

The colour patterns on the wings of Lepidoptera reside primarily in the cover scales, each the product of a single cell; the under scales are less intensively coloured and the wing cuticle is colourless, brownish or transparent. The overall wing pattern is a finely tiled mosaic of monochrome scales. Gradients in hue are accomplished by variations in the ratios of the limited numbers of scale colours, usually three to five, characteristic of a species. Colours are either structural in origin, due to interference, diffraction or scattering of light, or are chemical pigments (zoochromes). Most whites, blues and greens in butterflies and all iridescent (metallic) colours are structural, which additionally accounts for high ultra-violet reflectance (e.g., *Morpho* spp.). Transparency is achieved by different structural means in different families – the scales can be missing, translucent or arranged orthogonal to the wing surface. Zoochromes belong to four main categories (i.e., melanins, pterins, flavonoids and ommachromes) though several bile pigments have been found (e.g., in the ventral green of *Nessaea* spp.; Vane-Wright, 1979) and a number of zoochromes remain unidentified (Nijhout, 1981; 1985). A fundamental unity underlies the tremendous diversity in wing patterns; homologies between wing cell features, different wings, surfaces and taxa can be accounted for in relatively few pattern elements of the archetypal 'nymphalid' groundplan (Schwanwitsch, 1924; Suffert, 1927; Nijhout, 1991; Fig. 3.1a) and its derivative basic wing cell attributes (Nijhout, 1985; Fig. 3.1b). This emerges from studies of morphoclines of pattern elements (e.g., *Charaxes* spp., Nijhout & Wray, 1986), phenocopies induced by sub-lethal traumas (e.g., *Vanessa* spp., Nijhout, 1984) and occurrences of homeosis (Sibatani, 1983). Extant colour patterns in different Rhopalocera taxa may be derived from elements of this groundplan by distortion, elaboration, enlargement, reduction, loss, displacements, colour changes and fusion, although supernumerary elements may occur (e.g., leaf midrib in *Charaxes* spp., Nijhout & Wray, 1986). The complexity of wing patterns is due in large part to substantial wing cell autonomy in pattern formation. The validity of the model is suggested by the reconstruction of pattern homologies in *Heliconius*, in which red and black features are pattern elements, and by their fundamentally simple segregation as Mendelian loci (Nijhout *et al.*, 1990; see additional examples in Nijhout, 1991). As suggested by the theory of positional information (Wolpert, 1969), the determinants of simple to complex patterns may be few, including: (i) foci on the wing cell midline, which vary in number, precise position and shape; (ii) the characteristics of signals (morphogens) emitted from foci, which can differ in strength, timing and type; (iii) the ways in which signals are considered to interact from various foci; (iv) the function of wing veins which commonly act as sinks but also have no function or are sources; and (v) the interpretation landscape of scale-building cells (Nijhout, 1981; 1985). Pattern determination is responsive to both global and compartment-restricted determinants on the wings and within wing cells (Dennis *et al.*, 1984).

B *Factors in wing patterning*

Butterfly wing colour patterns must largely serve a biophysical function because they are located outside the metabolic pool in the dead cellular skeleton of wing scales (Silberglied, 1984). They are primarily shaped by the physical environment (i.e., thermoregulation) and by interactions between species within communities (i.e., defence mechanisms; social signalling; Gilbert, L.E., 1984), though can be influenced by genes pleiotropic for biochemical and physiological performance (section 3.4B). A colour pattern may serve more than one function, the outcome a compromise between conflicting selection pressures. However, generalizations as to the functional significance of particular patterns are hazardous because the variety of microhabitats, predators, competitors and biological attributes (i.e., size, robustness, flight characteristics, chemical defences) makes interactions potentially highly complex. Determining evolutionary primacy in co-ordinated structures, ultimate from proximate causes, is frustrated by feedbacks between pattern development and selection regimes over long time periods, which increases the likelihood of alternative evolutionary pathways and equifinality in pattern formation (Guilford, 1988). Numerous outlets for diversity in response are provided by the relative independence, in behavioural use and probably genetically, of pattern elements on different wings, surfaces and wing cells, and by contrasts between predator and butterfly visual systems, particularly regarding ultra-violet reflectance (Silberglied, 1984). The form that wing patterning takes depends on whether selection pressures for thermoregulation, anti-predator signalling and social communication conflict or are congruent. In turn, this hinges on whether anti-predator strategies have been primarily for conspicuousness or crypsis. Social signalling is enhanced by individual apparency but conspicuousness to predators requires aposematism reinforced by unpalatalibity or pretence (Turner, 1984). Apparency may be also influenced by thermoregulatory strategies according to basking mechanisms (i.e., reflectance *versus* absorbance). The outcome of such interactions is not necessarily the same for males and females, as the sexes usually differ in size, microhabitat use and behaviour and thus share different risks.

Predators of adults are commonly vertebrates (i.e., birds, lizards) but also invertebrates (e.g., dragonflies, spiders) (Moore, 1987) as indicated by direct observation, wing damage and wing remnants and predator excretions (Ford, 1975; Bowers & Wiernasz, 1979; Bengtson, 1981; Brower & Calvert, 1985; Chai, 1986; Melling, 1987). Evidence exists for selective predation in relation to prey size, colour pattern and palatability (Brakefield, 1984; Brower, 1984; Bowers *et al.*, 1985). Predatory birds display learned and unlearned responses to prey and tend to have divergent and labile tastes and hunt at specific times and in particular habitats; they also differ in their aversion to distasteful prey items (Rothschild, 1981; Endler, 1981; DeVries, 1987; Guilford, 1990). Adult butterflies are probably most susceptible to predation when settled but may also be successfully predated upon in flight (Young, 1980; Chai & Srygley, 1990). Defences against predation may be primary (i.e., avoidance of detection) or secondary (i.e., enhance escape when detected) but may also have a dual function (see

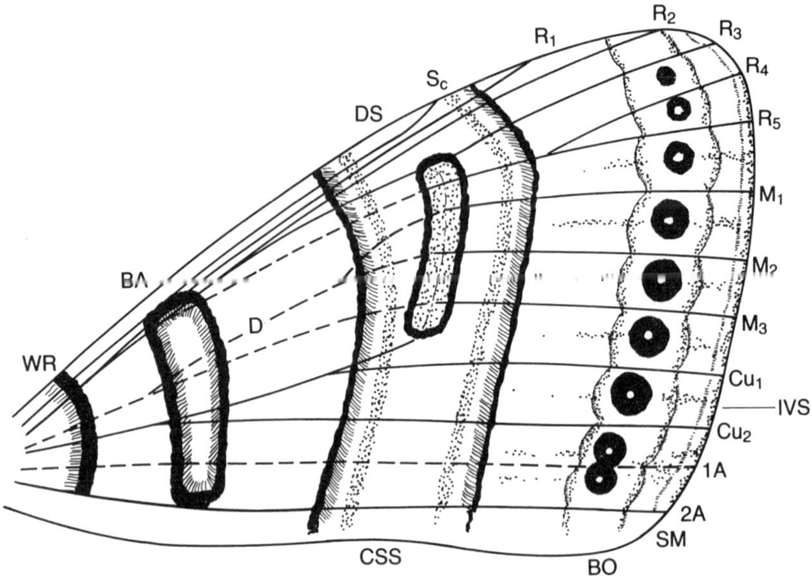
(a)
R2
R3
R1
R4
Sc
R5
DS
M1
BA
M2
D
M3
WR
Cu1
IVS
Cu2
1A
2A
SM
CSS
BO

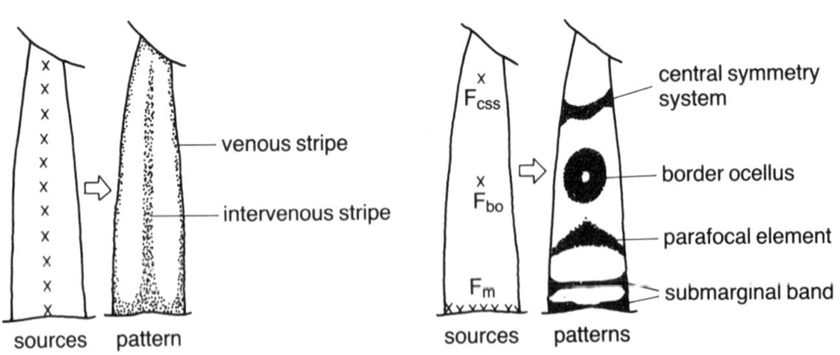
(b)
(i) Sources and their patterns
venous stripe
intervenous stripe
sources
pattern
Fcss
Fbo
Fm
central symmetry system
border ocellus
parafocal element
submarginal band
sources
patterns

3.1 Wing pattern structures in butterflies. (a) Elements of the nymphalid groundplan. BA, basal symmetry system; BO, border ocelli; CSS, central symmetry system; D, Discal wing cell; DS, discal spot; IVS, intervenous stripe; SM, submarginal bands; WR, wing root band. Venation forms are: C, costa; Sc subcosta; R, radius; M, media; Cu, cubitus; A, anal. (After Schwanwitsch, 1924; Suffert, 1927; with terminology from Nijhout, 1985.) (b) Summary of basic wing cell patterns: (i) two arrangements of sources and the patterns determined by these sources. All, some or none may be present in any one wing cell and the precise positions of sources along the wing cell midline is species-specific. (ii) Major themes of morphological diversity for each of the pattern elements. Diversity in shapes may come about through the interaction of signals from various sources. The horizontal dashed line through the border ocelli themes shows that this pattern element is sometimes represented only by its proximal or distal outline. (iii) Variations on one of the parafocal elements themes; each of the other major themes shows a similar diversity of variations. The parafocal elements often serve as a boundary between fields of different background coloration. (iv) Patterns formed by superposition of different elements (e.g., many Danainae and *Hypolimnas* spp.). Elements of pattern superposition appear to be restricted to the submarginal bands. Other pattern elements may fuse, but do not overlap. (After Nijhout, 1985 by courtesy of Academic Press Ltd.)

(ii) Major themes

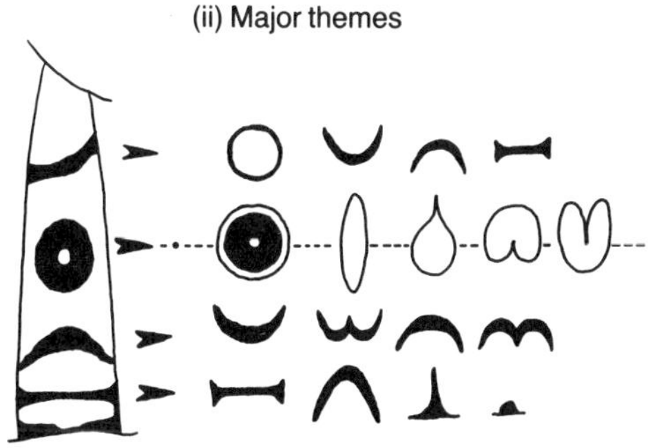

(iii) Variations on a theme

(iv) Pattern super position

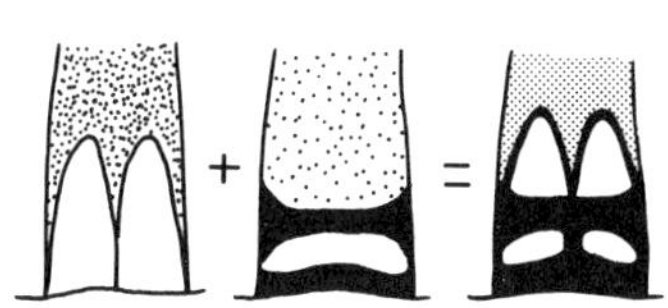

below). The former mainly involves crypsis (Endler, 1978) and masquerade of substrate as well as a number of behavioural responses; the latter incorporates evasive flights (Chai & Srygley, 1990), image blurring, aposematism and numerous categories of mimicry – including Batesian, Müllerian and additive mimicry, and possibly startling devices, decoys and deflective markings as in flash coloration, threat configurations, marginal ocellation and 'false heads' (Vane-Wright, 1976; Robbins, 1980, 1981; Endler, 1981; Turner, 1984). A number of these anti-predator systems, primarily aposematism, Müllerian and Batesian mimicry, are associated with chemical defences and are evolutionarily linked to hostplant allelochemics (see Fig. 3.2; Ehrlich & Raven, 1965; Brower, 1984). Defence strategies often involve an integrated repertoire of mechanisms and are closely tuned to ecology, particularly habitat and hostplant chemistry (Chai & Srygley, 1990; see section 3.2).

The butterfly visual system is characterized by having the broadest spectrum in the animal kingdom, low spatial discrimination and high sensitivity to motion and flicker. Social signals should range from ultraviolet through red, there is potential for long-distance signalling but fine details of wing pattern are resolved only during close encounters (Silberglied, 1984). Substantial sexual dimorphism exists for colour patterns and scent organs (androconial structures) on the wings and for tactile signals and secretions (Silberglied, 1984; Boppré, 1984). Variation between species is extensive, analogous patterns and structures having evolved convergently many times (Müller, in Longstaff, 1912). Although the majority of butterflies are monomorphic or weakly sexually dimorphic, a considerable number of species are strongly sexually dimorphic; in these cases, males are typically brighter and less polymorphic than females though occasionally this is reversed (Vane-Wright, 1979). The mating system of most butterfly species with male polygyny but restricted, often monogamous, mating in sperm-storing females makes sexual selection vastly stronger in males than in females (Trivers, 1972; Turner, 1984). Virgin females are infrequently found in the field and thus sexual selection for female colour by males is probably weak (Silberglied, 1984; but see Petersen, 1952; Magnus, 1958; Gilchrist & Rutowski, 1986). Intersexual selection has probably been the main driving force in sexual differentiation of male scent organs, tactile signals and secretions (Müller, *ibid.*; Boppré, 1984), but colour pattern dimorphism, specifically male 'brightness', is attributed variously to intersexual selection (Darwin, 1874), to predation and interspecific selection (Wallace, 1889) and recently to intrasexual selection (Silberglied, 1984; Vane-Wright, 1980; 1984). Dismissal of intersexual selection as an ultimate cause for 'bright' male coloration may be premature. It is based on experimental demonstration for lack of preferences and discrimination by females of male coloration in some species, though such experiments on females are notoriously difficult (Silberglied, 1984). Not only may there be selection for wing colour triggered by condition and investment potential during courtship (e.g., *Nathalis iole*, Rutowski, 1981), but there may be selection for long distance apparency-directed female solicitation of male wing colour as there has been for male scent (*Euploea core*; Latter in Boppré, 1984). Female solicitation has been demonstrated in a number of species (Rutowski; 1980, 1982a, b; Rutowski *et al.*, 1981; cf. Silberglied, 1984) and is cued into either habitat structure or male apparency (Wickman & Wiklund, 1983; Dennis & Shreeve, 1988). This may explain the dominance of ultra-violet in female discrimination and brilliant male colours which exceed requirements for short-distance signalling. A number of plausible evolutionary pathways exist for the coincidence of sex and colour pattern as for aposematism and unpalatability (Guilford, 1988), but intraspecific male – male selection can have arisen only as the primary cause for male 'brightness' in species in which males are territorial and vary for resource-holding power. Predation and defence are certainly significant parameters as is suggested by reduced sexual dimorphism and uniformity of colour patterns on dorsa and ventra in unpalatable compared to palatable species, and by greater selection for female than male crypsis (Brakefield, 1984; Landing, 1984; Chai & Srygley, 1990). Solutions set in a multivariate context, including population structure and community organization, may find that interspecific interactions are significant particularly for species-recognition in tropical rain forest habitats (Gilbert, L.E., 1984: 47; see section 3.2).

3.2 **Climatic correlates of global diversity in adult morphology**

Corresponding with the increased diversity from poles to equator is the increased variation in butterfly morphology. Significant increases probably occur in mean body size, wing expanse, wing structures (e.g., tail length; tail numbers) and in wing pattern elements (e.g., ocellus diameter). A greater diversity of shapes occurs for both the forewings and hindwings. Yet the most dramatic increases in variation are found in wing patterns, for different pattern elements and colour fields (D'Abrera, 1981; 1984; 1987; Smart, 1985). Quantification of this complex regional variation is extremely difficult but, despite this, gross pattern type may change systematically with latitude (Landing, 1984; see below).

A *Habitat stratification, the light environment and wing colour patterns*

The vertical stratification of mimicry complexes in Peru (Papageorgis, 1975) provides some evidence for the ecological compartmentalization of dorsal wing pattern variation in tropical rain forests. Five mimicry 'rings' were found (i) to occur as discrete groups at different height levels, (ii) with pattern intermediates flying at elevations intermediate between corresponding complexes, (iii) to comprise similarly patterned different taxa (viz., models and mimics) at any one level, but (iv) differently patterned

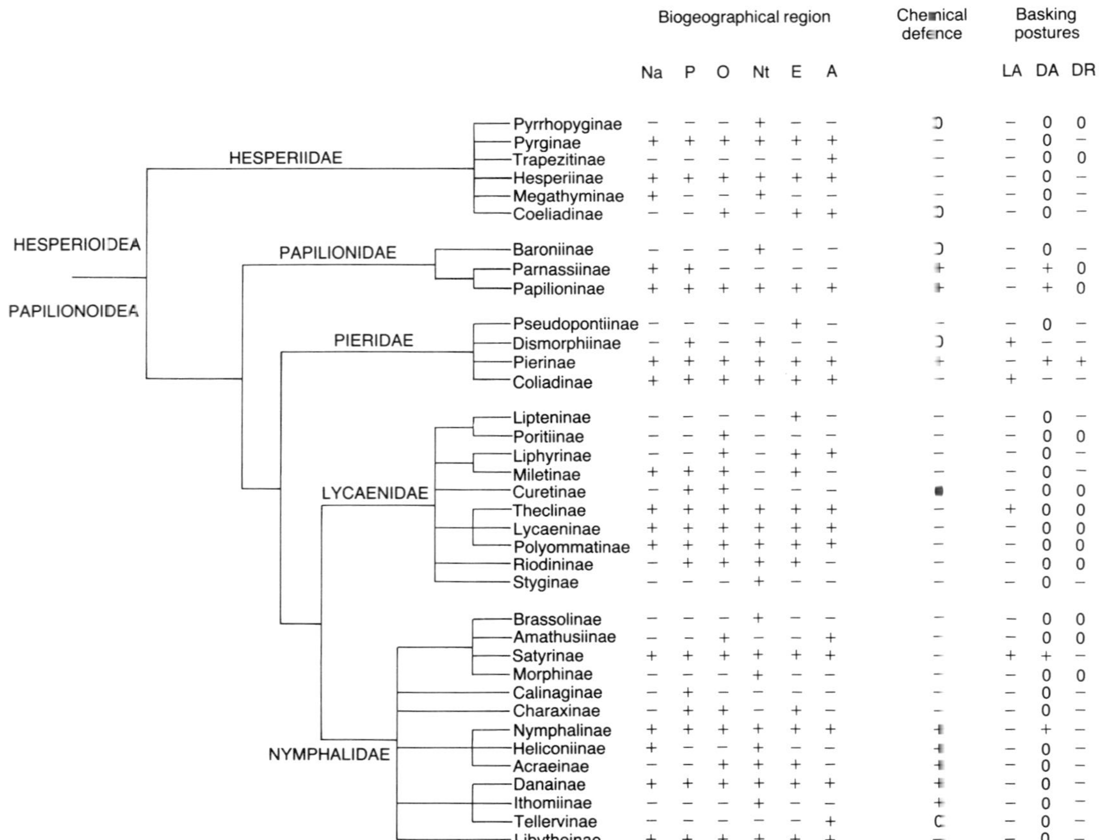
Biogeographical region
Chemical defence
Basking postures
Na P O Nt E A
LA DA DR
HESPERIOIDEA
PAPILIONOIDEA
HESPERIIDAE
PAPILIONIDAE
PIERIDAE
LYCAENIDAE
NYMPHALIDAE
Pyrrhopyginae
Pyrginae
Trapezitinae
Hesperiinae
Megathyminae
Coeliadinae
Baroniinae
Parnassiinae
Papilioninae
Pseudopontiinae
Dismorphiinae
Pierinae
Coliadinae
Lipteninae
Poritiinae
Liphyrinae
Miletinae
Curetinae
Theclinae
Lycaeninae
Polyommatinae
Riodininae
Styginae
Brassolinae
Amathusiinae
Satyrinae
Morphinae
Calinaginae
Charaxinae
Nymphalinae
Heliconiinae
Acraeinae
Danainae
Ithomiinae
Tellervinae
Libytheinae

3.2 A classification of butterfly families and subfamilies, their distribution in biogeographical regions, resting postures and chemical defence. Biogeographical regions: Na, Nearctic; P, Palearctic; O, Oriental; Nt, Neotropical; E, Ethiopian (African); A, Australian. Only presence (+) or absence (–) in a region is recorded, not bias in species diversity. Chemical defence: subfamilies including species believed to be chemically protected in the adult stage (unpalatable to birds) by their hostplants or from adult feeding (+, present; 0, probably present as the hostplant is the same or is closely related to hostplants butterfly taxa use which are chemically defended; –, not known to be present and only possibly absent). Basking postures: DA, dorsal absorbance; DR, dorsal reflectance; LA, lateral absorbance (+, present; 0, probably present; –, probably absent). Data on basking postures are very incomplete. Lateral absorbance identifies subfamilies known to include species that thermoregulate exclusively using a lateral basking posture with closed wings. Many species may heat up opportunistically when feeding, resting or roosting with their wings closed. The dorsal absorbance category also includes taxa that heat up primarily using substrate microclimates rather than direct sunlight. Hostplants are listed in Ackery, 1984; 1988; 1991. (Adapted from Ackery, 1984; Brower, 1984 and Shreeve & Dennis, 1991.)

conspecifics at different levels (Fig. 3.3). Under the canopy, relatively dull colours (blacks and browns) predominate, with increasing amounts of black pigment from the forest floor to the upper canopy. Supra-canopy species are distinguished by being relatively concolorous and brighter (i.e., white, yellow, orange, iridescent). Under the canopy, non-mimetic palatable and distasteful species share the same primary pigments, but the former group have cryptic ventra (Papageorgis, 1975; Chai & Srygley, 1990). The scheme is argued to be applicable to the Hesperioidea though not to the 'browns', which are regarded as a distinct class (Landing, 1984). Dimorphic sexes are alleged to fly at appropriate height levels (i.e., females usually at lower levels) and larger species in any class higher than smaller ones. Landing's novel claim is that these classes 'drop out', from floor to ceiling as forest height is reduced with increasing latitude or into drier regions, and adopt lower levels in angiosperm forests; thus transparents reach their limit in south Mexico, tiger-stripes at the USA border, red/blacks at the Canadian boundary, blue/blacks at the arctic circle, with browns and concolorous 'bright' species extending to the polar deserts. Flying height distributions of colour morphs is argued to relate to changing light conditions but not to thermal efficiency or predator stratification (Papageorgis, 1975); each pattern class is said to be maximally cryptic at different levels. This reasoning is based on the similarity in dorsal pigments between cryptic strictly non-mimetic species and aposematic species in each class, and the assumed premium placed on the former group not to be conspicuous. Thus, 'a warning pattern that also confers flight crypticity would give a distasteful butterfly an added advantage' (Papageorgis, 1975: 530). Gilbert, L.E. (1984) discusses how competition for enemy-free space, reproductive isolation, allopatry and predation may have led to these stratification classes.

Although the unexpectedly dull colour of aposematic models supports crypsis as the proximal cause of colour pattern stratification (Papageorgis, 1975), other factors require consideration, especially the balances between them. The contrast in colour patterns between day-flying (e.g., Uraniidae)

3.3 The relative frequency of individuals belonging to different wing pattern complexes flying at different height intervals in the virgin rain forest at Rio Liullapichis, Peru. The members of the orange complex, and presumably species with pure colours (i.e., white, yellow, blue, iridescent), fly above the canopy. N, number of observations; C, Morisita measure of overlap between adjacent complexes, where $C = (2\sum_{i=1}^{n} x_i y_i)/(\sum_{i=1}^{n} x_i + \sum_{i=1}^{n} y_i)$, in which x and y are the percentages of members of two adjacent complexes flying in the height zone i, and n is the number of flight zones. The overlaps are sufficiently low to consider the complexes vertically segregated. Note: uneven class intervals used. (From Papageorgis, 1975; by courtesy of the Scientific Research Society, USA.)

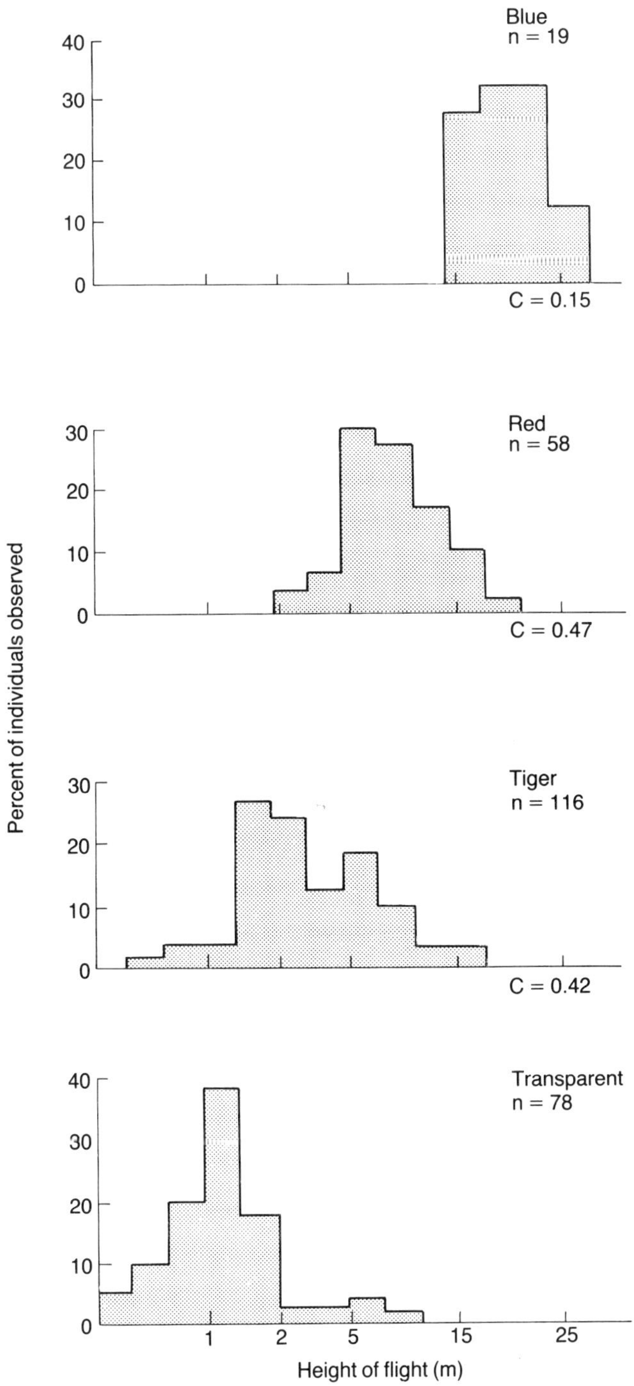
Blue
n = 19
C = 0.15
Red
n = 58
C = 0.47
Tiger
n = 116
C = 0.42
Transparent
n = 78
Percent of individuals observed
40
30
20
10
0
1
2
5
15
25
Height of flight (m)

and night-flying moths, and of butterfly species in exposed and shaded habitats, implies strong selection for bright colours in light environments – aposematism according to Turner (1984). The patterns may be maximally conspicuous to predators (Benson, 1978). If so, then bright pigmentary markers on the dorsa of ventrally cryptic species may have some mimetic (flicker) function – inexact mimicry for the dorsal pattern when exposed in stationary individuals, but effective enough to cause predator-evasion when in motion. Colour pattern classes may also have relevance for social communication, particularly *vis-à-vis* mate-location, strategies about which little is known for tropical species. Dismissal of thermoregulation as a factor has ignored flight speed relative to defence mechanisms, insect size, T_a variance within the canopy and different thermal strategies (see section 1.2; 3.3). Unpalatable species can generally occupy relatively closed and cooler habitats than palatable species which require higher T_b for predator evasion (Chai & Srygley, 1990). Unfortunately, observations on the geographical distribution of pattern classes (Landing, 1984) are deficient in several respects; fundamentally, they lack quantification but are also arbitrary and artificial (Murphy, 1986). A precise mechanism was not described though competitive exclusion for light environment was insinuated, but as this depends on parameters other than tree height and can vary as dramatically within temperate forests (Geiger, 1965), this does not explain the loss in pattern classes. A number of ecological constraints have been ignored that may explain wider aspects of decline in pattern diversity than Landing's pattern classes, which is only a small subset (i.e., colour field) of the variation system, and would be needed to account for the occurrence of sub-canopy pattern types in open habitats (e.g., partially transparent *Parnassius mnemosyne*).

B *Ecological and taxonomic constraints on wing morphology distribution*

Global variation in wing pattern diversity may largely reflect taxonomic diversity and the influence of historical factors and climate on resources, habitat complexity and structure (section 2.3A). Taxonomically limited wing patterns are restricted by butterfly – hostplant affiliations and the climatic limits of hostplant distributions (Ackery, 1984). Thus, many groups of higher taxa (e.g., Heliconiinae, Acraeinae, Ithomiinae, Danainae) and lower taxa (e.g., Troidini) are virtually restricted to the tropics (Fig. 3.2). As aposematic patterns of unpalatable species form 'resources' to be colonized by mimics (Gilbert, L.E., 1984), range limits for aposematic models cause pattern 'loss' amongst mimicry specialists (e.g., the dependence of Dismorphiinae mimics on Ithomiinae models in Costa Rica, DeVries, 1987; increasing monomorphism from core to periphery in Batesian mimics; Vane-Wright, 1979). The divergence of wing patterns from the Schwanwitsch groundplan may be promoted by co-evolutionary feedbacks on the models to escape 'parasitism' by mimics, even in largely so-called Müllerian rings (DeVries, 1987: 28). Thus mimicry complexes result in reduced pattern diversity by convergence but distinctive patterns by divergence and advergence (Turner, 1984; 1987 but see 1971). The inference is that taxonomically distinctive wing patterns are evolutionary markers of adaptive radiation and life history adjustments to novel hostplant taxa and their habitats.

Increasing autotroph species diversity results in greater habitat structural complexity, thus more microniches and potential for resource partitioning and specialization within distinct signalling zones (e.g., large 'eyespotted' *Morpho achilles* 'group' of rotting fruit and soupy fungi ground-feeders; conspicuous aposematic epiphytic mistletoe canopy feeders lacking eyespots; Young, 1980; Landing, 1984; Gilbert, 1984). Niche and signalling specialization may be enhanced in complex tropical habitats if populations are sparser and potential competitors and predators are more abundant than in temperate regions (Sbordoni & Forestiero, 1984). As this could theoretically lead to increased contacts with non-conspecifics, conspecific competitors and predators, relative to conspecific mates, there would be a premium on selection for visual markers to reduce harassment and to maximize anti-predator defences and for mate-seeking devices which do not detract from these adaptations (e.g., scent; habitat cues; Gilbert, 1984; Boppré, 1984). Taxonomic diversity, thus pattern diversity, declines with structural complexity at altitude, with latitude or disturbance (Papageorgis, 1975; DeVries, 1987; Thomas, 1991).

Regional heterogeneity for wing pattern diversity may be caused, partially, by differential species dynamics – taxa may vary for rates of speciation, dispersal, colonization and extinction – and habitat preferences generated by climatic changes, such as Pleistocene polyglaciation, in regions varying in physiognomic diversity (Dennis *et al.*, 1991). This is partly suggested by the number of replicates (i.e., species per genus), the differences between regional and 'global' frequencies in replicates and spatial bias for taxa. For example European Nymphalinae have fewer replicates, more divergent pattern forms and lower endemicity compared to Satyrinae (Dennis, unpub. data). This supports other evidence (e.g., for migration) that they are able to disperse over wider regions. Taxa with wide dispersal capacity tend to have wider habitat preferences and to be subject to more uniform selection pressures than those that do not. Therefore they are less likely to speciate or to become extinct; moreover, the wing patterns are likely to be conservative and adapted to a wider range of habitats rather than to specific ones (section 4.4). Conversely, many Satyrinae (e.g., *Erebia* spp. in montane southern Europe) have numerous replicates, displaying similar pattern forms, and higher regional fidelity. Taxa such as these, which are more stenoecious and with comparatively modest dispersal capabilities are more likely to become extinct or to speciate with climatic change (see section

4.3B). Depending on the precise nature of climatic change and regional physiognomy, a geographical bias can build up between such contrasting taxa (Dennis, unpub. data).

Geographical differences in wing pattern diversity may also be influenced by variation in constraints for development and activity. Higher temperatures in the tropics place fewer constraints on growth rates and development time, heating rates and attainment of adult T_b for OAR, thus on body size and wing expanse. Large insects are more apparent to predators, as well as to conspecifics, and possibly more sought after as prey (Bowers *et al.*, 1985). Thus, secondary defences (e.g., flash coloration, eyespots; false heads) should be enhanced. Yet this does not entirely account for commonly exaggerated patterns and features (e.g., brightness, tails, ocelli) in butterflies similar in size to those in temperate habitats. Instead, it is suggested that higher stable T_a in tropical habitats influences selection balances for thermoregulation and activity. As T_a is closer to T_b for OAR, there will be reduced selection for pattern devices enhancing thermoregulation (see section 1.2) which could compromise the evolution of social and anti-predator signalling (section 3.3). More importantly, attainment of optimal T_b values results in higher activity, increased apparency to predators and consequently increased selection for secondary rather than primary defences (section 3.3). Secondary defences tend to be brighter (more apparent) than primary defences, as indicated by the predominant crypsis in night flying moths, which are immobile in daylight hours. Enhanced secondary defences relating to increased potential for activity and predator-evasion reinforces any increased 'signalling' to predators due to increased predation in tropical habitats and may not conflict with selection for interspecific and conspecific communication, especially in open habitats. However, in closed habitats, at different levels under the canopy, selection balances change. T_a is lower than T_b for OAR, thermoregulation is less efficient with the lower radiation levels and butterflies may be less able to evade predators in the lower light conditions (Dennis *et al.*, 1986). Pigmentary wing phenotypes for different sub-canopy microhabitats may represent balances between long range crypsis and short-range secondary defences and non-ultra-violet social signals, thus between rates of attack and of successful counters. Although the unpalatability of models must be effectively advertised and sampled, there may be a distinct disadvantage for even chemically-defended aposematic species to be too apparent. An increased attack rate encouraged by a distant but indeterminate highly apparent prey could result, stochastically, in higher mortality, especially if predators vary in their tolerance of prey items. Patterns which are cryptic at a distance and at distinct flight levels in the rain forest canopy would reduce predation. These observations on the stratification of pattern in rain forest habitats amount to a greater diversity of microhabitats producing a greater diversity of pattern types.

3.3 Climate and microevolution of adult morphology

Microevolutionary variation in wing morphology should be more amenable to analysis than variation between higher taxa. It has been impeded by several factors: (i) the type and degree of changes in phenotype are not consistent for different species, (ii) trends vary in space (i.e., orientation; rate of change) and are often disrupted by local environmental diversity, (iii) quantification of the variation, genetic analyses and taxonomy are often inadequate or lacking, as are (iv) ecological and behavioural data, essential for interpreting changes in morphology (see Wcislo, 1989). As a result, much variation is regarded either as environmentally induced and non-adaptive (Bowden, 1979), as a pleiotropic biproduct of selection during development of early stages or due to uncertain historical influences (Ford, 1945; Beirne, 1943, 1947; Heslop-Harrison, 1947). While many phenotypic transformations in adult butterflies may lack obvious adaptive functions (i.e., phylogenetic inertia) but not be maladaptive, this section illustrates how relatively simple explanations, based on the interaction of exogenous agents and behavioural and physical differences in temperate and arctic butterflies, may account for part of the variation (Dennis & Shreeve, 1989).

A *Patterns and correlates of morphological variation*

Thirty-three of 55 British-resident, non-migratory butterflies vary significantly in size, colour or pattern from those on adjacent parts of the European continent or regionally within the British islands; nine of the phenetically uniform remainder have very restricted ranges (Heath *et al.*, 1984). The relative frequency and degree of phenetic differentiation increase systematically to the north and west (Fig. 3.4); the rate of increase is maximal in central and northern Scotland and between the British mainland and Ireland. Variation may be compounded by 'island' effects, but the latitudinal gradient is self-evident; localized variants occur on small islands or on disruptive ecological units which may have been subjected to genetic 'bottlenecking' (Dennis, 1977; Berry, 1979). Larger island morphs generally reveal a continuation and intensification of mainland patterns; they have more numerous and larger populations which are less subject to drift and receive gene flow from adjacent islands and the mainland (Dennis, 1977). The nature of the variation conflicts substantially for the 33 variable species (Table 3.1). Some increase in size, others are smaller; many are more melanic, others less so. Some are more cryptic in overall ground colour and less contrasting in pattern (e.g., *Gonepteryx rhamni, Coenonympha tullia, C. pamphilus*); others are the reverse of this (e.g., *Argynnis aglaja* ventra, *Hipparchia semele, Maniola jurtina*). Ocellation and marginal wing lunulation are more pronounced in some species (e.g., *Anthocharis cardamines, Polyommatus icarus*); in others they are diminished or absent (e.g., *C. tullia; Aphantopus hyperantus*). Common patterns of variation are

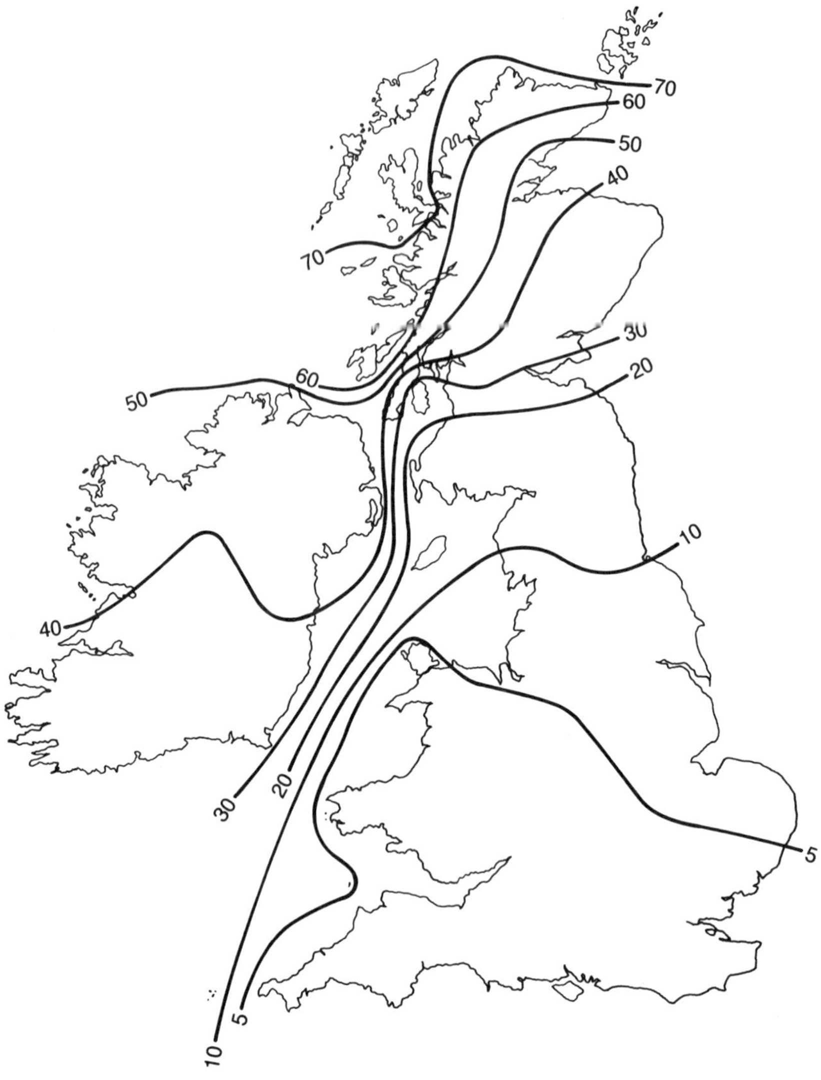

3.4 Phenetic variation among British butterflies. Isophenes for wing morphology variation plotted as percentage variation of the total fauna for 100 km square intersections and island centroids, from a base of zero in south-east England. (From Dennis & Shreeve, 1989; by courtesy of Academic Press Ltd.)

Table 3.1 Changes of wing morphology of variable species within the British Isles north and west to Scotland and Ireland (From Dennis and Shreeve, 1987, by courtesy of Academic Press Ltd)

Species	Major changes of wing characters to north and Ireland	Melanism D	Melanism V	Contrast and brightness D	Contrast and brightness V	Secondary defence D	Secondary defence V	Size: wing expanse	Basking method
Carterocephalus palaemon	D: contrasting black and orange V: darker ground colour with brighter spots and greener tone	+	+	+	+				D
Ochlodes venata	D: darker	+							D
Papilio machaon	D: darker with primrose ground colour, sub-marginal band more extensive	+		+?					D
Leptidea sinapis	D: black suffusion extends proximally from apex along costa and veins V: greenish-grey suffusion	+ (FW)			–				L
Gonepteryx rhamni	D: ♂ FW paler and HW lighter and greener. ♀ FW margin greenish-yellow especially at apex and HW yellower V: ♂. FW greenish-white			–	–				L
Pieris napi	D & V: black suffusion especially at base and along veins but also between veins D: yellow ground colour more frequent V: yellow suffusion deeper	+	+	+	+ (HW)				R+D
Anthocharis cardamines	D: black areas more extensive at FW apex, discoidal spot, HW marginal areas and possibly base. ♀ HW yellower V: HW possibly yellower WE: smaller	+		+ (♀)	+	+		–	R
Lycaena dispar	V: HW red band wider and groundcolour paler blue-grey. Spots larger WE: larger	–		+	–GC +P		+	+	R

Table 3.1 (cont.)

Species	*Major changes of wing characters to north and Ireland*	*Melanism* D	*Melanism* V	*Contrast and brightness* D	*Contrast and brightness* V	*Secondary defence* D	*Secondary defence* V	*Size: wing expanse*	*Basking method*
Lycaena phlaeas	D: HW copper band broader, ground colour greyer and red marginal band brighter	–		+	–GC +P	–			R
Aricia-agestis-artaxerxes	D: darker with smaller marginal lunules, white discal spot on artaxerxes V: greyer with smaller black spots and white dots WE: smaller	+			–	– (Lu) + (Di)	–	–	D(+R?)
Polyommatus icarus	D: ♂ blue not violet, median black spots more evident; ♀ very blue, lunules brighter V: ♀ distinctly marked, red lunules very bright WE: larger	– (♀)		+ (♀)	+ (♀)	+	+	+	R
Plebejus argus	D: more blue in (extinct) northern moss colonies WE: larger except *caernensis*	– (♀)		+ (♀)				+	R
Celastrina argiolus	D: brighter blue V: more expansive blue and more prominent spots	–		+			+		R
Maculinea arion	Devon D: brighter blue V: larger but fewer spots	–		+			+	±	R
Lysandra bellargus	D: ♀ more extensive blue and lunules duller and smaller V: ♂ darker with duller ground colour and smaller spots	– (♀)		+ (♀)	– (♂)	–	–		R
Aglais urticae	D: black suffusion with broader black margin, smaller marginal blue spots	+		–		–			D

Boloria selene	D: black suffusion especially of basal area and wing markings, brighter V: more contrasting	+		+	+	+?			D
Boloria euphrosyne	D: black suffusion V: HW deeper ground colour and black suffusion	+	+ (HW)	–	–				D
Argynnis aglaja	D: black suffusion, occasionally entirely black V: dark-green, prominent, large, silver spots, heavier black markings WE: generally larger	+	+ (FW)	–	+			+	D
Argynnis paphia	loss of f. *valezina*	+?	+?						D
Eurodryas aurinia	D: black suffusion and increase in pattern contrast (terracota versus pale yellow or white)	+		+?					D
Mellicta athalia	Cornwall: darker and duller WE: smaller in Cornwall than in Kent	+		–				–	D
Pararge aegeria	D: darker with larger and brighter pale patches, spots larger V: darker, lilac marginal band larger and purple, spots larger WE: larger	+	+	+	–	+		+	D
Lasiommata megera	D & V: darker suffusion and heavier patterning	+	+	–?	–?				D
Erebia epiphron	D & V: red band wider WE: larger in Scotland than Lake District			+ (FW)	+ (FW)			+	D
Erebia aethiops	D: blacker suffusion, smaller fulvous band and fewer spots V: fewer spots WE: smaller	+		–	– (HW)	–	–	–	D
Hipparchia semele	D: FW medial band wider and paler, highlighting apical spot V: FW more orange and brighter; HW black suffusion WE: smaller		+ (HW)	+ (FW)	+ (FW)		+	–	L

Table 3.1 (cont.)

Species	Major changes of wing characters to north and Ireland	Melanism D	Melanism V	Contrast and brightness D	Contrast and brightness V	Secondary defence D	Secondary defence V	Size: wing expanse	Basking method
Pyronia tithonus	D: darker suffusion especially of basal area, marginal band broader and deeper fulvous V: more contrasting and pronounced spotting	+		+	+		+		D
Maniola jurtina	D: blacker, fulvous band larger; FW apical spot larger V: darker; FW apical spot larger and bi-pupilled; HW spots fewer WE: larger and smaller	+	+	+		+	+ (FW) (HW)	±	D
Coenonympha pamphilus	D: pale ochreous with smaller apical spot V: paler grey; spots smaller with silver pupils			–	–	–	–		L
Coenonympha tullia	D: pale ochreous with small apical spot V: paler grey; spots absent or small with silver pupils		+? (HW)	–	–	–	–	±	L
Aphantopus hyperantus	D: possibly darker with smaller spots V: greyer; ♀ lighter yellow; smaller spots in both sexes WE: smaller	+?			–	–	–	–	D

D, Dorsal surface; V, ventral surface; FW, forewing; HW, hindwing; WE, wing expanse; GC, ground colour; P, pattern; Di, discal; Lu, lunules. Basking method—R, reflectance; L, lateral; D, dorsal. +, Increase; –, decrease.

evident within lower taxa (*Coenonympha* spp; closely related *Pararge* and *Lasiommata*) and within some higher taxonomic groups (e.g., tendency for Nymphalidae: Argynnini to become more heavily melanized with latitude and for brighter dorsa amongst lycaenid females). This may implicate responses common to particular phylogenetic groups. However, there are also related taxa which do not show parallel modifications in phenotype (e.g., Lycaenidae: *P. icarus* and *Aricia artaxerxes*; Satyrinae: Maniolini, *M. jurtina* and *A. hyperantus*).

Similar changes in butterfly morphology occur in Scandinavia, for which a number of clines have been mapped by Petersen (1947). Of 120 species (Henriksen & Kreutzer, 1982), 81 are variable. Some 30 of the 39 species that do not vary either are migrants or have very restricted ranges in Scandinavia. The dominant trends, into the Norwegian mountains or arctic Lapland, are a reduction in wing expanse, melanized and duller dorsa and duller less-patterned ventra, features which are characteristic of arctic endemic butterflies. However, a number are larger, with brighter dorsa and ventra and have enlarged ocellation (Table 3.2). Butterflies which occur in both Scandinavia and Britain vary in much the same manner although exceptions in northern trends do occur (e.g., *Gonepteryx rhamni*) and more significantly between populations of species occupying both montane and arctic Scandinavian environments (e.g., *Polyommatus icarus*).

The latitudinal trend in phenetic pattern changes coincides with the decline in numbers of species (section 2.3B); it also corresponds with changes in population status, phenological, migrational, physiological and biochemical gradients, as well as with less evident morphologial changes. Populations tend to be smaller, with higher variance, and are more isolated northwards (Warren, 1992a); the seasonal orientation of migrants is meridional (e.g., *Pieris rapae*; Baker, 1978, 1984). Voltinism declines (Fig. 2.2), broods tend to be more synchronized (e.g., *Maniola jurtina, Pyronia tithonus*: Brakefield, 1987a), and systematic trends occur in thresholds and sensitivity to token stimuli influencing diapause induction (e.g., *Aglais urticae, Inachis io*; Pullin, 1986b, 1988; Fig. 2.5). North – south trends in allozyme variation occurs (e.g., *M. jurtina*; Thomson, 1987b). Geographical variation also occurs in other aspects of wing morphology (e.g., *Pieris napi* androconial scales; Thomson, 1970a; Bowden, 1985) including polymorphisms (e.g., *Argynnis paphia valezina; Aricia artaxerxes* white discal ocellation; *Aphantopus hyperantus arete*; Robinson, 1971; Dennis, 1977). As much of this variation has been shown to have a genetic base (Lees, 1965; 1969; Lees & Archer, 1980; Pullin, 1986b; Brakefield, 1984; 1987a), there is strong evidence to support the theory that phenotypes of insect species are modified to suit local conditions. Some wing morph clines have also been shown to have a substantial genetic foundation; such are the latitudinal decreases in mean ventral hindwing ocellation and costality in *M. jurtina* (Brakefield, 1984), melanin and pteridine pigments in *Pieris* and

Table 3.2 Summary frequencies for the transformations in wing morphology northwards in Scandinavian and British butterflies (From Dennis & Shreeve, 1989)

Variables and attribute states		Scandinavia Increase	Scandinavia Decrease	Britain Increase	Britain Decrease
Melanism (base and lower veins)	Dorsal	34	7	20	7
	Ventral	8	0	10	0
Contrast/brightness	Dorsal	17	39	18	9
	Ventral	12	32	11	13
Secondary defence (marginal wing spots)	Dorsal	12	14	5	8
	Ventral	9	19	7	6
Wing expanse/size		7	43	9	10

Based on 120 Scandinavian species and 59 British species. Of the 39 Scandinavian species that do not vary, nine have widespread distributions, 22 have restricted distributions in Scandinavia, and eight are migrants. Twenty-six British species do not vary; nine of these have restricted distributions and 4 are wholly migrants.

in *Anthocharis cardamines* (Watt & Bowden, 1966; Shapiro, 1984b; Majerus, 1979) and a range of attributes (e.g., ground colour and ventral wing margin lunulation) in *Aricia artaxerxes* and *A. agestis* (Jarvis, 1966; Høegh-Guldberg, 1966; 1968; Høegh-Guldberg & Hansen, 1977).

Transformations in typical Müllerian and Batesian mimics indicate that a combination of selection and single gene switches can have a dramatic effect on the wing phenotype and wing shape (Clarke & Sheppard, 1960a, b; 1962b; Sheppard *et al.*, 1985; Nijhout *et al.*, 1990). Much of the clinal variation northwards in Britain and Scandinavia could be polygenic, thus equally responsive to selection (Brakefield, 1984). Some caution is required in making this assumption. Temperature and other climatic agents influence wing variation directly in some species (*Maniola jurtina, Erebia aethiops*; Thomson, 1973, 1980); every pattern element has its particular sensitive period to sublethal traumas (Brakefield, 1984; Nijhout, 1985). In *Aricia artaxerxes*, lower spot number can be produced by subjecting insects just before or during pupation to periods of cooling at 2–5°C for 9–12 h, thus simulating rare forms found in the field (Høegh-Guldberg & Hansen, 1977). Nevertheless, in all species there are genetic differences for phenocopy production and expressivity (Nijhout, 1984). The plasticity of phenotypes is not evidence for a lack of genetic variance nor of the irrelevance of evolutionary processes for attributes, since both the direction and degree of response to environmental cues may itself be under genetic control as in the case of polyphenisms (Clarke & Sheppard, 1972; Shapiro, 1976). Environmentally induced non-genetic effects on phenotypes, while not transmitted genetically to the next generation, can alter the strength and direction of selection affecting transmitted gene frequencies by markedly shifting the phenotypes exposed and expressed, and thereby subjected to selection and evolution. Selection can modify the form of attributes which are repeatedly produced by the same environmental conditions, even if there is a large non-genetic component for them (West-Eberhard, 1989) and phenotypic plasticity can be under genetic control (Thompson, 1991).

Two very different hypotheses are argued to account for the gradient in wing morphology transformations: (i) survival of isolated populations in small refugia associated with lower sea levels in western Britain during the Devensian glaciations and their subsequent colonization in the Flandrian (Beirne, 1943, 1947; Ford, 1945; Heslop-Harrison, 1947), and (ii) evolution in response to contemporary (Flandrian) conditions (Dennis, 1977; Dennis & Shreeve, 1989). The two hypotheses are not necessarily mutually exclusive (Dennis & Shreeve, 1989; see section 4.4); but interpret the variation very differently, the former as phylogenetic markers with little current relevance for survival (Bowden, 1979), the latter as of significance for individual fitness. Whereas resconstructions of climate and environment fail to support refugia in the vicinity of Britain for either the Devensian maximum or Loch Lomond Readvance (Fig. 4.3), the phenetic gradient correlates closely with climate (Dennis, 1977; Dennis & Shreeve, 1989; Table 3.3). Summer temperatures account for most of the variance in physical differences; significant amounts are also determined by sunshine, precipitation and windspeed. It is unlikely that winter temperatures are as important since the species affected occur at sea level and correlations with January sea level temperatures are low. The multiple correlations of morph transformations with several climatic variables and the extensive covariance among the latter are expected because of their causal association in pressure systems and air mass circulation, powerful meridional and zonal influences, and thus the marked regional bias in weather systems over Britain (Green, 1964; Hatch, 1973; White & Perry, 1989). These findings are very similar to the relationships between species richness and climatic attributes (section 2.3B). Thus it is plausible that as there is a causal link between climate and population status with increasing latitude, an indirect one may exist between population parameters and phenetic transformations, as well as direct links between climate and phenetics. A contemporary selectional hypothesis does not invalidate important historical processes, as appropriate environmental regimes and gradients have probably been perpetuated throughout the Pleistocene glaciations (section 4.3A).

As the impact of climate change may differ for open and closed habitats, disparities in phenetic transformations associated with habitat structure may be significant and have consequences for the type and direction of phenetic modifications (Dennis & Shreeve, 1989). Among ubiquitous British resident butterflies, significantly more that are distinctly variable are associated with early seres (70 per cent) than with later successional stages (25 per cent). Species of open habitats also extend further north of the Highland boundary fault (52 per cent) than butterflies associated with woodland (15 per cent) but, in doing so, probably experience threshold temperatures (critical conditions) more frequently. Early rather than later seres are more exposed and less buffered against fluctuating temperatures. Therefore, it is interesting that, of those species which occupy early seral habitats, significantly more extend beyond the Highland boundary fault that vary phenotypically (78 per cent) than those which do not (21 per cent). This supports other evidence that northern British environments are marginal for butterflies (see sections 2.2A; 2.3B).

B *Climate – behaviour interactions and wing morphology transformations*

Expectations are that in cooler climates there will be: (i) a reduction in adult flight time, (ii) lower activity rates (i.e., flight speed and evasion) and (iii) poorer responses to changing conditions. An inference from the negative correlations between summer climate and wing morphology transformations is that as summer conditions become cooler, less sunny

Table 3.3 Correlations between phenetic transformations in wing morphology (per cent species) and selected climatic variables in Britain (From Dennis and Shreeve 1989; by courtesy of Academic Press Ltd)

	Phenetic variants (% spp.)	Grid north	Grid east	July bright sunshine (hrs)	July mean temperatures (°C)	July maximum temperatures (°C)	January mean temperatures (°C)	January maximum temperatures (°C)	Precipitation total (mm)	Humidity (% time > 99%)
Grid north	0.90									
Grid east	−0.64	−0.41								
July bright sunshine (hrs)	−0.71	−0.86	0.30							
July mean temperature (°C)	−0.93	−0.92	0.67	0.76						
July maximum temperature (°C)	−0.85	−0.77	0.74	0.55	0.92					
January mean temperature (°C)	−0.12	0.39	−0.52	0.51	0.07	−0.21				
January maximum temperature (°C)	0.03	−0.25	−0.60	0.42	−0.06	−0.32	0.99			
Precipitation total (mm)	0.59	0.52	−0.69	−0.63	−0.61	−0.54	0.08	0.16		
Humidity (% time > 99%)	0.33	0.35	−0.44	−0.63	−0.43	−0.30	−0.08	−0.02	0.65	
Windspeed ($m.s^{-1}$ for 50% time)	0.67	0.43	−0.74	−0.12	−0.67	−0.83	0.53	0.62	0.36	0.04

Spearman correlations, r_s; significance levels: $> \pm 0.24$, $p < 0.05$; $> \pm 0.34$, $p < 0.01$; $> \pm 0.43$, $p < 0.001$. Sample size: N = 47, sampling at 100 km grid intersections and island centroids. Temperature data reduced to sea level. Climatic data from Meteorological Office, by courtesy of R. Whyman.

and increasingly unpredictable, selection for modifications in adult phenotype should enhance thermoregulation, mate advertisement, and predator-escape. The nature of modifications will be significantly affected by physical and behavioural differences between taxa, especially settling posture and basking mechanisms, but also by habitat structure – the context of thermal conditions, resources and enemies (Fig. 3.5). Behavioural differences often precede and stimulate morphological evolution (Wcislo, 1989). The model described here assumes: (i) that phenetic modifications have a significant genetic variance (see above); (ii) conservatism in basking and settling posture; and (iii) general spatial uniformity in species-specific behaviour patterns despite any changes in habitat associations. Butterflies 'settle' (i.e., bask, rest, roost) in very different ways and in different positions in their habitats; thus, they differ in their relative apparency to conspecifics and predators. Basking methods differ as to the surface exposed (i.e., ventra *versus* dorsa) and for dorsal baskers the mechanism employed (i.e., reflectance *versus* absorbance; see section 1.2). The majority of species are probably restricted as to thermoregulatory posture. Lateral absorbance baskers invariably use a ventral wing surface; dorsal absorbance baskers may combine with limited lateral basking (e.g., *Maniola jurtina*), whereas some dorsal reflectance baskers are known to absorb energy dorsally and laterally (e.g., *Pieris napi*; Kingsolver, 1987a). It is believed that these modes of behaviour developed in association with hostplant – habitat structures and are of some antiquity. Many species now exploit novel habitats produced by anthropogenic interference but similarities in behaviour in different habitats are usually maintained (Shreeve & Dennis, 1991).

The present model allows a number of predictions to be made as to modifications in phenotype from changes in thermal environment. Regarding thermoregulation, cooler, cloudier and windier conditions and lower irradiance levels result in reduced adult activity (i.e., rate, duration), which in turn influences their ability to locate mates, to court conspecifics successfully, to realize their potential fecundity and to colonize new habitats; depression of any of these biological functions adversely affects reproductive success. Modifications of phenotype in response to cool climates depends on species-specific basking modes. In lateral absorbance baskers, an advantage is obtained by increased melanization of the ventral hindwing surfaces particularly in the region overlapping the body (e.g., *Colias* spp.; Kingsolver, 1983; Kingsolver & Watt, 1983); in dorsal absorbance baskers, it is increased melanin of the dorsal forewings and hindwings (Wasserthal, 1975). In dorsal reflectance baskers, a reversal of this process is expected (less melanin), the dorsal median and distal surfaces tending to be brighter, enlarging the reflective area (Kingsolver, 1985a, b) and for the base and proximal venation to be melanized to increase energy absorption in the vicinity of the thorax (Kingsolver & Wiernasz, 1987). In *Melanargia*, the dorsal surfaces become blacker in warmer regions (Descimon & Renon,

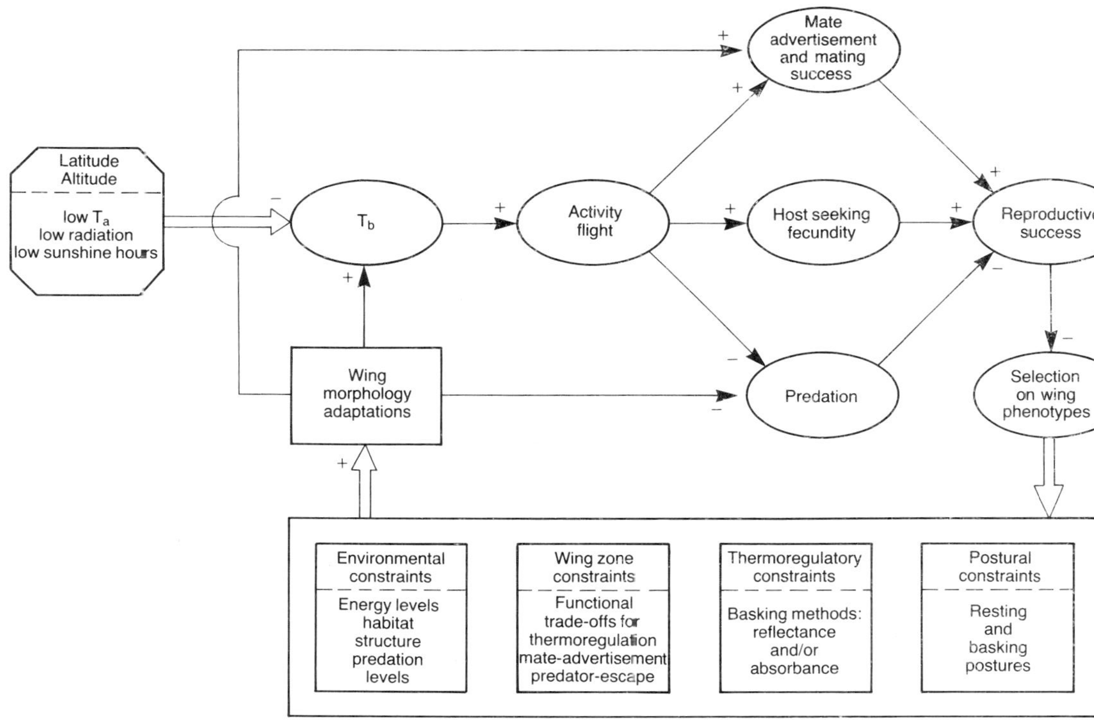

3.5 A generalized model for wing phenotypic transformations in butterflies within cooler climates. The type of transformations depends on a number of constraints, the most important of which determine posture and thermoregulatory mechanisms and have a phylogenetic basis. Octagon, zonal factors; rectangle, butterfly morphology; ellipses, biological processes; squares, constraints on morphological adaptations.

1975). However, in dorsal reflectance and dorsal absorbance baskers, there may also be melanization of the ventral hindwing base and venation, providing an advantage for insects which engage in periodic opportunist lateral basking. Many of the butterflies in north-west Britain follow these predictions (Table 3.1).

Although reduced energy levels at higher latitudes may be partly compensated by adaptations to maintain thermal efficiency, insects which are active for short periods of time will have fewer contacts with conspecifics. In cooler, cloudier and windier conditions, thresholds for courtship displays may be affected and courtship rituals be less successful. Energy levels also influence population size and density (section 2.2A) and this in turn may delay conspecific contacts. Thus, a second group of phenetic modifications may influence individual apparency. Brighter coloration (e.g., dorsal metallic blue in female *Polyommatus icarus*; ventral hindwing silver spotting in *Argynnis aglaja*; bright yellows/fulvous in *Maniola jurtina* and *Hipparchia semele*) in these circumstances may partially offset lower density, lower apparency and limited search times, especially for butterflies which lack effective habitat structures as predictable mate-location cues (Dennis & Shreeve, 1988). These modifications could also combine with secondary anti-predator devices when the insects are fully active.

The third group of phenetic modifications consists of those predicted to improve defences against predators, especially primary defences on the ventral hindwing. Reduced activity in adults associated with lower energy levels northwards results in their adopting passive postures (viz., nectaring, resting, basking and roosting) for longer periods. Immobile, apparent butterfly adults are more susceptible to predation. When butterflies are close to threshold T_bs for activity, secondary defences are less effective as they cannot make good their escape. In this situation, it is predicted that selection will enhance the primary defence function (i.e., crypsis) of the ventral hindwing, ensuring that adults are less apparent when immobile. Both *Gonepteryx rhamni* and *Leptidea sinapis* have greener ventral hindwings in Ireland, and *Aphantopus hyperantus, Coenonympha tullia* and *C. pamphilus* are paler on this surface. However, apparent secondary defences on the ventral hindwing will be less of a handicap to insects which conceal themselves in dense vegetation when inactive. On the other hand, secondary defences on the dorsal surfaces and ventral forewing, hidden when the individual is completely inactive, may be enhanced as a result of selection for signals which induce a larger shock in predators, increasing evasion time for active insects above the temperature threshold for activity (e.g., fulvous development and apical ocellus bipupillation in *M. jurtina* and *H. semele*; Dennis 1977; Thomson, 1980).

Alterations of the phenotype in response to constraints imposed by thermoregulation, mate advertisement and predation may not occur in isolation from one another. The nature of interactions or correlations among pattern elements on wing surfaces (viz., degree, sign) depends largely on basking posture and on the milieu in which insects exist, since this determines the circumstances in which wing surfaces are presented or exposed to conspecifics and enemies. It is possible to predict the correlations and changes in wing pattern elements; Kingsolver and Wiernasz (1987) have attempted this with some success for *Pieris* butterflies (i.e., *P. occidentalis* and *P. napi*) based solely on thermoregulatory adaptive hypotheses. Different basking postures place different emphases on particular wings and wing areas. In lateral basking species, the ventral surface is exposed for longer periods and virtually at all times the insect is not in flight. Therefore, compared to dorsal baskers, there will be: (i) a greater emphasis of all functions on the ventral hindwing surface, especially of thermoregulation which is concentrated basally, and therefore greater potential conflict between these functions on this surface; (ii) a greater trade-off between primary and secondary defence mechanisms on the ventral hindwing, as this surface must serve as defence when the insect is potentially active and when it is inert. The secondary defences of the ventral forewing, except the apex, can be independent of adjustments on the hindwing as most of this surface is concealed when the insect is at rest. In dorsal baskers, a greater emphasis on secondary defences may be expected for the dorsal surface in cooler climates, which is exposed when the butterfly is active, but this depends on time constraints and habitat structure.

In cooler, less sunny climates the ventral hindwings of lateral baskers will be exposed for longer periods of inactivity. Modifications for primary defence (crypsis) may then dominate any tendency towards enhanced mate advertisement, but need not be in conflict with thermoregulatory adjustments in the basal wing area. This seems to be the case in four out of the five lateral baskers which vary in Britain (i.e., *Gonepteryx rhamni, Leptidea sinapis, Coenonympha tullia* and *C. pamphilus*); the ventral hindwing is paler and cryptic. In *C. tullia*, both marginal wing pattern and ground colour of both wing surfaces are transformed northwards and with higher elevation (Dennis, 1977; Dennis *et al.*, 1986). Ocelli become smaller or absent and the white pupils replaced with silver; the wing colour is duller and more uniform in colour and pattern. There is a close correspondence between these phenotype changes and decline in energy (Fig. 3.6). Because of the lower temperatures and sunshine levels and the changeability of conditions, Scottish *C. tullia* are probably less active, immobile for longer periods and consequently less able to evade predator attacks or escape when captured. Prominent spots found in English lowland moss populations, which are used as deflectors, would in northern sites attract undue attention from predators. Thus, in cool, cloudy environments, selection favours a cryptic wing pattern; the smaller ocelli with their reflectant silver pupillation are likely to be evident only at close quarters and when, in sunlight, the insect can actively escape (Dennis *et al.*, 1986). The thermally

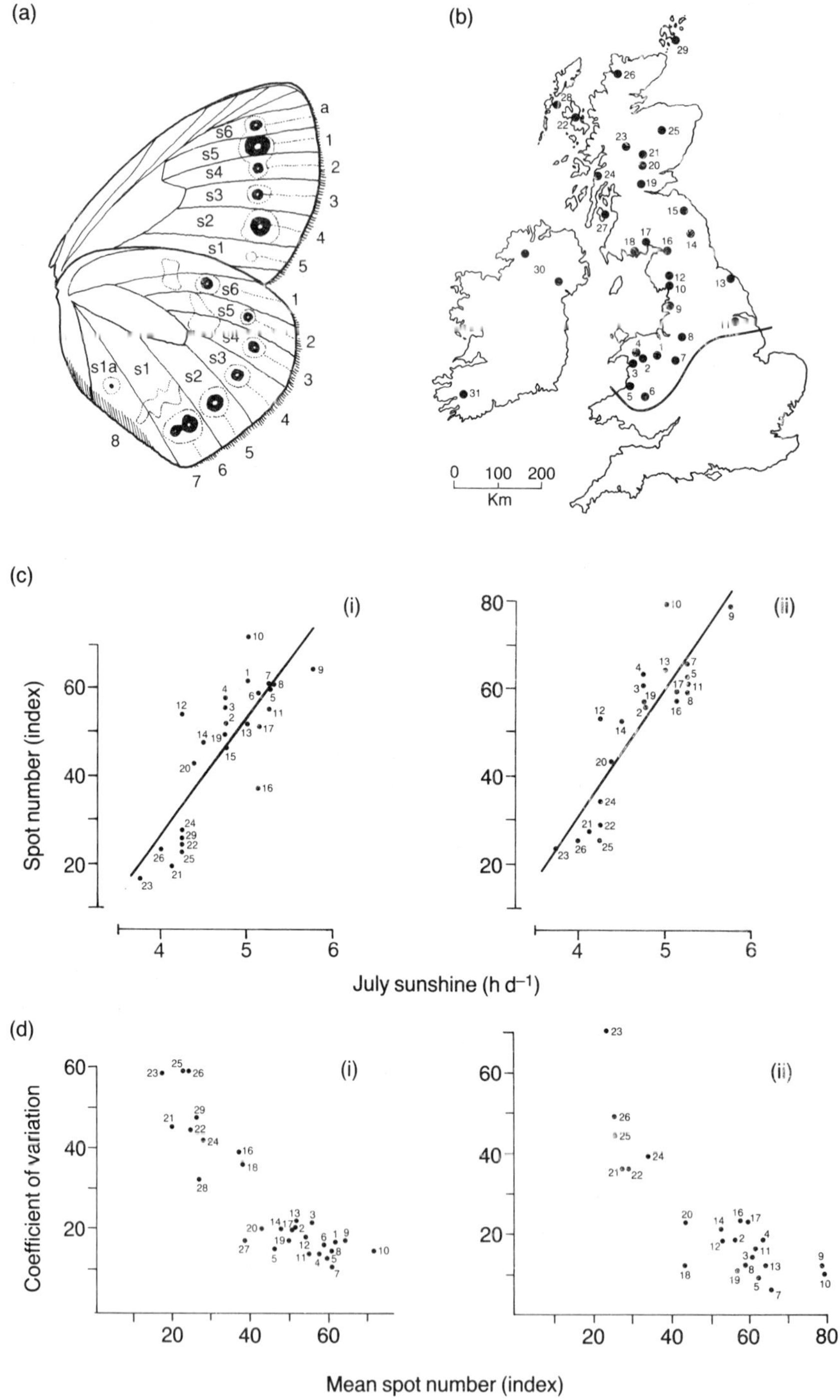
(a)
(b)
Km
(c)
(i)
(ii)
Spot number (index)
July sunshine (h d⁻¹)
(d)
Coefficient of variation
Mean spot number (index)

significant ventral hindwing basal coloration fits into the overall cryptic pattern. It is interesting that reduction in pattern and colour contrast extends to the ventral forewing and the dorsal surface, implying selection for crypsis in flight in cooler locations. The reduced apparency of the ventral apical ocellus also suggests that this functions more as a deflector than as a startling device in northern and high altitude populations of *C. tullia.* This observation is supported by components and factor analysis, which identified considerable potential for independent development of forewing and hindwing attributes, and the relative frequency of wing damage presumed to occur from attacks by birds, which is as high for the apical eyespot as for individual hindwing ocelli (Dennis *et al.*, 1984). Whether the apical ocellus acts as a deflector or as a startling device will depend on a number of biotic factors (e.g., mimicry, size, flight characteristics; predator perception) but possibly also on climate (i.e., activity).

In dorsal absorbance baskers there can be some uncoupling of functions inasmuch as dorsal wing surfaces are exposed when insects are active and ventral wing surfaces when they are inert. On the dorsa, the most important part of the wing for thermoregulation is the base, but for advertisement and anti-predator devices more distal zones are paramount. The dorsa of butterflies in cooler regions are often darker at the base and outer margins (e.g., *Boloria selene*), but have more extensive brighter areas in between highlighting larger apical ocelli when these occur (e.g., *Maniola jurtina, Pyronia tithonus, Pararge aegeria*). The form of phenetic modifications on ventra will depend mainly on their use in mate advertisement, courtship and concealment against predators, the latter influenced in turn by the nature of cover provided and used in the habitat. Brighter ventra may

---

3.6 Variation in wing spotting among British populations of *Coenonympha tullia* (Satyrinae). (a) Wing venation, neural space nomenclature, ocellation structure and spot coding (Turner, 1963) for a specimen of *C. tullia* illustrating the full array of marginal spots. As the wing spots become smaller, the wing spot components disappear in the order of white pupil, black spot and lastly the outer yellow halo. (b) Sites from which population samples have been obtained and the southern range boundary for the species. (c) Scattergram of mean spot number (index of spot development based on: 0, spot absence; 1, halo; 2, back spot and or white pupil) and mean daily hours of bright sunshine: (i) males, $Y = -80.1 + 26.7X$, $r_s = 0.80$; (ii) females, $Y = -85.1 + 29X$, $r_s = 0.86$. Correlations were high for July mean temperatures (males, $r_s = 0.73$; females, $r_s = 0.74$) and latitude (males, $r_s = -0.85$; females, $r_s = -0.78$), but were low for January temperatures and the annual range in temperatures ($r_s < 0.34$). (d) Scattergram of coefficients of variation against means for wing spot number (index of spot development): (i) males; (ii) females. Northern populations are more variable than southern populations but have lower mean spot numbers. (From Dennis *et al.*, 1986; courtesy of the *Entomologist's Gazette*.)

occur in response to selection for conspecific signalling: (i) for insects which are concealed at rest, or (ii) for insects whose pattern has the dual function of being apparent and possibly acts as a secondary defence in bright conditions, but is effectively cryptic when the weather is dull (e.g., green ground colour and enlarged silver maculation in *Argynnis aglaja*).

In dorsal reflectance baskers there is potential for modifications enhancing thermoregulation, mate advertisement and predator-escape in cool climates to reinforce a bright dorsal phenotype, as (i) enhanced thermoregulation involves expansion of the reflective area, and a larger reflective area would (ii) improve mate advertisement and (iii) heighten aposematism in species which have chemical defences and are unpalatable. The conjunction of these adaptations in Pierids (e.g., *Pieris napi*) has allowed them to exploit a scattered and unpredictable host resource, through an open population structure, patrolling mate-location behaviour and Müllerian mimicry (Dennis & Shreeve, 1988). However, any moderation of unpalatability or increased tolerance by predators of unpalatable prey would lead to selection for other defences; additional secondary ones (e.g., marginal wing deflectors) in species which can actively escape attacks during basking, but cryptic patterns in those species which cannot escape (e.g., arctic species; see below). This theoretical scenario may provide a plausible explanation for the striking phenotype of female *Polyommatus icarus mariscolore* from north west Scotland (Dennis, 1977; Fig. 3.7). The heightened blue reflective dorsa in female *P. icarus*, may not only improve thermal efficiency in a cooler climate but influence mate-location. Males preferentially approach their own phenotype rather than the subdued brown female forms found in England (Lundgren, 1977); this could provide a selective advantage for andromorph females (Vane-Wright, 1984) in cool climates, in which mate-location is influenced by reduced flight times and lower population density. However, blue females may be more vulnerable to predators especially as they spend less time than males flying and more time settled on the vegetation. To offset increased predation, larger marginal bright red lunules on the dorsa may have been selected to deflect predator attacks away from vital body organs. Ventral lunulation is also highly developed; this may function in the same way when females are ovipositing on low-growing Leguminosae amid bare ground, where the ambient temperatures are much higher and allow them to be active without resorting to reflectance basking and attracting the attention of potential mates and predators. (Dennis, 1984b). More prominent dorsal black marginal ocelli occur in males, and may provide them with increased evasion time when basking in sunny but cooler conditions. Northwards, the monomorphic rather than bimorphic state of *P. icarus* may result in predators developing a more efficient search image for the butterfly, especially as a larger proportion of the population will be blue and thus more apparent to predators; selection would be greater on the less mobile females. Similar changes may influence *Lycaena phlaeas*.

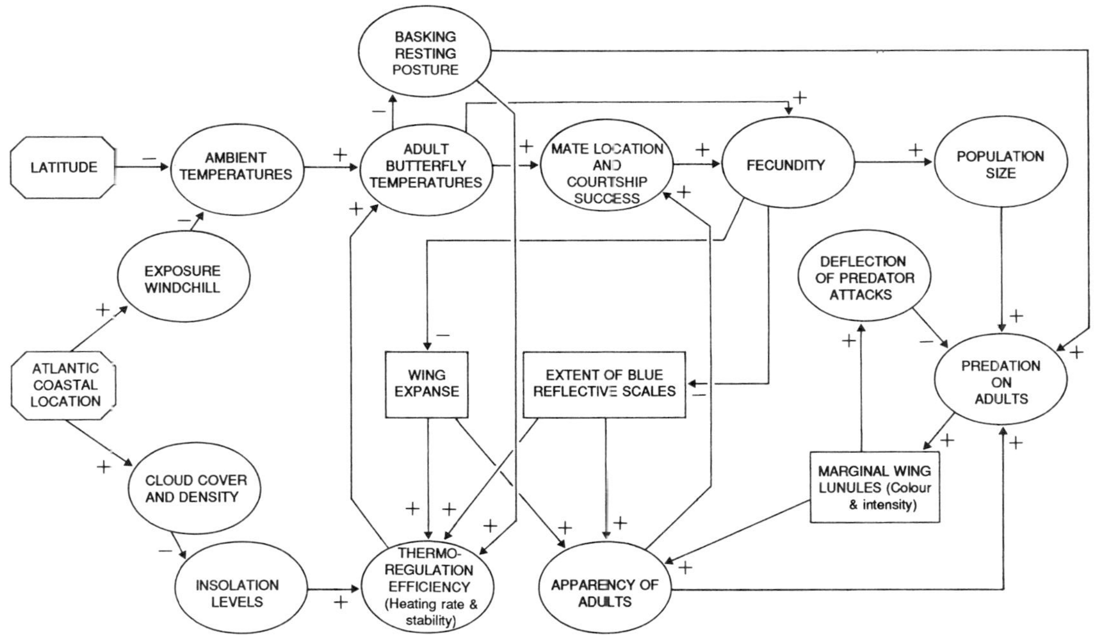

3.7 Potential factors that may influence wing morphology variation in northern British female *Polyommatus icarus*, a dorsal reflectance basker. Octagons, zonal factors; ellipsoids, factors affecting butterfly morphology; rectangles, changes in butterfly morphology. Signs (+, –) indicate corelations. Note that wing expanse in high latitude populations will also be influenced by switches in voltinism. In high latitude climates (arctic) where solar irradiance receipt is minimal, predation could select against apparent adults resulting in a switch to a small dull (melanic) phenotype that is cryptic and depends increasingly on absorbance basking and warm microclimates for thermal efficiency. (From Dennis & Shreeve, 1989; by courtesy of Academic Press Ltd.)

Although many butterfly wing pattern transformations in cooler climates can be accounted for simply on the basis of settling and basking postures and mechanisms, the model is incomplete without consideration of other factors which may influence trade-offs between the primary functions of thermoregulation, mate advertisement and predator-escape on different wings and surfaces. These include: (i) habitat structure and use, (ii) behavioural and activity profiles (e.g., mate-location systems), (iii) chemical defences and (iv) physical attributes (e.g., size, robustness, speed and mode of flight). Thus, rapid flight and evasiveness, territoriality and use of particularly warm microclimates for activity may account for the contrasting transformations in *Hipparchia semele* compared to other lateral baskers (Dennis & Shreeve, 1989: 338). Similarly, a difference in basking posture (i.e., wing angle) is believed to be linked to differences in melanization on the dorsa of *Pontia occidentalis* and *Pieris napi* (Kingsolver, 1987a, b).

The nature of trade-offs also depends on climatic constraints to activity; climatic variability is particularly important as it has the potential for influencing trends in both inter- and intra-population variability in butterfly morphology. It has been suggested that interpopulation variability in butterflies is more extensive in northern and western Britain (e.g., *Maniola jurtina*, Thomson, 1970b). Little information exists on intra-population variability. However, in *Coenonympha tullia*, the coefficient of variation for wing ocellation is significantly larger in Scottish populations than those from England (Fig. 3.6). Increased inter-population variability in northern and western Britain is largely to be expected, and is probably the outcome of selection, in isolated populations, to specific conditions (Ford, 1945; Brakefield & Shreeve, 1992) or to genetic drift in isolated populations (Brakefield, 1989). In both cases climatic trends and variability could influence the outcome by controlling population size, density and spacing, as well as by reducing gene flow among populations, and in the case of a selectional scenario alone by increasing the variance in selection pressures among populations occupying distinct habitats on different substrates or topography. Any increased intra-population variance in northern populations would be more difficult to explain. Rare alleles are expected to be lost, and average heterozygosity to decline, with 'bottlenecking' or 'founder events', though substantial genetic diversity at polymorphic gene loci tend to be lost only in association with more extreme stochasticity (references in Brakefield, 1989). In *C. tullia*, not only does the coefficient of variation increase significantly for wing ocellation in northern populations (Figure 3.6), but it does so as the mean value for ocellation decreases. As *C. tullia* populations tend to be more widely spaced in southern than in northern Britain, it is possible that 'bottlenecking' is less frequent in the north than in the south. But, if such intra-population variance is found to increase in other species which are known to occupy marginal habitats in Scotland, an alternative solution needs to be considered. Such increased intra-population variance northwards may reflect on an increase in the variability in climatic conditions, *relative* to thresholds for functions, such as thermoregulation, flight and dependent activities, in Scottish populations compared to those from further south. An increase in effective climatic variability would also select for greater phenotypic plasticity in northern populations, that is if the variation in the morphological attributes discussed in this section are important enough for survival. This would certainly go some way to explaining the extensive phenotypic variation between populations and between broods in different years, and also account for the direct influence of temperature on some morphological attributes in butterflies that have been reared under controlled environments (e.g., *Maniola jurtina*, Thomson, 1973; *Pieris* spp., Shapiro, 1976).

British environments lack the extremes of summer low radiation levels that occur at higher latitudes (Downes, 1965; Wallén, 1970). The evolution of wing morphology in arctic butterflies must involve differences in balance between the pressures imposed by the different biological functions. The dominant mode of transformation of species extending to the arctic from temperate regions and of the phenotypes of arctic endemics is towards small melanized or pale cryptic insects (Table 3.2). Transformations for thermoregulation (i.e., small size, melanization) are in phase with the requirement of primary defence (i.e., crypsis, dull phenotype) for the long periods of inertia when weather conditions are unsuitable for activity. In these circumstances, it is argued that any selection for increased individual apparency between conspecifics, as may occur for some species in northern Britain, would probably be countered by increased predation. Response times to insolation changes will be slower for the same species at arctic than at temperate latitudes for the same ambient temperatures because of lower radiation levels, and an insect closer to its lower thermal limit has a lower probability of escaping visually hunting predators. As dorsal cryptic patterns would interfere with phenotypes for reflectance basking, selection for enhanced thermoregulation may then switch to basal dorsal and ventral absorbance. Thus, selection for increased crypsis may account for the encroachment of melanin over the medial and distal wings and yellow suffusion in arctic *Pieris napi* females, as yellow pigment negates reflectance basking (Kingsolver, 1985a) and is either less apparent or less attractive to males (Petersen, 1952). This represents a shift towards crypsis away from aposematism in more sedentary high latitude *P. napi* females than males, which may depend more on substrate temperatures. Possibly this provides an explanation for the conundrum between predictions and observations based solely on thermoregulation (Kingsolver & Wiernasz, 1987).

The essential feature of the general model outlined above is that it is accessible to testing using variation in space and time (see section 3.4A). The next step is the formulation of specific hypotheses for particular functions in different taxa. Adaptive mechanisms for thermoregulation have already been shown to exist for wing patterns in Pieris (Kingsolver & Wiernasz, 1991a, b). For experimental designs and techniques readers are directed to the excellent work of Kingsolver and Wiernasz to whom theoretical and empirical advances in this field are largely due.

C *Climatic influences on adult size*

Distinctive patterns in adult size occur for different levels – across species and within species across populations – at different scales. The most obvious trend would seem to be the decrease in maximum size, mean size and variance with latitude, and possibly altitude, for multiple species populations though this has not been formally documented (e.g., 20°E meridian; cf. Afro-tropical regions *versus* West Palaearctic; D'Abrera, 1980; Higgins & Riley, 1983). Spatial patterns in size distribution are difficult to explain because of: (i) the variety of agents and constraints involved, direct and indirect, abiotic and biotic; (ii) the different life-history stages that can be affected, particularly adults and larvae; (iii) the array of functional correlates for size (e.g., developmental, physiological, physical); and (iv) the variety of ecological consequences (e.g., reproduction, predation, dispersal, foraging efficiency, competition and population dynamics) which may determine feedbacks and life-history trade-offs (cf. section 2.1; Scriber & Slansky, 1981; Blau, 1981). Size also has different components (i.e., wing expanse, body sub-components), correlations between which produce substantial residual variance; for instance, insects with similar body weights may have different thorax dimensions, wing areas, wing loading and wing shape. Climate has a number of potential influences as a proximate or ultimate agent. The increased variance at low latitudes implies a variety of trade-offs functioning in different micro-niches within complex tropical systems; conversely, at high latitudes the spread of development over several years in many species may act as a buffer against seasonal constraints on growth and size. Selection for large size in the tropics is likely to involve a complex nexus of factors, for instance, increased fecundity and egg size, resource acquisition, adult longevity, thermal stability and dispersal capacity. At higher latitudes in cooler climates, seasonal restrictions on development time and thresholds for heating and activity probably limit adult maximum size. Large insects develop and heat up slowly and are more apparent, vulnerable and eagerly pursued by predators. Moreover, as large insects tend to have smaller populations and a lower intrinsic rate of increase, r, than small insects, larger species may also be more vulnerable to extinction upon colonizing new sites than smaller species (Gaston & Lawton, 1988), especially in cooler environments.

Climatic differences may also influence geographical patterns in sexual dimorphism. To the extent that relatively small male size (i.e., male : female < 1) is the result of either male protandry or increased food intake by female larvae (Singer, 1982), any decline in seasonality towards the humid tropics may correlate with decreasing sexual dimorphism in size; there would be less advantage for males to emerge early and any increase in adult life span may encourage investments in adult resources compared to larval resources. Any behavioural differences between species in the tropics and high latitudes may also affect size ratios. For example, any tendency for territoriality to be linked with sexual selection for male size (Wickman & Wiklund, 1983; Wickman, 1985b), would lead to geographical differences in sexual dimorphism for size associated with shifts in the relative frequencies of perching and patrolling species influenced by climate (Wickman, 1988; Dennis & Shreeve, 1988).

At the infra-species level, very different spatial patterns in size emerge. Wing expanse may decrease (e.g., *Anthocharis cardamines*) or increase (e.g., *Polyommatus icarus*) with latitude and altitude, or there may be little obvious geographical trend (e.g., *Coenonympha tullia*); disruptive pockets of populations comprising smaller or larger individuals can also occur (e.g., *Hipparchia semele* on Great Orme's Head, North Wales; island populations in Western Isles of Scotland) (Dennis, 1977; Dennis & Shreeve, 1989; Dennis *et al.*, 1986). Important observations on latitudinal trends in wing expanse have recently been made on 16 Swedish butterflies by Nylin and Svard (1991). Among these species (12 Satyrinae and four Lycaeninae), 13 display constant voltinism (12 univoltine: *Lycaena helle, Heodes virgaureae, Palaeochrysophanus hippothoe, Coenonympha tullia, C. arcania, C. hero, Aphantopus hyperantus, Maniola jurtina, Erebia ligea, Hipparchia semele, Lasiommata maera, L. petropolitana*; 1 bivoltine: *Lasiommata megera*) and three have shifts in voltinism within the study region from a single generation in the north to two or more further south (*Lycaena phlaeas, Coenonympha pamphilus, Pararge aegeria*). Species without shifts in voltinism generally have clear size clines, approximating to a linear model of wing expanse on latitude, decreasing in size from south to north (Table 3.4). These species were also larger in continental Europe. Net negative trends with latitude for the three species with shifts in voltinism were much weaker, but indications of reversals in wing expanse with latitude appear to coincide at locations where shifts in voltinism occur (e.g., *Lycaena phlaeas* at c. 62°N). Similar observations had previously been made by Petersen (1947).

Although environmental agents can greatly complicate explanations at this scale, potential factors accounting for trends in adult size with latitude may be divided into two sets, those affecting development, especially larval growth, and those directly influencing the size of the adult stage. Among direct climatic influences on adults, size may theoretically affect warming rates and thermal stability (see section 1.2), but the relative advantage of differently sized individuals would depend on basking mode, behaviour and microhabitat use as well as on ambient temperatures (T_a) or solar

Table 3.4 Regressions of wing length on latitude for sixteen butterfly species with constant (C) or shifting (S) voltinism within Sweden (From Nylin & Svard, 1991; by courtesy of *Holarctic Ecology*)

Species and voltinism	*Sex*	*N*	*b* (mm degree)	r_p^2	*p*
Lycaeninae					
P. hippothoe	♂	92	−0.038	0.03	n.s
(C)	♀	101	−0.047	0.03	n.s
H. virgaureae	♂	136	−0.188	0.40	< 0.001
(C)	♀	105	−0.122	0.16	< 0.001
L. helle	♂	64	−0.054	0.05	n.s.
(C)	♀	47	−0.025	0.01	n.s.
L. phlaeas	♂	86	−0.021	0.01	n.s.
(S)	♀	61	0.053	0.04	n.s.
Satyrinae					
C. tullia	♂	139	−0.194	0.29	< 0.001
(C)	♀	39	−0.307	0.40	< 0.001
C. arcania	♂	79	−0.352	0.27	< 0.001
(C)	♀	41	−0.161	0.07	n.s.
C. hero	♂	67	−0.291	0.52	< 0.001
(C)	♀	27	−0.148	0.15	< 0.050
C. pamphilus	♂	206	−0.035	0.04	< 0.010
(S)	♀	91	−0.026	0.02	n.s.
A. hyperantus	♂	194	−0.214	0.21	< 0.001
(C)	♀	61	−0.155	0.12	n.s.
M. jurtina	♂	123	−0.185	0.03	n.s.
(C)	♀	72	0.073	0.00	n.s.
E. ligea	♂	190	−0.210	0.34	< 0.001
(C)	♀	47	−0.179	0.28	< 0.001
H. semele	♂	39	−0.088	0.01	n.s.
(C)	♀	22	−0.040	0.00	n.s.
L. megera	♂	40	−0.221	0.05	n.s.
(C)	♀	19	−0.300	0.17	n.s.
L. maera	♂	104	−0.227	0.23	< 0.001
(C)	♀	54	−0.362	0.21	< 0.001
L. petropolitana	♂	65	0.051	0.03	n.s.
(C)	♀	49	0.044	0.03	n.s.
P. aegeria	♂	82	−0.174	0.18	< 0.001
(S)	♀	55	−0.042	0.00	n.s.

irradiance (R). For example, although female *Euphydryas phaeton* are twice the size of males and have a much greater wing loading, their size and restriction to a warm sheltered microhabitat enables them to maintain elevated T_bs against short term thermal fluctuations and to reduce time spent basking (Gilchrist, 1990). For conditions close to the threshold for activity, selection for size will depend much on the premium for continuous activity (i.e., oviposition in females) as opposed to that for rapid heating (i.e., male territorial flights and predator-escape). In northern regions species occupying woodlands, which provide a buffer against temperature extremes (e.g., *Pararge aegeria*), may be larger and adapted to thermal stability whereas those in open habitats (e.g., *Hipparchia semele*) are perhaps more likely to be smaller and adapted to warming up rapidly. However, these relationships are likely to be complex as other factors, climatic and non-climatic, may influence adult size directly. In cool climates, larger winged dorsal reflectance baskers may be disadvantaged less than equally sized dorsal absorbance baskers, because of a larger thermal capacity (effective wing area: body size ratio). Larger insects may be more apparent and exposed to a wider range of predators, but this will depend on size ratios and predator visual acuity. Large insects could also be at a disadvantage in isolated habitats, especially those exposed to high winds (e.g., small *Argynnis aglaja* on Flodday, South Uist) (MacArthur & Wilson, 1967; Dempster *et al.*, 1976; Dennis, 1977).

Climatic effects on early stadia are inextricably bound up with seasonality, hostplant quality, intraspecific competition and life-history strategies (see section 2.1). Observations from the Swedish study (Nylin and Svard, 1991) suggest that an indirect influence on larval development may dominate any direct impact of climate on adult size *per se*. Firstly, the size differences are argued to have little thermoregulatory significance for adults (Kingsolver & Watt, 1983), although this needs to be tested. Secondly, trends are apparently identical for species having different basking methods. Even primarily reflective baskers (e.g., *Pieris napi, Lycaena phlaeas*; Petersen, 1947; Nylin & Svard, 1991) become smaller northwards, though such species may also be capable of dorsal and lateral absorbance and the nexus of selection pressures in arctic environments may cause a switch in emphasis as to basking modes (Dennis & Shreeve, 1989). Thirdly, and difficult to explain by any direct influence on adults, is an indication of 'saw-tooth' clines in size for switches in voltinism (Fig. 3.8). Nylin & Svard (1991) suggest that the observed size patterns with latitude and altitude are to a large extent governed by the length of the growing season, affecting selection for development time and size (see Fig. 2.1). A model maximizing the yearly rate of increase (r_a) predicts that final body size should be larger in southern populations of a species than in the north, as a result of the longer favourable season allowing a longer development time (Roff, 1980). This positive correlation between size and development time may then

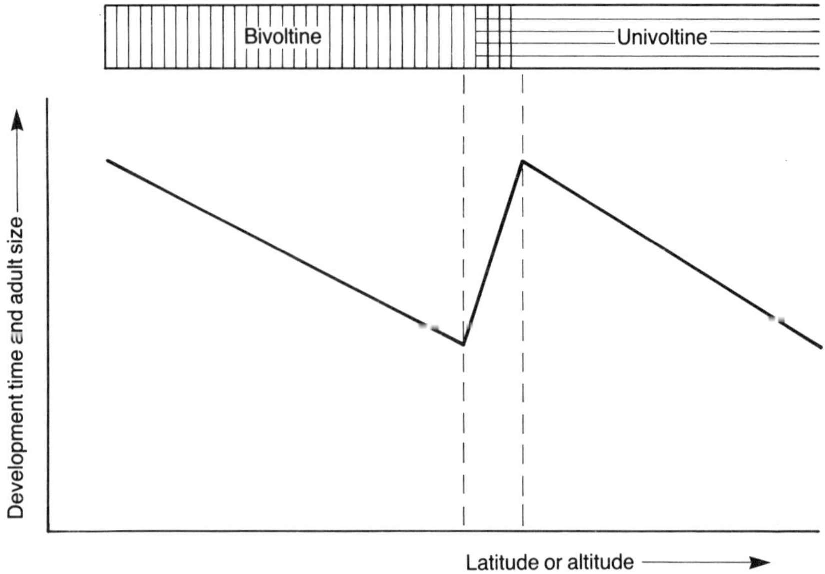

3.8 Schematic illustration of a 'saw-tooth' cline for size varying with latitude or altitude as an effect of the length of the developmental period on selection for developmental time. Size can be expected to increase from north (higher altitude) to south (lower altitude), but the size cline should 'reverse' at boundaries where shifts in generation number occur. (Adapted from Nylin & Svard, 1991, by courtesy of *Holarctic Ecology*.)

produce larger-sized individuals where cooler climates induce switches from multiple to decreased voltinism, generating a 'saw-tooth' cline at the point where voltinism changes (e.g., *Polyommatus icarus* in northern Britain; Masaki, 1973; Roff, 1980).

The consistent significant trend in wing expanse with latitude in Scandinavia may be reinforced by the positive correlation between latitude and altitude. Where this does not occur, the relationship between size and latitude may be less prominent: firstly, because differences in growing season associated with latitude will not be as large; secondly, because other factors, climatic and non-climatic ones, may also influence larval development and thus ultimately affect adult size. Although altitude is positively correlated with latitude in Britain, species in northern Britain for which comparisons in size can be made come from the western seaboard; in this region the effect of latitude on growing season is countered by warm prevailing westerlies and a warm ocean current. Thus, any increases in wing expanse among butterflies with latitude in Britain must necessarily first address developmental regimes. There is also potential for local and regional climatic factors to influence the growing season, development rates, development times and size. Steep lapse rates in temperatures for montane Britain may force biennial regimes on some species (e.g., *Coenonympha tullia*; Melling, 1984) affecting development rates and size. Some high altitude populations of Scottish *C. tullia* are among those with the largest mean wing expanses (Dennis *et al.*, 1986). The growing season for populations on northern aspects or in frost hollows may be shortened, as may that for populations affected by drought. On the Great Orme's Head, North Wales, a number of climatic and substrate factors are believed to combine to select for dwarfism and early emergence in two unrelated univoltine species (*Hipparchia semele*, Satyrinae; *Plebejus argus*, Lycaenidae) (Dennis, 1977; 1992). Finally, analysis of geographical trends can be frustrated by short-term environmental effects, such as seasonal weather. For example, in siliqua-feeding pierids, such as *Anthocharis cardamines*, dry springs can produce a food shortage and a preponderance of small adults from small final instar larvae (Wiklund & Åhrberg, 1978).

3.4 **Climate and temporal variation in wing morphology**

Temporal variation in wing morphology occurs at three main scales: inter-annual, seasonal and intra-brood. The focus of attention here is given to seasonal polyphenisms in adults which are closely linked to consistent switches in climate. Variation within emergence cohorts has been reviewed recently by Brakefield (1984). Long-term trends in pattern diversity are discussed in chapter 4.

A *Seasonal environments and adaptive polyphenisms*

Seasonal polyphenisms in wing attributes are part of an extensive array of co-ordinated adaptations in morphology and life-history traits to temporal switches in climate and habitats. The wing features affected are (i) colours and patterns on ventra and dorsa, (ii) wing shape, and (iii) wing size. This is typically accompanied by various changes in body dimensions, physiology, biochemistry, activity, behaviour and reproductive strategy (see section 2.1). As switches in seasonal environments present similar challenges as geographical changes in conditions, the resulting wing modifications should adhere to much the same model (Fig. 3.5). The nature of seasonal wing transformations depends on differences in (i) adult activity and longevity and (ii) habitat and substrate use. The two groups of factors carry implications for reproductive strategies and success, individual apparency and exposure to predation and, in turn, influence the relative exposure of different wing surfaces and areas. Two distinct seasonal environments (i.e., temperate and tropical regions) can be distinguished on the basis of how they influence activity and habitats. In temperate regions, seasonal contrasts in thermal conditions produce differences in habitat structure for adult butterflies

3.9 Seasonal polyphenism in *Junonia almana* (L.), Nymphalinae from India. Top row, wet season; bottom row, dry season; left, dorsal surface; right, ventral surface. (By courtesy of P.M. Brakefield & T.B. Larsen.)

(e.g., leaf cover). In tropical environments similar but larger differences in habitat structure are produced by seasonal differences in moisture.

There has been limited quantification of seasonal polyphenisms in both tropical and temperate regions (Robertson, 1980b; Brakefield, 1987b). In the tropics, dry season forms are generally more cryptic than wet season forms (Brakefield, 1984; Denlinger, 1986). Comparisons of two satyrinae (i.e., *Melanitis leda; Orsotrioena medus*) and one nymphalinae (i.e., *Junonia almana*) revealed that dry season forms have (i) reduced ventral ocellation, (ii) duller, less striking ventral coloration, (iii) similar or enlarged dorsal ocellation and patterning, (iv) more falcate wings, (v) more pronounced tails, and (vi) larger body and wing size. Ventra are especially more cryptic; in *Junonia almana* the effect is a very realistic dry leaf 'mimic', a miniature example of *Kallima inachus* (Brakefield & Larsen, 1984; Fig. 3.9). Similar seasonal changes have been described for five *Bicyclus* butterflies (Satyrinae) from Malawi (Brakefield & Reitsma, 1991). The seasonal differences are argued to represent shifts in balances between primary (dry season) and secondary (wet season) anti-predator defences. Reproductive success during the wet season is optimized by an active adult life involving relatively rapid mating and oviposition on available hostplants. In the dry season, delay in reproduction due to hostplant deficiencies places a greater premium on longer-term survival incorporating quiescence or aestivation (Brakefield & Larsen, 1984). The larger size of dry season forms is consistent with fat-body development associated with delayed egg maturation and an extended life-expectancy. The seasonal shifts in activity coincide with changes in habitat use and predators (Owen, 1980; McLeod, 1984). However, as in the case of the model for geographical variation, selection for pattern development on any surface depends on settling postures (i.e., basking, resting, aestivation) – thus on wing surface exposure – and potential for activity. Unlike the ventral hindwing, patterning on the dorsa and forewing ventra in tropical dry season butterflies may not be cryptic, as T_bs are high enough for evasive action, and these surfaces are concealed in settled insects. Instead, enhanced secondary defences (i.e., eyespots, false heads) may contribute to escape when settled insects are detected (e.g., *Melanitis leda*). The reasons for seasonal polyphenisms in other tropical groups (e.g., nymphalinae: *Precis* spp.; Pieridae: *Colotis* spp.) have yet to be determined; both distinctive aspects of dry season behavioural ecology (e.g., *Precis octavia*; anachoresis & aggregations; McLeod, 1984) as well as alternative wet season defences (i.e., aposematism and unpalatability; mimicry) need to be considered. A characteristic observation of dry season forms is that they have more variable ventral phenotypes (Owen, 1971); this contradicts the expectation of increased pattern uniformity amongst organisms subject to intense visual selection on the same background (Endler, 1978). This may suggest that greater variability enhances crypsis and frustrates the search images of predators (Brakefield & Larsen, 1984). It is possible too that the resting background of dry season forms, inactive during daylight in addition to night hours, is less uniform than that of inactive wet season forms.

In temperate species, relatively short adult lifespan places a premium on continued activity during both vernal/autumnal and aestival broods. Thus, selection should tend to enhance thermal efficiency. The phenotypic transformations will depend on basking postures and mechanisms, with trade-offs between heating rate and overheating (Kingsolver & Watt, 1983; Kingsolver & Wiernasz, 1987). Comparisons of the thermoregulatory behaviour and polyphenic response between pierine butterflies (family Pieridae, subfamily Pierinae) and their sister group (subfamily Coliadinae) suggest that seasonal polyphenism in pierines is adaptive for thermoregulation and involves both the ventral and dorsal wing surfaces (Kingsolver & Wiernasz, 1991b). During the cooler seasons, lateral absorbance baskers should develop darker basal ventra (*Leptidea sinapis; Colias* spp., Watt, 1969), dorsal absorbance baskers darker basal dorsa (e.g., *Polygonia c-album*) and dorsal reflectance baskers more reflectant dorsa with increased basal, but decreased medial, melanism (e.g., *Pontia* spp., Shapiro in Kingsolver, 1985b; *Celastrina argiolus; Pieris* spp. Kingsolver, 1987a). Smaller body size and wing expanse during the cooler seasons implies selection for rapid heating and lower threshold temperatures for flight especially as lower temperatures during the preceding larval stages usually produce larger individuals. As temperate weather conditions frequently enforce inactivity at any stage in the year, it is expected that wing patterns for resting and basking postures will be selected for crypsis against their seasonal substrates (e.g., dark ventra in hibernating phenotype of *P. c-album*, Nylin, 1989; larger light dorsal patches in early spring *Pararge aegeria* from overwintering pupae, Robertson, 1980b), but as warmer conditions are more predictable in summer, secondary defences should be better developed in aestival broods (e.g., longer tails in *P. c-aureum*; Fukuda & Endo, 1966; Hidaka & Takahashi, 1967). Reasons for the marked transformations of *Arashnia levana* (i.e., vernal *levana* fritillary-pattern; aestival *prorsa* admiral pattern) remain obscure but could involve mimicry of *Liminitis* in summer broods (Shapiro, 1976). Some taxa lack seasonal variation, though demonstrating a latent capacity for producing it. This could indicate the existence of approximately similar conditions in spring and summer (e.g., *Pieris napi mogollon*); typically, such species tend to have long-lived adults and adapt to seasonal conditions by migration (e.g., *Inachis, Aglais; Cynthia*; Shapiro, 1976).

Distinctive polyphenisms occur widely in early stadia (see section 2.1B) These are mainly colour morphs which are cryptic against their substrates (Wiklund, 1975; Smith, 1978; 1980; Hazel & West, 1979; 1983; West & Hazel, 1985), but include a variety of adaptations to adverse conditions (e.g., cocoon structure; wax secretion, cuticle thickness; larval instar number; Scott, 1979; Tauber *et al.*, 1986). Many of these are not strictly seasonal polyphenisms inasmuch as proximate triggers are often typically colour, texture or type of substrate, though climatic attributes can be directly involved (e.g., light, humidity, temperature in pink/green pupal morphs of *Danaus chrysippus*; Smith *et al.*, 1988) and the distribution of substrates is not seasonally random (Shapiro, 1976). Colour and structural morphs are also often integrated with diapause (e.g., pupal colour in *Papilio xuthus*; Ishizaki & Kato, 1956; larval colour in *Plebejus icariodes*; Scott, 1979).

B *Developmental correlates of variation within broods*

A number of morphological attributes have been found to correlate with emergence time. In *Maniola jurtina*, mean spotting significantly declines during the emergence period (Creed *et al.*, 1959; Dowdeswell, 1962; Beaufoy *et al.*, 1970; Brakefield, 1984; Shreeve, 1989) and post-aestivation (Scali, 1972; Brockie, 1972; Scali & Masetti, 1975); female bipupillation frequency increases later in the season (Shreeve, 1989), and intra-seasonal differences occur in ventral hindwing spot placement (Brakefield, 1984). In *Pararge aegeria*, increased dorsal hindwing ocellation (i.e., four spots) occurs in mid-summer (Parker, 1984; Shreeve, 1987). Intra-seasonal increases in size and wing expanse may also be expected, especially in directly developing species.

Some of these correlations with the emergence period may correspond to temporal differences in selection pressures on the adult phenotype. For instance, predation against less cryptic phenotypes during aestivation would account for reduced ocellation in late season Mediterranean *M. jurtina* females, and accord with Brakefield's (1984) model for satyrid spotting. Increased predation or vulnerability may also account for increased ocellation in territorial male *P. aegeria* which have biased distributions to dense woodland in mid-summer (Shreeve, 1989; Dennis *et al.*, 1986). However, much morphological variation may be influenced by developmental processes rather than by selection directly for pattern elements (Nijhout, 1981; 1984; 1985). Genes pleiotropic for development rate in *M. jurtina* influence spot number and placement (Brakefield, 1984). There is evidence that powerful selection against high spotted genotypes in late pre-adult development also occurs in *M. jurtina* (Dowdeswell, 1961; 1962; McWhirter, 1965; Brakefield, 1984), though as individuals with increased spotting are known to move greater distances, it is possible that lower spotting in the late season could reflect differential emigration of spot classes (Brooke *et al.*, 1985). Intra-seasonal variation implies endocyclic selection or pleiotropic antagonism for the attributes at different stages in the life-cycle (McWhirter, 1967; Ford, 1975).

Climatic conditions which influence development rates and exposure to selection agents hold significance for intra-seasonal, interseasonal and geographical variation in attributes correlated with development. Thus, hindwing ventral anality in *M. jurtina* ocellation increases with altitude and latitude (Brakefield, 1984). As wing pattern variants can be produced by sublethal traumas on pre-adult stages (Kuhn, 1926; Nijhout, 1985), weather conditions, such as frosts and excessive diurnal maxima, may cause sporadic outbursts of variation within populations (e.g., the numerous aberrations of *Argynnis paphia* in six years during twentieth century UK, Robertson, 1980a), though this phenomenon may also be associated with large population increases (Ford, 1945, 1975).

4

Past climates and evolutionary history

Climatic change can lead to alterations in abiotic and biotic components of natural selection and to perturbations through stochastic agencies. Species may respond by changing in abundance or distributions, by evolving or becoming extinct. Whether there is any response and the nature of that response depends on intrinsic features of the gene pool and on extrinsic aspects of environmental change: type, severity, rate and persistence (Holt, 1990). Changes, when they occur, are the result of historical events working through ecological and genetic processes (Myers & Giller, 1988). The focus of this chapter is the exploration of such processes. Unique events to illustrate these phenomena are set in a European frame. Failure to evoke links at finer levels in low-resolution historical reconstructions, specifically ecological processes, leads to an overloading of assumptions and to unfounded, incohesive and inconsistent biogeographical models. As a sporadic fossil record detracts from the substantial value of Lepidoptera as tools for testing evolutionary theory, particularly evolutionary rates, attention is also given to methods for improving retrodictions.

4.1 **Climate and distribution changes in recent history**

Britain and adjoining regions of western Europe have the longest period of climatic and butterfly records in the world but, despite this, links between the two and other environmental factors have generally encouraged little more than casual observations. In this section the first part reviews changes in ranges and distributions while the second attends to causal modelling for spatial changes.

A *Climatic correlates of distribution changes*

The close association between species richness and climatic variables in Britain (section 2.3) implies that changes in climate would result in the expansion or contraction of butterfly ranges. Values for regression coefficients predict a loss or gain of five species per spatial unit (10–30 km^2) for every 1°C change in mean summer temperatures and six species for every 1 h in the mean duration in daily summer bright sunshine (Table 2.6b). The indications are much the same for sampling frames based on weather stations and systematic intervals founded on grid intersections and for different periods of climatic data (Turner, 1986; Dennis & Williams, 1986). Such enticing predictions carry with them cautionary notes and qualifications: (i) changes in temperature and sunshine persist long enough to permit a biotic response; (ii) there are no obstacles to dispersal; and (iii) climatic changes influence all areas uniformly. Sufficient attention also needs to be given to standard errors for the regression coefficients (Table 2.6) and potentially non-linear influences by predictors on essential resources such as hostplants.

In Britain, records for temperatures (from 1659; Manley, 1974) and rainfall (from 1766; Wigley *et al.*, 1984) date much earlier than any effective systematic observations on population changes in the Lepidoptera, which have been reviewed back to 1864 (Beirne, 1955). Records for other climatic variables (e.g., sunshine and snow cover) exist from the beginning of the twentieth century, whereas observations on pressure and winds, thus atmospheric circulation, have been documented back to 1750 (Lamb & Johnson, 1966). The close agreement between Mossman's (1896) Edinburgh temperature data (1764–1896) and Manley's central England temperature series (r_p, 0.8; differences of means, 0.8°C) may validate the application of the central England series to larger parts of Britain (Duncan, 1991). These are illustrated as 10-year running means by Manley (*ibid.*) and 20-year running means by Parker (1988). The rise in mean annual temperature from 8.8°C in the 1890s to 9.5°C in the 1990s suggests a potentially significant influence on Lepidopteral distributions. However, this linear trend is not consistently matched by the monthly mean figures. Only for October is there a systematic rise in temperature over this period. Nevertheless, shorter periods since 1900 exist when monthly temperatures have generally risen (June to September from 1930 to 1950; July and August in the 1980s) or fallen (March, June to September from 1950 to 1970; April and May from 1960 to 1987). Regarding the period of Lepidopteral records, summer temperatures were especially low during the last quarter of the nineteenth century. As summer warmth, in particular, is implicated in affecting butterfly populations (see section 2.2), distributions could have been influenced by these changes. But any relationship is unlikely to be simple for at least two reasons. Firstly, conditions for summer months are not always in phase with one another, especially early and late summer conditions (e.g., 1980s, Parker, 1988); secondly, running means conceal large variations in annual statistics which may counter systematic trends. Although annual totals of rainfall have not changed systematically since 1800, January rainfall has increased, whereas summer rainfall has decreased. July and August rainfall has significantly declined since 1970; the period has been influenced more by anticyclonic conditions and includes four of the driest summers in England and Wales since 1766 (Parker, 1988). Only north-west Britain received as much summer rainfall as the mean for the period 1951 to 1981, much of southern and eastern Britain recording less than 85 per cent of the average (Mayes, 1991). As this directly reflects on cloud cover and sunshine totals, this too could have implications for butterfly distributions.

Long-term trends in butterflies do not match the climatic statistics. In Britain, over the past 150 years, half of the 59 resident species have undergone major contractions in range and four species have become extinct (i.e., *Lycaena dispar* 1851, *Cyaniris semiargus* 1877, *Aporia crataegi* c. 1925, *Maculinea arion* 1979). Only nine species have periodically expanded in range and only three of these (i.e., *Anthocharis cardamines*, *Inachis io*, *Ladoga camilla*) have recovered ground lost since the 1880s (Table 5.3; Warren, 1992b). Figures for Holland from 1901 are similar: of 63 species, 29 have declined or are extinct and only seven have expanded in range (Swaay, 1991). Autecological surveys have been carried out on 21 species that have declined in Britain; in all but two, the causes are attributed entirely to habitat despoliation or loss (Warren, 1992b). However, climatic changes may have been responsible for shorter-term range changes. Turner (1986) points to the coincidence in the extinction (i.e., *C. semiargus, A. crataegi*) and contraction (i.e., *L. camilla, Polygonia c-album*) of four Midland species with predictions based on regression coefficients for the climatic downturn between 1870 and 1920. A number of butterflies also retreated in Scotland and north-east England at this time (e.g., *Pararge aegeria*, *Lasiommata megera, Anthocharis cardamines*; Dennis, 1977; Thomson, 1980). Expansion in the ranges of some butterflies in the 1930s and 1940s (e.g., *L. camilla, P. c-album*) coincided with warmer, sunnier summers. The most recent period of range expansions, since 1960 and especially during 1988–92, may well be caused largely by the current warmer, drier conditions (Brugge, 1992). At least six species have 'advanced' in north-east England (i.e., *Ochlodes venata*, *Thymelicus sylvestris*, *L. megera*, *Lycaena phlaeas*, *Polyommatus icarus* and *A. cardamines*; Dunn & Parrack, 1986; Dunn, T.C., pers. comm.), *P. aegeria* and *A. cardamines* have increased their ranges in Scotland (Barbour, 1986; Long, 1980; Thomson, G., pers. comm.), and *T. sylvestris*, *P. aegeria*, *Pyronia tithonus*, *Celastrina argiolus* and *P. c-album* in the north Midlands and north Cheshire (Warren, R.G., 1984; Hardy, P. and Hind, S.H., pers. comm.; Dennis, pers. obs.). A number of other local range extensions have been reported in the regional literature (Emmet & Heath, 1989).

Although these observations are suggestive of a causal link between changes in climate and distribution, they point to serious problems in constructing historical explanations as well as raising a number of specific questions. The problems relate to distribution records and their processing and to the interpretation of correlations. Distributions and patterns are all too often treated as reality, whereas in fact they are hypothetical representations of what we hope may exist (Rosen, 1988). The Biological Records Centre and regional data bases have never had the resources to attain an even geographical coverage and unbiased sampling. Analysis of distribution changes depends substantially on the spatial resolution of mapping and the nature of records, which is usually restricted to binary information on presence or absence. Reported changes in distributions can also trigger modified behavioural responses in recording (Pollard, 1979). Consequently, data records require some standardization, for instance on the basis of numbers of records or spatial units surveyed per time period (Swaay, 1990), but even these procedures do not automatically justify the data base; they cannot account for idiosyncrasies in recording behaviour (e.g., timing; biased observations; inducements to search hitherto unexplored units), which occur due to the sparse distribution of reliable recorders. Causation of a range change is typically given as a statement tied to a 'correlation', or more usually from a single coincidence. Rarely are attempts made to quantify the changes, either distributional or climatic in space or time, to determine covariances or to model properly the functional connections in ecological processes (see Pratt, 1986–87; Aspinall, 1987; 1988).

B *Modelling climatic determinants of range changes*

Testing reconstructions of historical distributions based on climatic data is not generally feasible for the Lepidoptera (Dennis, 1977) because of the paucity or absence of records. But it is possible to test predictions, for instance linking summer climate with range changes, by analysing the relationships between butterfly populations, distributions and climatic change as an analogue for processes in the remoter past. Suitable data are facilitated by synoptic recording as under the Butterfly Monitoring Scheme (Pollard & Yates, 1992). Parry (1981) has successfully modelled such a retrodictive procedure to test the relationship between upland farming and climatic change in northern Europe (see Fig. 4.1a). Several premises are implicit: (i) the more precisely the processes can be modelled the more probable is it that appropriate tests can be constructed; (ii) effective tests depend largely on the quality of the data, particularly on the way in which distribution records are processed. Quantitative data allow more rigorous testing of relationships. The most appropriate regions for study are those for which a number of species collectively attain their geographical limits, either along latitudinal gradients (e.g., north-east England) or for altitude (e.g., Pennines). Not only are these 'boundary' regions more likely to be sensitive to changes, but they facilitate monitoring and mapping more than do larger, more uniform regions. This methodology carries with it several provisos (see Hengeveld, 1990: 159). Ecological data on a species need to

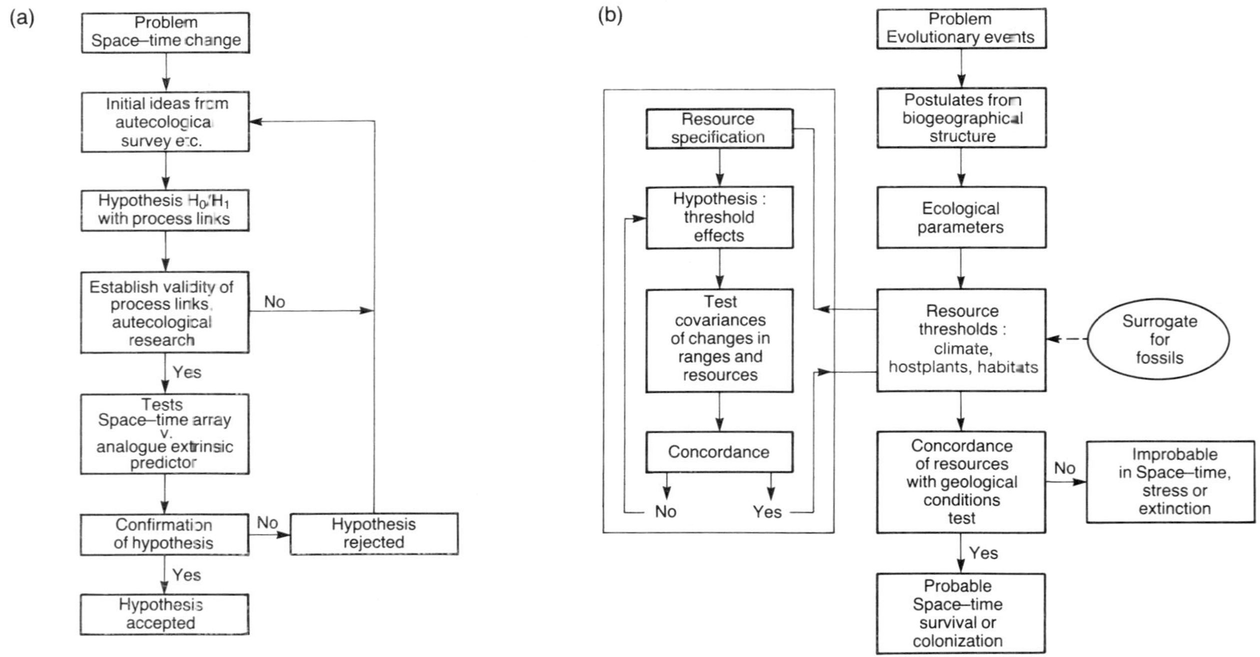

4.1 Retrodictive modelling of butterfly distributions. (a) Scheme for determining the biological significance of factors in distribution changes ascertained from observational records. Termination of the process is determined by the amount of variance accounted for by the test procedures. (b) Scheme for postulating space–time distributions of organisms and evolutionary events without adequate fossil records. Circularity of argument is avoided if past conditions are determined other than from biogenic data (i.e., global circulation models). Resource thresholds for species should be tested by autecological survey. See text for discussion.

be collected from a wider area than the boundary region as extreme and unusual conditions at the boundary may result in only a subset of the ecological requirements of a species being adequately determined, especially if such studies are short-term. Confirmation of a hypothesis allows retrodiction of events governed only by the same conditions, as larger changes may trigger feedbacks in ecosystems (e.g., further temperature increases may promote droughts adversely influencing hostplants). To interpret the effect of larger climatic changes, more sophisticated modelling is required based on suitable regional analogues. Testing contractions in range against climate has the disadvantage of the need to sift out numerous anthropogenic influences; human disturbances are less likely to cause wholesale range expansions, but this cannot be overlooked (Pollard, 1979; Aspinall, 1987, 1988). On the other hand, testing expansions in range against climate change may be hindered by the difficulties organisms have in dispersing over inappropriate biotopes (Thomas, 1984; but see Shreeve & Dennis, 1991). Both test scenarios are frustrated by artificial introductions.

'Correlation' and regression parameters are often not interpreted with sufficient caution, especially in multivariate designs (Vincent, 1981). Although a high significant correlation may confirm a hypothesis, it is not proof of the process modelled: other factors or links may have been omitted. Similar distribution changes in the past cannot simply be interpreted as having emanated from the same causes; equifinality is potentially rife in biotic systems. Conversely, low correlations are not proof of the absence of a relationship say between changes in heat and distribution changes. Analyses based on coarse data intervals (monthly means) may be 'blind' to compensatory high-resolution weather events in summer or climatic conditions affecting life-histories during other months of the year that are not considered. Coarse mapping units may also miss large 'population' responses; this reason, threshold effects from different climatic base figures and regional variability in climatic spectra, can explain why not all species are affected in 'boundary' areas and why one species is not similarly affected in different regions.

Analysis of distribution changes for individual species provides a more rigorous test of climatic influences; it requires high quality data and more precise modelling of processes than for a regional biota. This approach is well illustrated by the analysis of distribution changes in *Ladoga camilla* and *Pyronia tithonus* (Pollard, 1979; 1991a). Autecological research on *L. camilla* from 1972 to 1977 identified a key factor operating on late larval and pupal mortality in June, probably caused by bird predation which increased with development time in cooler Junes (Fig. 4.2). The subsequent accumulation of records from the entomological journals against mean monthly temperatures for May, June and July identified a close fit for new site records from 1930 to 1942, indicating a spread of the butterfly, with the longest period of warm Junes in the twentieth century. A number of

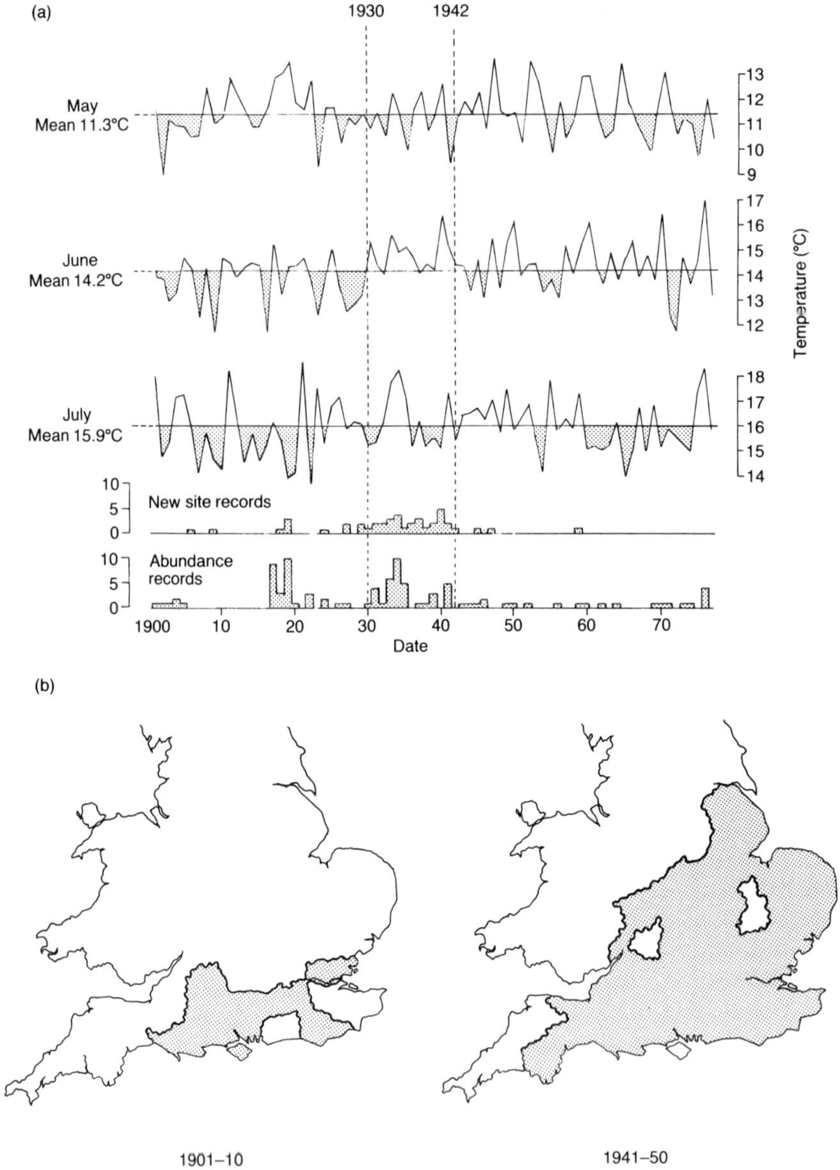

4.2 Range changes in British *Ladoga camilla*. (a) Mean monthly 'central England' temperatures for May, June and July 1900–77, and records from entomological journals indicating a spread to a new area and unusual abundances. (b) Distribution on a vice-county basis as shown by records in the entomological journals 1901–10 and 1941–50, to indicate the change in the range during the period. (From Pollard, 1979 by courtesy of Blackwell Scientific Publications Ltd.)

valuable points are illustrated by this study. Firstly, clear modelling of the chain of causation is shown to be crucial as is the need to assess potentially important subsidiary factors to range changes. Fecundity in *L. camilla* was shown to relate closely to flight period temperatures and sunshine (Fig. 1.8), which were high in the early 1930s. Moreover, the decline in coppicing during the twentieth century was shown to provide suitable conditions for the hostplant, *Lonicera periclymenum*, and egglaying enabling the insect to colonize new sites in the 1930s. Secondly, it exposed vulnerable links in the research design. Key factor programmes are labour-intensive and it is rarely possible to apply them to many sites so as to assess regional variability in population dynamics. The key factor in this peripheral locality does not preclude the existence of different key factors towards the core of the distribution, such as parasitization in earlier instars (Shaw, 1981; Pollard, E., pers comm.). Despite this disadvantage, this methodology is theoretically more sound than *a posteriori* 'shotgun' techniques of scanning large covariance matrices for possible predictors.

Expansion in the range of *Pyronia tithonus* in northern Britain has been shown to correspond closely to changes in phenology. The length of the flight period, which is significantly shorter close to the edge of the range, has become significantly longer during the years of recording by the Butterfly Monitoring Scheme and the mean flight date has become earlier. Though neither relates significantly to distance from the range edge, a causal link may occur between flight period extension and range expansion (Pollard, 1991), but autecological research is required to expose any chain of extrinsic factors involved. More recently, Pollard and Yates (1992) have shown how the Butterfly Monitoring Scheme data can be exploited to analyzse extinctions and foundations of populations at sites covered by the scheme. Such extinctions and foundations, arbitrarily defined as presence and absence respectively of species at sites over a four-year period, have been shown to be significantly correlated with variability in population estimates (I; $r_p \geqslant 0.53$; $p < 0.01$). In the case of *Celastrina argiolus*, which typically has low mean transect counts, foundations and extinctions seem to occur periodically. The four years 1988–91 have witnessed a burgeoning of foundations at BMS sites in this species. Similarly, foundations in *Polygonia c-album* have occurred noticeably at the edge of its range in Britain. These studies underline how vital is a synoptic data base of the Butterfly Monitoring scheme type for generating new hypotheses as well as for recording and interpreting range changes in regional faunas.

4.2 Climate and environment in prehistory

Reconstructing the history of organisms lacking a fossil record is facilitated by three observations. Firstly, there is a close and logical relationship between butterfly distributions and climatic controls (sections 2.3 and 4.1).

Secondly, the recent revolution in palaeoclimatic research, improvements in drilling techniques, climatic proxy data and geochronological methods allows environmental changes and organic evolution to be studied at greater spatial and temporal resolutions (Bradley, 1985; Cronin & Schneider, 1990 for references). Thirdly, climatic oscillations on the geological timescale have been large compared to those over the period of instrumental recording, sufficiently large to impinge on all biota. This section briefly introduces some basic concepts pertinent to a historical reconstruction of the European butterfly fauna.

A Cenozoic climatic trends

Continuous climatic records, at least for large parts of the Quaternary (c. 1.6 ma BP), are now available from deep sea sediments, terrestrial polar ice sheets and increasingly from terrestrial sediments (Bartlein & Prentice, 1989; Cronin & Schneider, 1990; Harland *et al.*, 1990) One climatic curve from Atlantic sediments extends back to 70 ma BP (Miller *et al.*, 1987; Fig. 4.7). Proxy data now include a variety of indicators – $\delta^{18}O$ isotope ratios from ice cores (Imbrie *et al.*, 1984) and benthic foraminifera (Miller *et al.*, *ibid*), $CaCO_3$ spectra from ocean sediments and deuterium changes in ice cores and wood (Jouzel *et al.*, 1987), calcite deposits (Winograd *et al.*, 1992; Ludwig *et al.*, 1992), pollen from terrestrial (Hammen *et al.*, 1972; de Beaulieu & Reille, 1984; Kershaw, 1986; Follieri *et al.*, 1988; Adam, 1988; Guiot *et al.*, 1989; Hooghiemstra, 1984) and marine deposits (Rossignol-Strick & Planchais, 1989) and terrestrial mollusca (Rousseau & Puisségur, 1990). Other organisms (e.g., Coleoptera, Coope, 1979; Atkinson *et al.*, 1987) provide shorter sequences through the last glacial – interglacial cycle, whereas numerous proxy sources have been used to extend the sequence of historical climatic data back to the early Holocene (e.g., harvest yields, dendrochronology, drift ice, glacier movements; Parry, 1978; Lamb, 1982). Techniques for extracting climatic signals from biotic indicators are now far advanced (Atkinson *et al.*, 1987), often involving multivariate procedures such as correspondence analysis, allowing reconstruction of moisture as well as thermal conditions (Rousseau & Puisségur, 1990). Retrodictions of climatic changes have also been made using general circulation models (GCMs) (see Kutzbach & Wright, 1985; Kutzbach & Guetter, 1986; Kutzbach & Gallimore, 1988) allowing data-model comparisons (Ruddiman, 1990).

A variety of forcing mechanisms operate on the complex earth–atmosphere–ocean system to produce climatic changes on different time-scales – viz., fluctuations in solar output, changes in the Earth's geography and orbital parameters, in surface ice cover, heat storage and circulation in oceans, in volcanic dust and so on (see Lamb, 1977; Musk, 1985). Despite the apparent complexity, evolutionary history amongst the Lepidoptera has been dominated by three fundamental climatic patterns – the glacial–

interglacial cycles, mass extinction episodes and a Cenozoic cooling trend. The glacial–interglacial cycles have been deterministically modelled by three Milankovitch elements: (i) eccentricity of the Earth's orbit (periodicity 100 ka); (ii) oscillation of the Earth's axial tilt (periodicity 41 ka) and (iii) precession of the equinoxes or perihelion timing (periodicity 23 ka) (Berger, 1978; 1984; Imbrie & Imbrie, 1979). Some 46 cycles have been identified from $\delta^{18}O$ isotope ratios over the past 2.5 ma, but only three over the past 300 ka (Ruddiman *et al.*, 1986; Shackleton in Harland *et al.*, 1990); the 100 ka eccentricity cycle has dominated the late Quaternary and the 41 ka tilt cycle the early Quaternary (Ruddiman & Raymo, 1988). These elements, which produce variations in solar radiation receipt and effect latitudinal and seasonal energy distributions, have existed throughout Earth history but have been too fine-grained to identify beyond 3 ma BP (Shackleton *et al.*, 1988; see Kerr, 1992 for interaction with other factors). Both CO_2 and CH_4 signals covary closely with modelled insolation values and ice volume ($\delta^{18}O$), implicating them in reinforcement of climatic cycles (Shackleton & Pisias, 1985; Barnola *et al.*, 1987; Genthon *et al.*, 1987; Davis, 1990; Chappellaz *et al.*, 1990). GCMs of the parameters and their effects (i.e., ice sheets, sea level changes) have permitted 'snapshot' reconstructions of high latitude climates for specific times, such as the Devensian maximum (18 ka BP; Musk, 1985; Kutzbach & Guetter, 1986; COHMAP, 1988; Guetter & Kutzbach, 1990), but difficulties of interpretation remain for tropical regions (Rind, 1990) and sub-Milankovitch events such as the Younger Dryas (Shackleton, 1989; see Fig. 4.3, Late Devensian Zone III) which are widely confirmed from data (Denton & Hughes, 1981; Gasse *et al.*, 1990).

Mass extinctions reputed to have a periodicity of c. 26 ma (Raup & Sepkoski, 1984, 1988; Sepkoski, 1989) almost certainly involve a climatic component (Wolfe, 1990), but it is still unclear whether these catastrophes (e.g., K–T boundary at 66.5 ma BP) are truly periodic rather than random and whether the cause is an extraterrestrial impact or terrestrial volcanicity (Alvarez *et al.*, 1982; Officer *et al.*, 1987; Courtillot *et al.*, 1988; Clube, 1989). The K–T changes are best explained by a short cold period (Collinson, 1990). More certain is the Cenozoic trend towards lower global mean temperatures punctuated by sharp declines as at 37 ma BP and 2.5 ma BP (see Fig. 4.7; Miller *et al.*, 1987). Although initial GCM experiments suggest that such climatic changes are too large to be explained by plate tectonics (i.e., Atlantic widening or orogenesis) (Barron, 1985), the inferred decrease in atmospheric CO_2 throughout the Cenozoic can be accounted for in renewed denudation of silicate bedrock associated with rejuvenation and orogenesis (Berner *et al.*, 1983; Raymo *et al.*, 1988; Ruddiman & Kutzbach, 1991).

B *Environmental correlates of climatic cycles*

Climate may affect butterfly evolution indirectly through other environmental factors such as vegetation and soils, water table levels, ice cover, relative sea levels and human agencies. These factors, too, can have a reciprocal influence on climate having an immediate relevance for butterfly biota. Such interactions are increasingly incorporated as feedbacks within the natural systems of GCMs. Other environmental consequences of climatic change may have no causal link with butterfly evolution, but can provide useful markers of biogeographical processes (e.g., fossil insects) or of regional climatic severity (e.g., periglacial landforms) (Coope, 1987; Williams, 1975; Watson, 1977). Stratigraphic techniques used to date and reconstruct climatic changes provide parallel space–time data on these environmental components (Flint, 1971; Godwin, 1975; West, 1977; Denton & Hughes, 1981).

Palynology and the identification of macroscopic remains have permitted the detailed mapping of specific plant types and vegetation patterns at regional and continental scales (Godwin, 1975; Huntley & Birks, 1983; Huntley & Webb, 1988; Huntley, 1990). In turn, this provides insights into the stochastics, lags, rates and processes of plant migration and colonization, succession, 'community' structures, shifts in tree limits, refugia, extinction and other responses to environmental changes. These data are of particular value for reconstructing palaeohabitats. Equivalent fossil material, valuable corroboratory evidence, exists for some fauna (e.g., mammals, Coleoptera; Stuart, 1974; Coope, 1979; 1987). Owing to the impact of climatic change on the hydrological cycle, rhythmical fluctuations in sea levels and ice cover accompanied Milankovitch and subordinate periodicities; geomorphological data document glacial advances and retreats, timing and ice cover, and the shift in snow lines (Denton & Hughes, 1981; Burga, 1988) as well as oscillations in relative sea levels and isostatic adjustments (Greensmith & Tooley, 1982; Fairbanks, 1989). Eustatic sea level changes are positively correlated with glacial and interglacial stages, which differ by as much as 100 m between cycle troughs and peaks, whereas isostatic adjustments depend on regional disparities in the relative loads of ice and water (West, 1977). Palaeobotanical research and archaeology have accumulated substantial information on human 'artefacts' which can be linked to climate and environmental changes, for example, fires, clearances, pastoralism and cultivation (Evans, 1975; Simmons & Tooley, 1981). These data present a rapidly burgeoning kaleidoscopic backcloth of past biotopes for butterfly biogeography and evolution.

4.3 **Biogeographical inferences of climatic changes**

The relatively primitive state of biogeography compared to many other sciences is often attested to in the apparent lack of proof systems, the alleged absence of specifically biogeographical processes and the preponderance of narratives over hypotheses (Rosen, 1988; Peters, 1991). The virtual preclusion, but not impossibility, of engineering experiments has focused debate on the philosophical frame of various strategies, for instance, deductive or inductive approaches, investigative *versus* explanatory methodologies. Investigative biogeography seeks pattern in order to discover processes, whereas explanatory biogeography seeks to explain patterns by referring to current processes (Patterson, 1983). Even when prodigious fossil material is available to test or support a hypothesis, determination of cause is probabilistic not objective and it is impossible not to invoke an explanatory mode of reasoning, especially as long-term phenomena can be viewed as a sequential action of short-term ecological processes (Schoener, 1988). When the fossil record is fragmentary or absent, dependence on an explanatory strategy increases, but this does not exempt researchers from formulating testable ideas. This section briefly outlines a methodology for testing biogeographical and evolutionary hypotheses for organisms without fossils and illustrates how this may be extended to probabilistic reconstructions of faunal history.

A *Faunal structures, biogeographical postulates and historical templates*
Retrodictions can be made using resource ranges and thresholds for taxa determined from autecological surveys, as surrogates for fossil material, to test postulates derived from the analysis of faunal spatial structures. Acceptance or rejection of a hypothesis depends on the concordance of resource criteria with geological conditions (Dennis, 1977; Fig. 4.1b). This procedure has some validity. Specific resources can be shown to be limiting to species and the influence of changes in resources on species' ranges can be tested using historical data sets or synoptic monitoring attached to autecological surveys (see section 4.1B). Data for species' thresholds and geological conditions have independent sources and rates of change are not specified by the conditions. A fundamental assumption is that resource thresholds have not changed substantially during the period investigated. Most hypotheses can be corroborated using analogous organisms which have an adequate fossil record. This methodology should avoid direct speculations of origins from distribution patterns and of evolutionary rates of morphological attributes from the presumed antiquity of taxonomic distinctions based on those attributes (see Ford, 1945; Beirne, 1947).

The fauna of a region consists of elements which differ in origin, age and adaptations (Mayr, 1965). Correct identification of elemental patterns is as essential for the adequate formulation of postulates as is the choice of methodologies for their testing. It is important to understand that state-of-the-art technology for determining pattern (i.e., cladistics; multivariate analyses) does not strictly produce objective 'patterns' (Sneath & Sokal, 1973; Myers & Giller, 1988; Dennis *et al.*, 1991). Determining postulates from faunal structures has its limitations. Palaeoecological data (*viz.*, non-analogue assemblages and heterogeneous dispersal rates) contest the existence of fixed 'communities' (Graham & Grimm, 1990). Significant pattern elements derived from multivariate techniques realistically indicate affinities only in a time-slice. Awareness is required of species (or areas) that share identical spatial patterns (or species) but have different histories (e.g., recent endemics and relicts on the same island). Conversely, species may resolve to significantly different patterns but share basically the same long-term history; for instance, boundaries may not mark ecological limits but simply be established at tension zones in different locations because of different dispersal rates (Barton & Hewitt, 1989).

Butterflies are limited largely by a number of habitat components (for egglaying, larval feeding, mate-location, nectaring, roosting etc.) and climatic conditions, particularly temperature. Although it is not yet possible to translate all these requirements to the geological timescale, data on host-plants, general habitat types, summer and winter warmth account for a high proportion of the variance for successful colonization and survival (Dennis, 1977). Macroscopic remains and empirical limits for pollen furnish increasingly extensive data on hostplants and habitat types. Reconstructions of climate are, as yet, restricted by the nature of geological data, which are limited to estimates of means and extreme values, and knowledge of climatic effects on butterfly populations, which is in its infancy. A simple but nevertheless effective system, if applied cautiously, would be to compare the ranges of seasonal mean conditions (January, July means) for taxa with the conditions for specified regions and times. As this is becoming a familiar procedure (Dennis, 1977; Atkinson *et al.*, 1987), it is important to understand what conclusions can be drawn from it as well as the limitations. Seasonal temperature means at the edge of a taxa's range are not *ipso facto* limits for species nor unqualified markers for their presence or absence. Stations recording screen temperatures are merely samples; butterfly species are capable of surviving in very small areas for long periods (< 1 ha; Warren, 1992a) and may occupy warmer or colder local habitats within the body of the range, relating to aspect or topography. Screen temperatures and their summary statistics are not biologically relevant measures for butterflies; they act not only as surrogates for microclimates but also for different influences of thermal conditions during the life-cycle (e.g., T_0min; T_b for FAR). Habitat and hostplants, especially if influenced by man, may act to limit species at the range, not climate. If a distinction can be made between realized and potential niche then logically the concept expands into that of realized and potential range. One derivative of this is that widespread habitats which allow the build-up of larger metapopulations may permit the survival of a species, stochastically, in a wider array of microniches and microclimates (Bennett *et al.*, 1991). Edge of range distribution records may also resolve more to extreme conditions (i.e., warm years) than to climatic averages over a standard period. When these means are compared with climatic curves, awareness is needed of the generalization of such records (means of means), the failure of biotic markers to reflect extremes in the geological record and the potential effects of past non-analogue habitat types (Guiot *et al.*, 1989).

Apart from the limitations of geological and present data, there is further justification for the use of comparative summary statistics. Autecological research and relationships between species richness and thermal environment implicate increased 'stress' levels with climatic extremes at the edge of ranges; thus with heightened probabilities of extinction. Comparative data on the climatic ranges of a number of currently sympatric species for a geological time series may then be interpreted much as a probability spectra. This approach has been applied here to reconstruct tolerance, colonization and extinction curves for the British butterfly fauna for the last glacial–interglacial hemicycle (Fig. 4.3); columns F, H and I are based on the limits of species taken individually, whereas column G illustrates limits for summer temperatures based on the mutual climatic ranges of species resident in lowland England for the western palearctic. Multivariate analyses justify their treatment as a single faunal unit; they represent extreme examples of the European 'extent' group (Dennis *et al.*, 1991).

B *Extinction, colonization and refuges*

Biogeographical changes (i.e., range expansions, contractions and shifts, species extinction, distribution fragmentation or consolidation) for specific taxa are ultimately the result of population extinction and colonization processes; the latter may be affected by climate directly, for instance through the impact of seasonal climatic conditions on survival, fecundity and dispersal, or indirectly as through the medium of hostplants and habitats by influencing natural and anthropogenically induced vegetation changes (i.e., succession, retrograde succession, plagioseres, managed landscapes) and land losses and gains (i.e., ice cover, marine transgressions). Historical reconstructions for butterflies indicate that extinction and colonization events have varied in magnitude, regionally and temporally and in the faunal elements affected, in response to the magnitude and type of climatic changes (section 4.4B).

True extinction involves the loss of a taxon without descendants. This is the result of population losses, referred to here as population extinction. Loss of all populations over a specified area is here termed regional extinction. Regional extinction and species extinction have occurred often enough in historic times to indicate that they can be caused by apparently small, subtle but fundamental changes in habitat quality, in butterflies (e.g., *Maculinea arion*, Thomas, 1984) as well as in other organisms (birds and mammals, Marshall, 1988). Mass extinctions and extinction events in the geologic record mark catastrophes for large numbers of unrelated taxa. Prominent environmental changes in Britain during the last glacial–interglacial hemicycle point to a number of regional extinction events. A more detailed chronological narrative for individual species is given

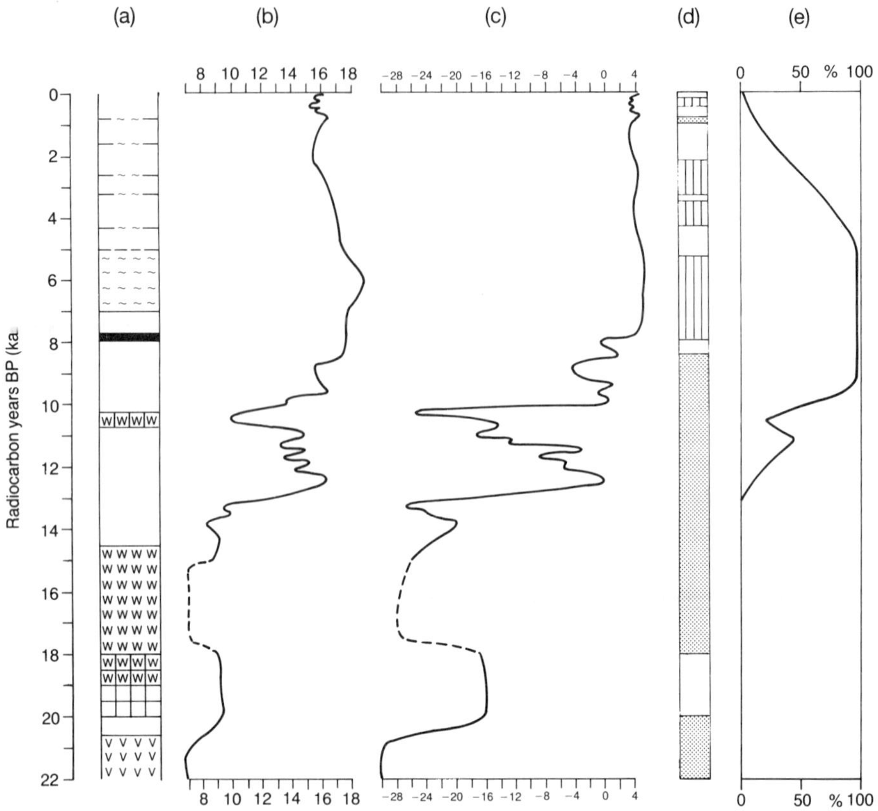

4.3 Colonization, extinction and evolution of British butterflies during the last glacial–interglacial hemicycle. (a) Geomorphological events: cross-hatched, glacial advances in Britain; V, large ice wedges; W, small ice wedges, involutions and pingos (Williams, 1975; Kolstrup and Seddon in Atkinson *et al.*, 1987; Watson, 1977); thick bar, severance of land connection with continent (Kolp, 1976); ~~~, development and expansion of ombrogenous peat and recurrence surfaces (Godwin, 1975). (b) Mean temperature of the warmest month for central England (Lamb, 1966; Lamb *et al.*, 1966; Manley, 1974; Parry, 1978; Atkinson *et al.*, 1987; Huntley & Prentice, 1988). (c) Mean temperature of the coldest month for central England (Lamb, 1966; Lamb *et al.*, 1966; Atkinson *et al.*, 1987; Walker & Lowe, 1990). (d) Annual precipitation: grey shading, drier than period 1940–70; vertical shading, wetter than period 1940–70 (Lamb, 1966; Parry, 1978; Lockwood in Musk, 1985; Bridge *et al.*, 1990; Blackford & Chambers, 1991). Precipitation may have been some 14 per cent lower than today at 18 ka BP outside the main ice accumulation centres. (e) Forest cover (Godwin, 1975; Bennett, 1989). (f) Tolerance curves for known resident British butterflies of mean screen temperatures for the warmest month in central England: lower curve, for lowest temperatures per unit time; higher curve, for highest temperatures per

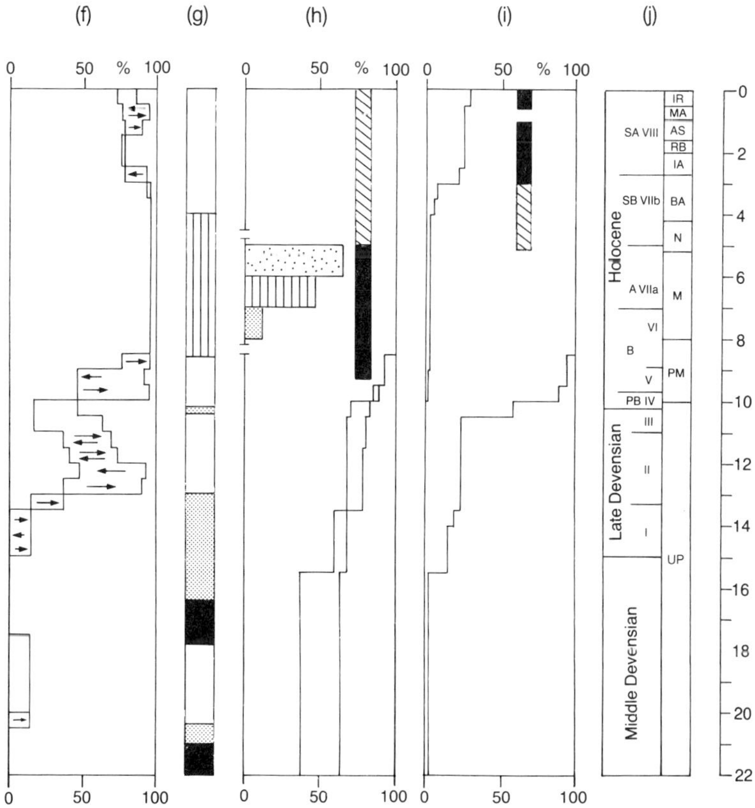

unit time (resolution 500 a based on 59 species). Arrows indicate direction of changes, increases or decreases in butterfly fauna. (Data on mean temperature tolerances extracted from distributions in Antonova, 1980; Higgins & Riley, 1983; Higgins & Hargreaves, 1983; Emmet & Heath, 1989.) (g) Thermal limits for British butterflies: vertical shading, tolerance of conditions in central England based on the mutual climatic range of species currently resident in lowland Britain (based on 55 species); grey shading, mean conditions for the coolest month probably critical for > 50 per cent of the current British fauna; black, mean conditions for coolest month critical for > 90 per cent of the British fauna. (Data on mutual range and tolerances extracted from Antonova, 1980; Higgins & Riley, 1983; Higgins & Hargreaves, 1983; Emmet & Heath, 1989.) (h) Hostplant and habitat limits for the current resident British fauna within Britain: lower curve, presence of habitat structural components (not symbionts), as known, for resident British species in central England (see Table 4.1); upper curve, presence of hostplant components for resident British species within Britain. Hostplants (macroscopic fossils and empirical pollen limits) are assumed to occur for the whole pollen zone in which they are found (Godwin, 1975; Birks, 1980). Ten species are excluded (i.e., *Carterocephalus palaemon*, *Thymelicus*

4.3 (cont.)
lineola, *Thymelicus acteon*, *Ochlodes venata*, *Cupido minimus*, *Lysandra coridon*, *Lysandra bellargus*, *Hamearis lucina*, *Erebia epiphron* and *Erebia aethiops*) due to lack of adequate fossil material. (Hostplant and habitat constraints based on Henriksen & Kreutzer, 1982; Kudrna, 1986 and Dennis, 1992.) Bar chart, species vulnerable to forest climax development: stippled, butterfly species with hostplants not typical part of the field layer or high strata in forest and requiring open habitats; vertical shading, species whose hosplants attain a suitable quality in the regeneration cycle of forests but require the continuous production of open spaces; grey shading, species requiring both arboreal or liana hosplants and habitats (see Table 4.1F). Divided bar, constraints for specific groups: black, period of isolation, range contraction and microevolution for species dependent entirely on open conditions; shaded, population extinction and contraction for species (N = 29) dependent on forest habitats *sensu lat.* (i) Arrival sequence and climatic constraints: upper curve, proposed cumulative earliest date for continued presence of British butterfly fauna north of the southern English coastline based on climate, hostplants and habitats, calculated for 500 a periods (see Table 4.1; current isothermal gradients assumed; mean screen temperatures for the warmest month < +1°C than those for central England). Lower curve, cumulative curve for species experiencing upper and lower critical thresholds for mean screen temperatures of the warmest month in central England during the late Holocene (N = 59). Black bar, period of microevolutionary and ecological changes relating to lower critical threshold summer thermal limits; this period may have started earlier (shaded zone) depending on the extent to which local and regional climates were influenced by anthropogenic agents (i.e., deforestation, grazing, loss of cover, shelter and windchill). (j) Chronology and cultures: left part, Blytt & Sernander Post Glacial periods and Godwin pollen zones (PB, Pre Boreal; B, Boreal; A, Atlantic; SB, Sub Boreal; SA, Sub Atlantic); right part, cultures (UP, Upper Palaeolithic; PM, Proto Maglemosian; M, Mesolithic; N, Neolithic; BA, Bronze Age; IA, Iron Age; RB; Romano-British; AS, Anglo-Saxon; MA, Medieval age; IR, Industrial Revolution). Note: the conventional ^{14}C chronology is subject to a correction of + 720 ± 200 a at 10 ka BP (Dansgaard *et al.*, 1989; Becker *et al.*, 1991). None of the tolerance curves for climate considers local climate contrasts (i.e., aspect, slope angle). Records are attributed to the entire 500 a period in which they occur.

elsewhere (Dennis, 1977; 1992); here, the emphasis is on different processes affecting regional population extinctions and the part played by climate.

The climatic tolerance curve (Fig. 4.3f) indicates that there are three episodes during the last 22 ka when climatic conditions *per se* may have been extreme enough to cause some loss of species from the present land area of the British Isles: (i) the late Devensian 22–13 ka BP; (ii) Younger Dryas (Zone III) 11–10 ka BP and (iii) periods during the recent Holocene 4.0–0.2 ka BP. The Devensian maximum glaciation (c. 18 ka BP) may well have been a *tabula rasa* for the current butterfly fauna. Ice extended south from Scandinavia and from regional ice caps to cover much of northern Britain, Wales and Ireland (see Dennis, 1977: 12; 1992). Summer mean temperatures (T_{sc}) lay outside the threshold for all species, even it seems *Erebia epiphron*, although conditions south of the ice sheets may have been drier than today. Winter mean temperatures (T_{sc}) of – 30°C indicate that mean monthly minima were well below – 40°C and absolute minima below – 50°C, very probably beyond the tolerance of diapausing stages of all current British butterflies (see section 2.1B). Hostplants and habitats of a substantial fraction of the fauna were absent from Devensian Britain. There is corroboratory evidence of the severity of the environmental and climatic conditions at 18–14.5 ka BP from the Coleoptera, in the apparent removal of even the hardiest north-eastern Siberian Carabidae at that time (Coope, 1987). Although temporary populations may have colonized southern Britain during occasional warmer and sunnier summers, these data point to the Devensian maximum as a datum line for the entry of the present fauna. The Devensian climate had a direct impact on butterfly survival, fecundity and dispersal and an indirect one on their biotopes and hostplants.

The Younger Dryas was less severe climatically. At its nadir summer screen conditions indicate probable survival for less than 20 per cent of the fauna and winters probably less than 50 per cent. However, as hostplants and hostplant-habitats were available for 82 per cent and 76 per cent of the present fauna respectively, it is quite possible that a larger fraction of it was able to survive in warmer local habitats with southerly aspects (Table 4.1D). Periodic climatic deterioration after 4 ka BP may well have led to population extinctions in northern and western Britain, as is implied by a cumulative curve for species experiencing summer critical screen temperature thresholds in central England during the latter Holocene (Fig. 4.3i). This period is alleged to have been cool and damp, thus cloudy, and population extinctions would have been geographically biased to higher, oceanic, thus wetter, cloudier and cooler regions. However, considerable caution is required when interpreting the plant palaeoclimatic indicators as such changes may have an anthropogenic basis. Even so, any climatic changes would have been reinforced by extensive anthropogenic modifications in uplands to cause widespread regional extinctions at modest altitudes (see below).

Climatic change in the Holocene has also been permissive of regional population extinctions by triggering environmental transformations – initially through vegetation succession to climax forest closure between 7 and 5 ka BP and subsequently through retrogressive succession by providing opportunites for the invasion of human cultures and advancing technology (Fig. 4.3) (Simmons & Tooley, 1981; Bennett, 1989).

The influence of forest closure on butterfly populations depends much on the hostplant-habitats of species. Some 39 per cent of the present British butterflies use hostplants and habitats that are not typical groundflora of the closed forest or part of early regeneration complexes (Tables 4.1F and 5.3). As the mid-Holocene tree line extended well above 600 m in England and Wales (Taylor, 1980), and locally to higher levels (750 m, Turner & Hodgson, 1979; 900 m, Turner, J., 1984), such species depended for their survival on the maintenance of open ground. A mosaic of open spaces at different stages of development probably existed in the lowland forests, caused by senescence, tree fall, natural fires, mass movements, flooding and activities of animals such as beavers, sufficient to ensure the widespread distribution of species characteristic of the forest clearings and forest floor (e.g., *Pararge aegeria; Argynnis paphia*). Simmons (1990) judges that lightning fires associated with increased convection and higher temperatures between 7 and 5 ka BP may have created more frequent and larger clearings but the climatic conditions require confirmation. However, the survival of species whose hostplants were not part of the forest assemblage would have depended on immature disclimaxes associated with topography (e.g., early seres, scars and scarps), lithology (e.g., extreme nutrient status, either siliceous or calcareous, as a dominant edaphic control) and altitude (i.e., exposed mountain tops). These species would have experienced extensive regional extinctions in the lowland forest belt. Evidence is steadily accumulating for the long-term intentional maintenance by early man of open spaces in uplands, by firing, to ensure a suitable plagioclimax for deer (Simmons & Tooley, 1981; Simmons, 1990). As these upland spaces may well have provided the bulk of suitable habitats for species requiring extensive early successional conditions, it is ironic that man may have been largely responsible for their prehistoric refuges.

Holocene climatic change produced enriched environments suitable for man to exploit, but the degree to which 'invading' human cultures had a beneficial or detrimental impact on butterfly populations depended on ecological balances, which differed regionally, and on increasing human population pressure, thus on developing land-use intensity. Initially, the onset of the Neolithic in Britain and succeeding Bronze and Iron Age cultures must have enhanced survival for open-dwelling butterfly species, which could then colonize lower elevations and the expanding, increasingly complex mosaic of vegetation communities associated with clearances (see below). But, the continuing exploitation of northern uplands would have led to the loss of most butterflies of open habitats from them and the deforestation of lowlands to the curtailment of species reliant on forest biotopes.

The exploitation by man of uplands in highland Britain set in motion a chain of processes that produced nutrient-deficient, degraded, quasi-tundra ecosystems. These processes are generally characteristic of oceanic uplands during the telocratic phases of interglacials (Godwin, 1975). However, in the Holocene, retrogressive succession, podsolization and declining soil infiltration capacity were triggered much earlier by Mesolithic settlers (10,250–7,500 a BP) (Simmons, 1990). Modelling the processes indicates their intensification through edaphic and climatic feedbacks (Dennis, 1992; Fig. 4.4). Climatically, dwarfed vegetation communities lack shelter and experience excessive windchill and low ground temperatures; the latter are exacerbated by surplus surface water caused by the removal of trees, lower evapotranspiration rates and decreased infiltration capacity. These conditions encourage continued heath and moss associations and can deter tree growth even with relaxation of grazing pressure. As a consequence, populations of all but a small number of butterfly species specializing on heath, acid-tolerant grassland and moss and tolerant of cool summer conditions (e.g., *Coenonympha tullia*; *Erebia epiphron*; *Callophrys rubi*) would have become regionally extinct at higher elevations, particularly if summer warmth declined after 3 ka BP. The lower elevations and permeable, calcareous substrates of scarplands in warmer and drier lowland Britain have resisted this chain of processes, even though the base-rich soils have encouraged continued use for both cultivation and pastoralism to the present day.

The deforestation of British lowlands has had a profound effect on the ranges and distributions not just of species dependent on arboreal and liana hostplants (e.g., *Ladoga camilla*; *Quercusia quercus*) and on trees as habitat components (e.g., *Apatura iris*), but also on those species whose hostplants are habitual components of the forest groundflora (e.g., *Mellicta athalia*; *Hamearis lucina*) and regeneration complex (e.g., *Gonepteryx rhamni*). More than 90 per cent of the islands were covered with forest before the elm decline at 5 ka BP; the forest varied regionally as to its dominant taxa and in altitudinal zonation (Bennett, 1989). Little deciduous cover now remains (< 4 per cent) and a substantial number of species dependent on these habitats are restricted to the few surviving remnants (e.g., *Mellicta athalia*; *Satyrium pruni*; Thomas, 1974; Warren, 1984). Ironically, their continuing existence during the period of maximum deforestation depended on traditional woodland management practices, some of which may date back to the Neolithic but certainly to post-Roman times (Dennis, 1992). Regional extinctions of such species in Britain would have

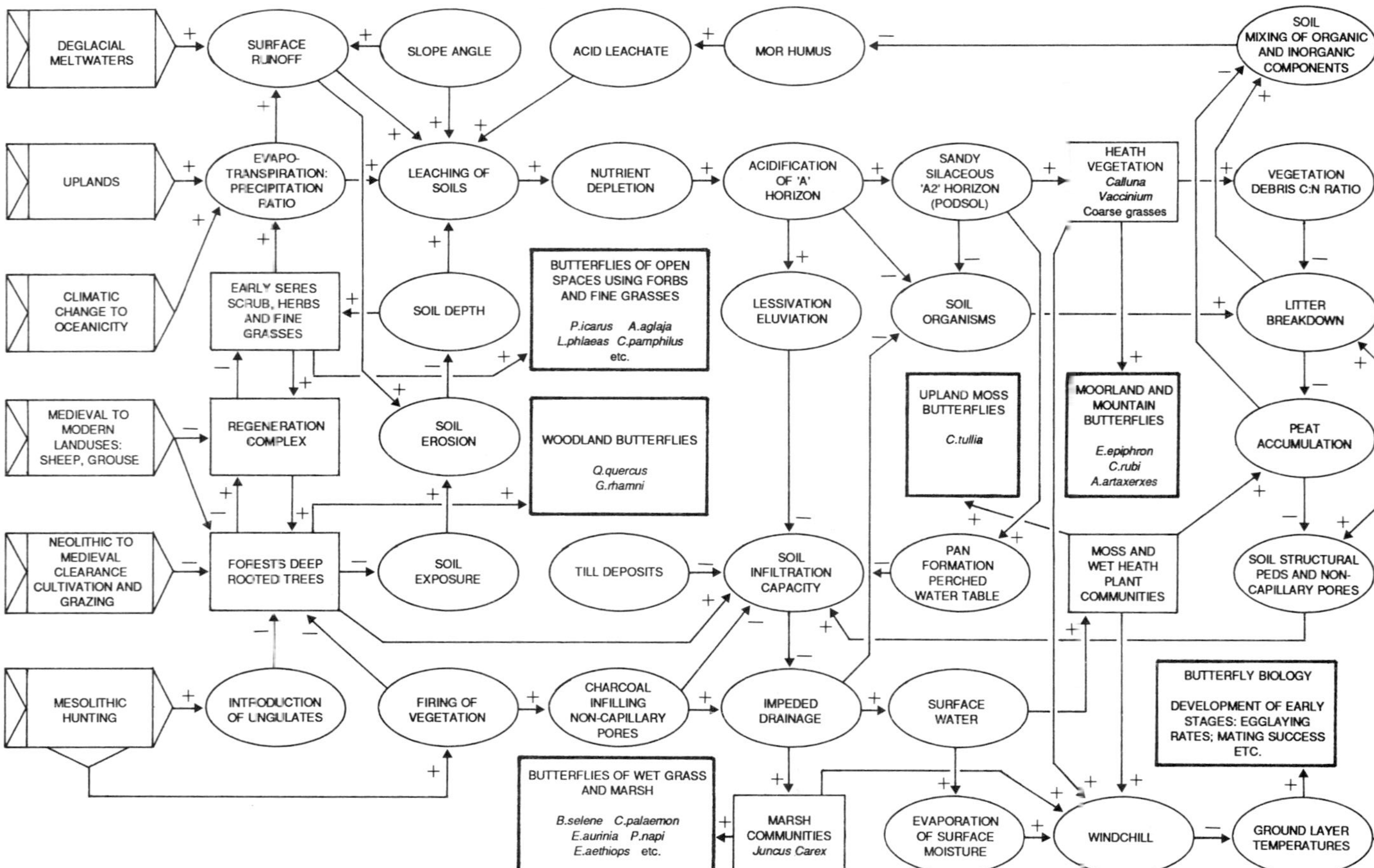
DEGLACIAL MELTWATERS
UPLANDS
CLIMATIC CHANGE TO OCEANICITY
MEDIEVAL TO MODERN LANDUSES: SHEEP, GROUSE
NEOLITHIC TO MEDIEVAL CLEARANCE CULTIVATION AND GRAZING
MESOLITHIC HUNTING
SURFACE RUNOFF
EVAPO-TRANSPIRATION: PRECIPITATION RATIO
EARLY SERES SCRUB, HERBS AND FINE GRASSES
REGENERATION COMPLEX
FORESTS DEEP ROOTED TREES
INTRODUCTION OF UNGULATES
SLOPE ANGLE
LEACHING OF SOILS
SOIL DEPTH
SOIL EROSION
SOIL EXPOSURE
FIRING OF VEGETATION
ACID LEACHATE
NUTRIENT DEPLETION
BUTTERFLIES OF OPEN SPACES USING FORBS AND FINE GRASSES
P.icarus A.aglaja
L.phlaeas C.pamphilus
etc.
WOODLAND BUTTERFLIES
Q.quercus
G.rhamni
TILL DEPOSITS
CHARCOAL INFILLING NON-CAPILLARY PORES
BUTTERFLIES OF WET GRASS AND MARSH
B.selene C.palaemon
E.aurinia P.napi
E.aethiops etc.
MOR HUMUS
ACIDIFICATION OF 'A' HORIZON
LESSIVATION ELUVIATION
SOIL INFILTRATION CAPACITY
IMPEDED DRAINAGE
MARSH COMMUNITIES
Juncus Carex
SANDY SILACEOUS 'A2' HORIZON (PODSOL)
SOIL ORGANISMS
UPLAND MOSS BUTTERFLIES
C.tullia
PAN FORMATION PERCHED WATER TABLE
SURFACE WATER
EVAPORATION OF SURFACE MOISTURE
HEATH VEGETATION
Calluna
Vaccinium
Coarse grasses
MOORLAND AND MOUNTAIN BUTTERFLIES
E.epiphron
C.rubi
A.artaxerxes
MOSS AND WET HEATH PLANT COMMUNITIES
WINDCHILL
SOIL MIXING OF ORGANIC AND INORGANIC COMPONENTS
VEGETATION DEBRIS C:N RATIO
LITTER BREAKDOWN
PEAT ACCUMULATION
SOIL STRUCTURAL PEDS AND NON-CAPILLARY PORES
BUTTERFLY BIOLOGY
DEVELOPMENT OF EARLY STAGES: EGGLAYING RATES; MATING SUCCESS ETC.
GROUND LAYER TEMPERATURES

been more sporadic than upland losses, perhaps resulting in unusual distribution patterns (e.g., 'tonsure model' of upland forest distribution; Simmons, 1990), but overall extinction has been biased to northern Britain (Dennis, 1992).

The tolerance data for species, based on both climatic and resource parameters, indicate several potential colonization events. As with regional extinction, climate may affect the colonization process directly or indirectly via habitat changes. The most prominent colonization episodes occurred in the Late Devensian and early Holocene. The tolerance data allow the construction of a cumulative arrival curve (Fig. 4.3i; Table 4.1). This may represent a conservative estimate for arrival for several reasons: (i) the fossil record may underscore the arrival of hostplants, though this is not proof of the existence of dependent herbivores; (ii) although screen temperatures lie beyond thresholds for many species during Zone III, those experiencing marginal conditions may well have found suitable local habitats; (iii) migration rates may have been extremely rapid (see below). Even so, there is reason for confidence in the basic features of the arrival curve. Changes in thermal conditions when they occurred (c. 13 ka; 10.2 ka) were enormous and extremely rapid (Coope, 1987; Atkinson *et al.,* 1987) and the coarse 500 a intervals used absorb large variations in response (Coope, 1991). Early arrivals were generally limited by climate, but later ones increasingly by habitat resources (> 15 per cent of the fauna; Table 4.1E). By 8,600 a BP, climatic conditions in central England corresponded with the mutual climatic range of all species now found in southern England (Fig. 4.3g); these conditions terminated c. 4 ka BP. Subsequently, the tolerance curve points to potentially extensive regional colonization of butterflies in northern Britain during the medieval period and after the 'Little Ice Age'. In the later Holocene, other colonization events would have been encouraged by environmental changes (see above); regional extinctions caused by upland degradation must have led to the spread of

---

4.4 Process-response and feedback of climatic and human agencies on natural environments of British uplands and their effect on butterfly populations. Boxes, vegetation; highlighted boxes, butterflies; ellipses, environmental conditions and factors influencing butterflies. Signs are correlations. Environmental changes in upland environments around 7.5–7.0 ka BP (wetter climate, Mesolithic hunters) shifted the soil water balance significantly towards paludification. The major spread of peat in England and Wales occurred after 4 ka with Neolithic and Bronze Age forest clearance and a possible deterioration of climate (Tallis, 1991b). The diagram excludes more recent influences of industrial pollution which have seriously affected the ecology of some uplands (e.g., Pennines). (From Dennis, 1992; by courtesy of Oxford University Press.)

Table 4.1 Historical data for resident British butterflies for the last deglacial hemicycle

	A	B	C	D	E	F	G	H	I
Carterocephalus palaemon	(LG)	13.0	+	–	10.2, C	–[1]	–	12.4	–30
Thymelicus sylvestris	10.0*	15.0	+	–	9.7, C	–[1]	0.5	15.3	–20
Thymelicus lineola	(LG)	15.0	+	–	9.7, C	–[1]	2.5	16.1	–20
Thymelicus acteon	(EH)	15.0	–C, F?	–	9.7 C, F?	–[1]	3.5	16.5	–10
Hesperia comma	22.0	18.0	+	*S	13.3, C	–[2]	–	10.7	–30
Ochlodes venata	(LG)	15.0	+	–	9.8, C	–[1]	–	15.0	–30
Erynnis tages	22.0	18.0	+	–	9.8, C	–[2]	–	14.9	–30
Pyrgus malvae	22.0	15.0	+	–	10.0, C	–[1]	–	14.7	–20
Papilio machaon	22.0	15.0	+	*S	13.6, C	–[2]	–	10.0	–30
Leptidea sinapis	22.0	13.0	+	*S	13.2, C	–[1]	–	10.7	–30
Gonepteryx rhamni	9.7	13.0	–F	–	9.7, F	–[1]	–	14.7	–30
Aporia crataegi	8.0	13.0	–F	–	8.0, F	+, –[1]	2.5	15.6	–30
Pieris brassicae	22.0	15.0	+	–	10.2, C	–[2]	–	12.0	–30
Pieris rapae	22.0	15.0	+	–	10.2, C	–[2]	–	13.3	–30
Pieris napi	22.0	18.0	+	+	15.0, C	–[2]	–	8.9	–30
Anthocharis cardamines	22.0	15.0	+	–	10.2, C	–[2]	–	12.4	–20
Callophrys rubi	22.0	15.0	+	+	10.2, C	–[1]	–	8.9	–30
Thecla betulae	8.0	13.0	–F	–	8.0, F	–[1]	–	14.9	–20
Quercusia quercus	9.0	13.0	–F	–	9.0, F	+	–	15.1	–20
Satyrium w-album	9.0	13.0	–F	–	9.0, F	+	–	14.9	–20
Satyrium pruni	8.0	13.0	–F	–	8.0, F	–[1]	3.5	16.4	–20

Table 4.1 (cont.)

	A	B	C	D	E	F	G	H	I
Lycaena phlaeas	22.0	18.0	+	+	15.0, C	–[2]	–	8.9	–30
Lycaena dispar	13.4*	15.0	+	–	9.8, C	–[2]	3.0	16.2	–20
Cupido minimus	(LG)	15.0	+F?	–	10.2, C, F?	–[2]	–	12.4	–20
Plebejus argus	22.0	15.0	+	–	10.1, C	–[1]	–	13.7	–30
Aricia agestis	22.0	18.0	+	–	9.8, C	–[2]	2.0	15.4	–20
Aricia artaxerxes	22.0	18.0	+	⋆S	13.2, C	–[2]	–	10.7	–30
Polyommatus icarus	22.0	18.0	+	+	15.0, C	–[2]	–	8.9	–30
Lysandra coridon	(EH)	15.0	–F?	–	9.7, C	–[2]	3.0	16.0	–20
Lysandra bellargus	(EH)	15.0	–F?	–	9.7, C	–[2]	3.0	16.0	–20
Cyaniris semiargus	11.0*	15.0	+	–	10.1, C	–[2]	–	13.0	–30
Celastrina argiolus	9.7	13.0	–F	–	9.7, F?	+, –[1]	–	13.7	–20
Maculinea arion	13.3	15.0	+	–	9.8, C	–[2]	2.5	15.9	–20
Hamearis lucina	(EH)	13.0	–F?	–	9.8, C	–[1]	2.0	15.5	–20
Ladoga camilla	9.7	13.0	–F	–	9.7, C, F	+	2.5	15.9	–20
Apatura iris	13.0	9.0	–H	–	9.0, H	+	2.5	15.9	–20
Aglais urticae	22.0	13.0	+	+	13.0, H	–[1, 2]	–	8.9	–30
Nymphalis polychloros	13.0	13.0	+	–	9.8, C	+	2.5	15.9	–20
Inachis io	22.0	13.0	+	–	10.0, C	–[1,2]	–	14.2	–20
Polygonia c-album	22.0	13.0	+	–	10.0, C	+	–	14.7	–30
Boloria selene	22.0	15.0	+	+	15.0, C, H	–[1]	–	8.9	–30
Boloria euphrosyne	22.0	13.0	+	⋆W	10.2, C	–[1]	–	8.9	–20
Argynnis adippe	22.0	13.0	+	–	10.0, C	–[1]	–	14.7	–20

Table 4.1 (cont.)

	A	B	C	D	E	F	G	H	I
Argynnis aglaja	22.0	18.0	+	–	10.2, C	–[1]	–	12.4	–30
Argynnis paphia	22.0	13.0	+	–	9.9, C	–[1]	–	14.7	–20
Eurodryas aurinia	22.0	15.0	+	–	10.1, C	–[1,2]	–	13.9	–20
Melitaea cinxia	15.0	15.0	+	–	9.7, C	–[2]	3.0	16.0	–30
Mellicta athalia	22.0	13.0	+	+	13.0, H	–[1]	–	8.9	–30
Pararge aegeria	15.0	13.0	+	–	10.0, C	–[1]	–	14.2	–20
Lasiommata megera	22.0	18.0	+	–	10.0, C	–[2]	–	14.6	–20
Erebia epiphron	(D)	18.0	+	+	15.0, C	–[2]	9.7	9.3	–30
Erebia aethiops	(LG)	13.0	+	–	10.2, C	–[1,2]	–	13.6	–20
Melanargia galathea	22.0	15.0	+	–	9.8, C	–[1,2]	0.5	15.4	–20
Hipparchia semele	22.0	18.0	+	–	10.2, C	–[2]	–	13.3	–20
Pyronia tithonus	22.0	15.0	+	–	9.8, C	–[1,2]	–	14.8	–10
Maniola jurtina	22.0	15.0	+	–	10.1, C	–[1,2]	–	13.0	–20
Aphantopus hyperantus	22.0	15.0	+	–	10.1, C	–[1,2]	–	14.0	–30
Coenonympha pamphilus	22.0	18.0	+	–	10.2, C	–[2]	–	12.4	–20
Coenonympha tullia	13.3*	15.0	+	*S	13.5, C	–[2]	–	10.5	–20

A Hostplant foundation date. Date (^{14}C ka BP) from which hostplants are known to have been present in the British Isles from macroscopic remains and empirical pollen limits. Maximum date is 22 ka and dates in parentheses are estimates; *, almost certainly a minimum date. D, Devensian 22–15 ka; LG, Late Glacial 15–11 ka; EH, early Holocene > 10.2 ka. Geological and current hostplant data as in Fig. 4.3h.

B Habitat foundation date. Date (^{14}C ka BP) from which basic habitat structures required by species have been present in central England. < 18 ka, ice contact and polar desert; < 15 ka, polar desert, bare ground and patches of short herbs;

< 13 ka, tall herb grassland; < 9 ka shrubs and woodland; > 9 ka, climax forest components. Geological and current habitat data as in Fig. 4.3h.

C Late Glacial residence. Temporary and permanent residence in Britain of butterfly species during the Late Glacial c. 15–11 ^{14}C ka BP. +, present and –, absent on the basis of hostplant, habitat structure, summer and winter mean screen temperatures. The limiting factor(s) for species which were probably absent is given by F, hostplant; H, habitat; C, climate.

D Loch Lomond stadial (pollen Zone III) residence. Residence in Britain of butterfly species during the whole period from c. 11–10.2 ^{14}C ka BP; +, presence and –, absence based on the same conditions as in C above. *, species subject to marginal climatic conditions, but probably finding suitable local climates associated with topography. Limiting conditions for marginal species: W, winter temperatures; S, summer temperatures.

E Butterfly foundation date. Date (^{14}C ka BP) from which butterfly species have probably been resident in Britain; based on the suitability of resources (hostplants, habitat structure) and climatic conditions (summer and winter mean screen temperatures). Limiting condition influencing this date: C, climate, F, hostplant; H, habitat structural component. Mean screen temperature assumed to be ≤ + 1°C higher in southern England than in the central Midlands.

F Hostplant constraints during the forest climax development 9–5 ka BP. +, none, species utilize arboreal hostplant and habitat components; $^{-1}$, herbs and shrub hostplants attain a suitable quality for butterfly species in the forest regeneration cycle and depend on the continuous production of open spaces in the forest; $^{-2}$, species use habitat and hostplants not typically associated with forest areas and require open spaces outside the forest cover (e.g., riparian, scree, cliff, scar, scarp, limestone grasslands, dunes etc.).

G Climatic constraints on butterfly species in the later Holocene. Date from which climatic conditions (summer mean screen temperatures) first become marginal for resident British butterflies in central England following the climatic climax development in forest cover (to nearest 500 a in ^{14}C ka BP).

H July mean screen temperatures (°C) at the northern range limit of resident British butterflies in the western Palearctic to 60° east. Values interpolated from climatic data for 254 climatic stations (Wallén, 1970).

I January mean screen temperatures (°C) at the northern limit of resident British butterflies in the western palearctic to 60° east. Values as above for H above, but given to the lower value for 10°C intervals, in view of the uncertainty regarding the influence of low winter temperatures. Limited data are available on the supercooling point of overwintering phases in some butterflies; T_c: *T. acteon*, – 28°C; *P. brassicae*, – 23°C; *L. dispar*, – 27°C; *A. urticae*, – 20°C; *I. io*, – 18°C (McNeil & Fields, 1985; Pullin, 1992; see section 2.1B).

---

some species, such as *Coenonympha tullia* with the expansion of highland *Eriophorum vaginatum* and *Sphagnum* mosses, whilst lowland clearances and deforestation exposed larger areas for butterflies requiring open conditions (e.g., Satyrinae; grassland Hesperiidae). Early landnam-type clearances which encouraged forbs such as *Melampyrum pratense* and *Rumex* species would have enhanced the spread of butterflies such as *Mellicta athalia, Melitaea cinxia* and *Lycaena phlaeas*. The intensity and expanse of later, medieval, clearances was probably reflected in increasingly vibrant metapopulation dynamics of species dependent on early seral vegetation associations.

Observations on Devensian and Holocene colonizations and extinctions, corroborated by the insect fossil record, illustrate that as scale increases (temporal, spatial, taxonomic), the patterns and processes inferred extend beyond those observed in ecological time (Myers & Giller, 1988). Indications are that measures for attributes of the migration process (i.e., lifetime track, linear range) underlying colonization would have exceeded those from contemporary populations. Migration is subject to selection by the relative costs of leaving or staying within habitats (Baker, 1978; 1984). During the Late Devensian and early Holocene return of temperate fauna from Mediterranean latitudes, the cost of leaving a current habitat would have been much less than today (*viz*., shorter migration distances to suitable habitats; abundant nectar sources en route) and the benefits of moving greater (*viz*., hostplants occurred in Late Glacial Britain for > 70 per cent of the present fauna; absence of parasites), whereas the cost of staying in a recently colonized site would have steadily increased (*viz*., resources increasingly utilized by expanding populations of fauna in the ameliorating climate; resource decline with vegetation succession; growing predator and parasite populations). To the extent that lifetime tracks of individuals are under genetic control, it is feasible that these episodes of northward species migration would have led to the selection for greater directionality and thus for longer movements towards more suitable habitats at higher latitudes (Dennis, 1977).

It is possible to speculate only on absolute values for lifetime movement attributes as these are generally not well known for modern populations. Although very limited migration occurs for species with relatively closed population structures (Thomas, 1984; Thomas, 1985), mark–release–recapture (MRR) procedures grossly underestimate maximum distances moved (Shreeve & Dennis, 1991). Indirect methods of measuring genetic diffusion give higher values for dispersal than direct measurements which miss long distant migrants (e.g., *Heliconius erato*; Barton & Hewitt, 1989). Large numbers of relatively sedentary species have been recorded in transit across open fields or gardens (Baker, 1984) and movements > 1 km d^{-1} noted for them using MRR (Dennis, 1986b; Warren, 1987). The upper limit to individual migration is probably c. 300 land km $lifespan^{-1}$ (Baker, 1978); typical values of movement for *Pieris rapae* and *P. brassicae* (0.8–1.1 km h^{-1}) and extreme values for *Ascia monuste* (12–17 km h^{-1}) indicate that the maximal latitudinal shift in ranges required for species with an established hostplant-habitat resource could be well within 500 a, the time unit selected for response to changes in conditions and resources (Fig. 4.3).

Although Fig. 4.3 specifies conditions only for species occurring in Britain today, range shifts, colonization and regional extinction can be modelled for different taxa, for example arctic elements, which must have occurred in Britain during the late Devensian and Younger Dryas (Dennis, 1992). As faunal elements contrast for climatic tolerances (Fig. 4.5) and resources and the degree to which these resources have been influenced by geomorphological events (e.g., ice cover), they would have been affected very differently by changing conditions during the glacial–interglacial cycles, particularly high-latitude elements compared to those at lower latitudes (Dennis *et al.*, 1991; see section 4.4B). Faunal elements have

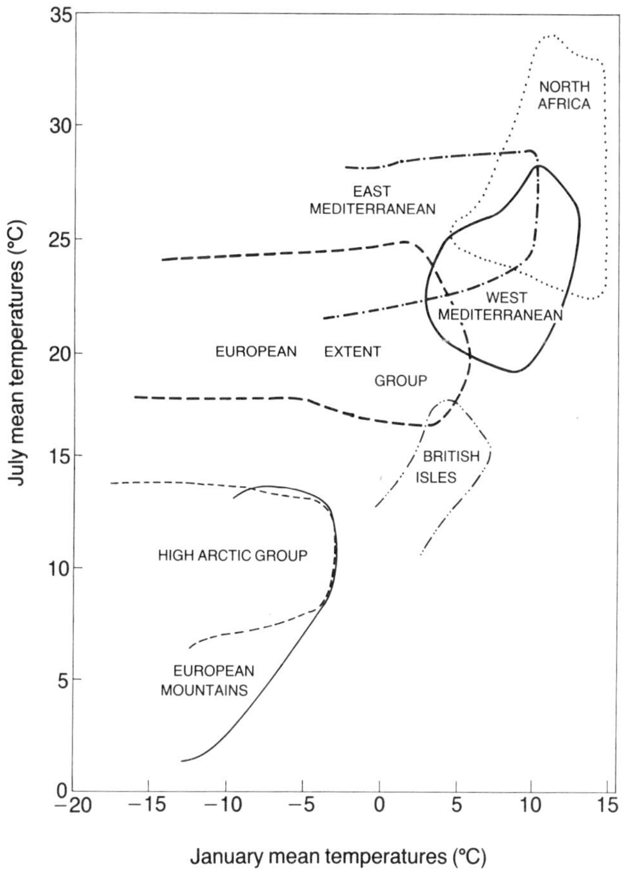

4.5 The characterization of selected European butterfly faunal units (see Fig. 4.8) on July and January mean screen temperatures for 254 locations over Europe and North Africa to 60° east. The lower altitudinal limit for the southern European montane unit is 1500 m which is generally the lowest residence level quoted for the majority of subalpine butterflies (Higgins & Riley, 1983). (From Dennis *et al.*, 1991; by courtesy of Academic Press Ltd.)

undergone systematic range expansion – refuge cycles, refuges representing the restriction of ranges to relatively small areas. Fossil plants and insects corroborate these massive range changes (Coope, 1979, 1987; Huntley, 1988). A variety of refuges have also existed (*viz*., montane, upland, fenland, cuesta, forest) at different times during the Holocene climatic changes for different taxa occurring in Britain today (Dennis, 1977; 1992). The main difference between these refuge–range expansion cycles in Britain and those for faunal elements on the continent is one of scale and 'lifespan'; that is each glacial cycle heralds a virtual *tabula rasa* for the altithermal interglacial British fauna, signalling the complete loss of genetic variation from the region of the British Isles. At its simplest, the lowering temperatures during the telocratic phase of a glacial–interglacial cycle adversely affects mortality, reproduction and dispersal in butterfly populations, prior to any significant influence on global ice budgets and sea levels, thus minimizing the probability of removal migration during retrogressive succession.

C *Evolution of the migratory habit*

Migration *sensu* Baker (1978) has a number of individual and group components (*viz*., directional bias, distance, speed, degree of return, regularity, frequency of irruption); this section explores the role of climate in the evolution of 'true' migrations, which involve regular, direction-biased, long-distance seasonally reversed transfers of fractional if not whole populations (e.g., *Danaus plexippus*; references in Ackery & Vane-Wright, 1984). Although migrational behaviour varies extensively between taxa, there are common elements that can modelled in three stages: (i) predisposition or exaptations, (ii) adaptations to a environmental gradient and (iii) protraction of the gradient and responses to spatial dislocation of resources.

The most important pre-adaptations for distant return migrations are multivoltinism, adult longevity and the capacity to evolve powerful flight. Multibrooded species (e.g., many Pierinae and Nymphalinae) have the greatest capacity for intraseasonal return flights; moreover, as they are invariably forced to use different hostplant resources between broods, and often different hostplant species in different biotopes, their survival depends on regular long-distance movements (Shapiro, 1980d). Nevertheless, multivoltinism is not an absolute prerequisite; sequential use of disjunct resources is sufficient for seasonal migrations to develop as in the case of species which are univoltine and hibernate in the adult stage (Larsen, 1976). Lack of predictable spatial variation in resource quality over short distances precludes the evolution of directionality and reversals. Only if an environmental gradient favours migration in one direction during one part of the year, but in a direction > 90° from this in another part of the year, can a bias for seasonal direction and thus reversals evolve (Baker, 1984).

Theoretically, the polarization in quality of resources used successionally, even if uniformly distributed, could result in selection for seasonal movements, thus a bias in direction ratio and return frequency. Lebanese *Gonepteryx rhamni* migrate between winter lowland hibernation quarters and summer upland breeding sites (Larsen, 1976). In Californian *Nymphalis californica*, a spring upland migration and antumn downslope migration follow a wave of hostplant quality, young fresh growth of *Ceanothus*, despite its omnipresence (Shapiro, 1975b). Gradients typifying return migration occur with altitude, latitude and into continental interiors (Larsen, 1976; Baker, 1978) and each has a climatic basis. Climate primarily controls host resource quality and availability, and suitability of hibernation sites, which in turn affects reproduction and survival through life-history traits such as developmental rate, potential for continuous development and somatic maintenance. As yet, the data are inadequate to determine which factors are relevant for particular species (see Baker, 1978: 440). Nevertheless, Calvert and Brower (1986) have identified a large number of adapations to hibernal climate among *Danaus plexippus* colonies in the Transvolcanic Belt of Mexico. Protraction of an environmental gradient and increasing isolation of seasonal resources imposes selection for increasingly distant movements by insects dependent on seasonal migrations. Selection for such movements would be enhanced, as they would for regularity and completeness of migrations and size of irruptions, as the seasonal variance in conditions at the termini of such gradients increases.

As distant return migrations are not restricted to a taxon, to one single geographical region or to a single climatic regime, different conditions and historical events have been propitious for the evolution and enhancement of the mechanism, probably at different geological times (section 4.4). Both regional and climatic bias in the frequency of and the variation among migrant species provide some insight into historical events and the variability of migration systems (Larsen, 1976). In the west palaearctic, the seasonal alternation of conditions with latitude permits continuous development of multivoltine species capable of migrating northwards in spring and southwards in autumn. Univoltine species are generally forced to diapause through the long winter of northern latitudes and through the long summer droughts at Mediterranean latitudes. The mountain systems of southern Europe greatly exaggerate the environmental differences between the Mediterranean region and northern Europe (Fig. 4.6). Primarily, they present steep gradients over short distances; this provides predictable seasonal changes in resources and predictable cues (zeitgebers) for insects to exploit these resources. This in itself would have been sufficient to trigger return migrations. Selection for migration parameters would have varied with climatic change: with the telescopy of biomes during each glacial stage, shorter removal and return migrations would have been favoured as compared to longer distances with their protraction in the Tertiary and during interglacials; these changes corresponded to shifts in the polar front (Ruddiman & McIntyre, 1981). The widespread distribution of butterfly hostplants, and the rapid responses migrant species potentially have to environmental changes, would have allowed them to exploit open areas during summers far north of the alpine mountain belt at times when brief warm periods occurred in the glacial phases. This emphasises the stochastic input of selection regimes on butterfly migration. Thus, the apparent extensive 'wastage' of individuals at extreme latitudes and altitudes is not evidence of maladaptation; it simply reflects the impact of varying selection pressures by seasonal regimes on long distant migrants which oviposit en route and which consequently maintain a stock of dispersal genes.

Data accumulating on species undergoing migrations increasingly point to variation in migration components between species (e.g., *Pieris brassicae* and *Vanessa atalanta*) and within species for different regions and different years. This has led to a reconsideration of the part played by contemporary variations in seasonal weather on migrations (cf. Johnson, 1969; Baker, 1978; Pollard, 1982b). Much of the variation between taxa is probably due to differences in physiology, ecology and morphology, but there is a considerable potential for contemporary extrinsic agents to influence migrations within taxa. Selection regimes are affected both by regional differences and by temporal variations in climate. The latter contributes a significant stochastic input both in the form of short-term fluctuations in weather systems and long-term climate changes, patterns which are never quite repeated and which include unpredictable extreme events. The contemporary influence of weather systems on migration is evidenced in the disparity of distances covered in different years, and in the differences in timing, return frequency, regularity, completeness and irruption magnitude, for a variety of species (e.g., *Colias croceus; Cynthia cardui; Vanessa atalanta; Pontia daplidice* in Europe); also, in the degree of synchroneity between different species and between different populations of the same species in different regions (references in Baker, 1984; Emmet & Heath, 1989). For example, global scale disturbances such as El Niño events can influence rainfall in widely separated semi-arid environments and the covariance of butterfly migrations on different continents (Drake & Farrow, 1988). Although migration direction is largely independent of wind direction (Williams, 1958; Baker, 1978), long-distance spring migrations from the Mediterranean are enhanced by warmer meridional air streams to the west of anticyclones and to the east of depressions over north-west Europe (Fig. 1.3). Insects generally fly higher downwind and thus faster than upwind (Gibo & Pallet, 1979). In other insects, migration is greatly influenced by meso- and local scale weather systems (Drake & Farrow, 1988; McManus, 1988). Altogether, despite the dearth of adequate data on responses of migrants to synoptic pulses, it is suggested that many butterfly migration systems are less deterministically controlled by mean climatic gradients than suggested by Baker (1978; 1984). Thus, the variability of long- and short-term atmospheric events may largely explain erratic movements in some species (*C. cardui*, Larsen, 1976), particularly as the stochastic nature of weather systems may preclude any region within the west palaearctic as providing a consistently adequate seasonal resource zone.

4.4 **Adaptations to palaeoclimates**

Recently, the gradualist notion of evolution has been challenged by two sets of 'observations': firstly, a non-uniform fossil record which suggests long periods of stasis interspersed with relatively short periods of rapid change; secondly, distinct periodicities of long-term climatic events with very different wavelengths. The alternative view is one of 'punctuational equilibria' (Gould & Eldridge, 1977; Calow, 1983), from which it has been proposed that climatic periodicities have the capacity for determining four different tiers of evolutionary processes (Gould, 1985; Bennett, 1990; Table 4.2). The fundamental maxim is that whatever 'progress' is made at one tier can be undone by a higher one. Thus, Milankovitch events of the second tier can erase microevolutionary changes of the first tier; similarly, accumulated trends in lineages at the third tier in geological time may be lost in mass extinctions (fourth tier). These tiers provide a useful frame for the formulation of questions linking climate and butterfly evolution. The first section explores the climatic forcing of butterfly macroevolution and the development of a cold fauna. Milankovitch cycles in the Quaternary caused extensive regional population extinctions and migrations, but are believed to have disrupted speciation events. The second section examines the potential for macroevolution and species extinction in a region affected by Pleistocene glaciations. Much microevolution of the first tier is caused by the direct and indirect influence of climatic change. The last section

---

4.6a (overleaf) Physical conditions during the last glacial maximum, 18 ka BP. The distribution of mountains, ice cover, coastline and polar front. Ice sheets (continuous lines) and ice caps (black areas), after West, 1977; Starkel, 1977; Frenzel, 1979; and Denton & Hughes, 1981. Light shading, land over 1000 m. Potential significant extensions of the coastline at 18 ka BP (thick continuous line) follow the 120 m bathymetric contour (Godwin, 1975). Polar front and winter sea ice limit (thick pecked line), after Ruddiman & McIntyre, 1981. Note that high mountain terrain is largely restricted to Europe south of 50° north; this divides Europe into distinctly different climatic zones, with hostile conditions for most butterflies north of alpine belt during the glacial maxima. Line x-y, transect in Fig. 4.6b.

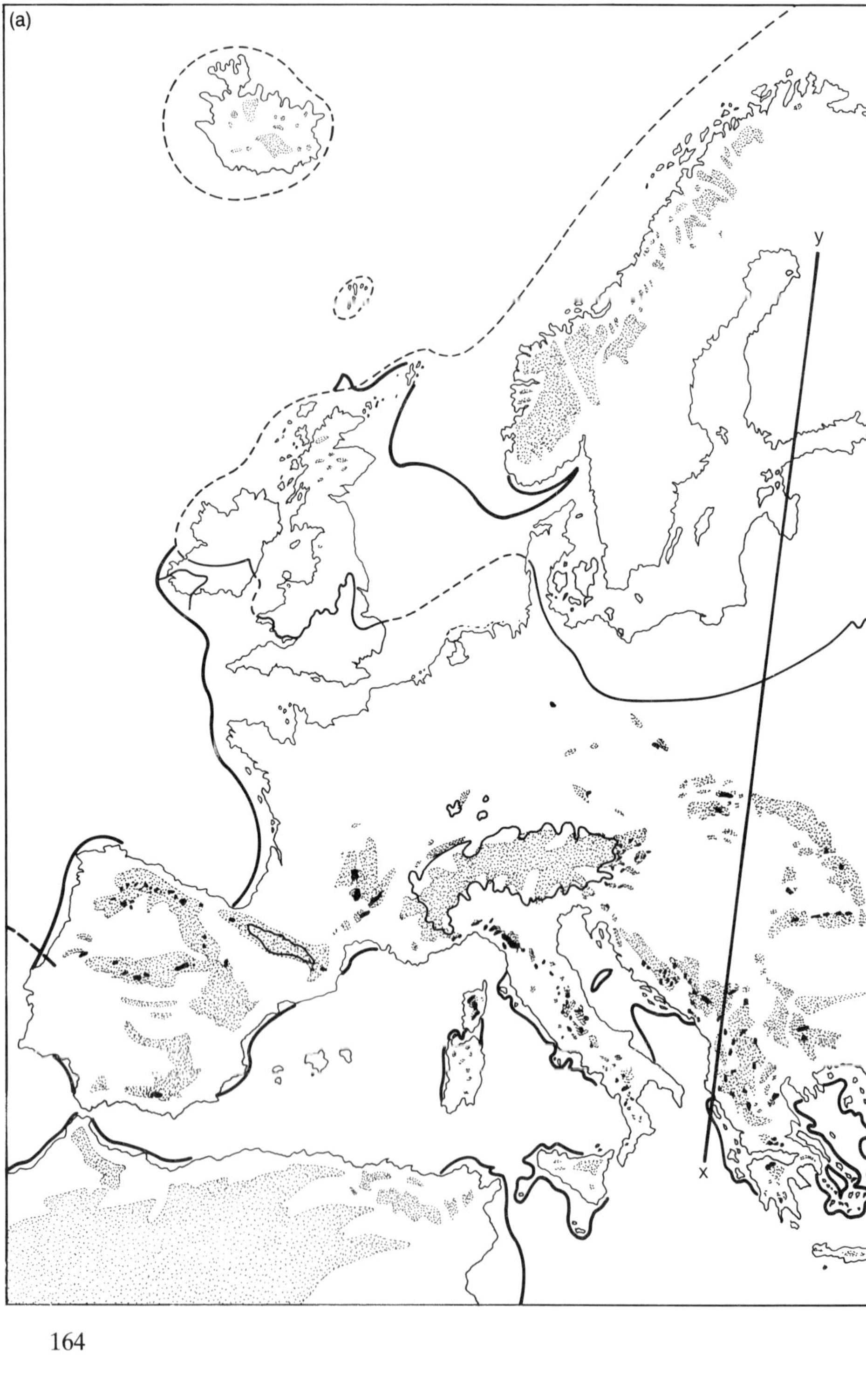
(a)
y
x

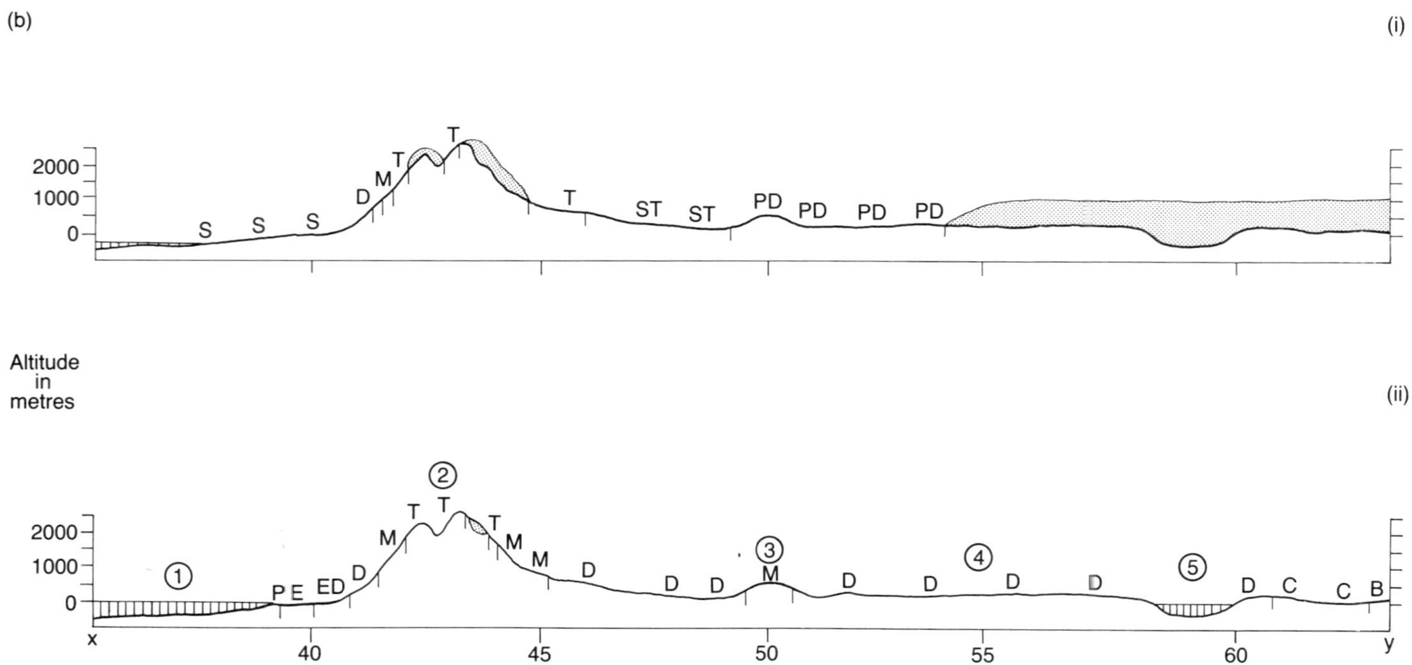

4.6b Schematic north–south transects of vegetation during (i) a glacial stage and (ii) an interglacial stage across Europe. Line of transect shown in (a). PD, polar desert; S, steppe; T, tundra; C, boreal coniferous forest; M, montane and subalpine forest (pine, fir, spruce and beech); D, deciduous forest; E, evergreen forest; P, pine; B, birch; stippled areas, ice and snow fields; 1, Mediterranean sea; 2, Dinaric alps; 3, Dukla Pass in Carpathian mountains; 4, north European plain; 5, Gulf of Finland.

Table 4.2 Temporal hierarchy of processes controlling evolutionary patterns seen in the geological record (From Bennett, 1990 by courtesy of the Palaeontological Society)

Tier	*Periodicity*[a]	*Disturbance*	*Evolutionary processes*
First	–	–	Microevolutionary change through natural selection among populations in more or less stable communities
Second	20–100 ka	Milankovitch cycles	Disruption of communities, leading to loss of accumulated change from the first tier
Third	–	–	Macroevolutionary change through speciation in conditions of geographic isolation created by environmental changes during the second tier
Fourth	c. 26 ma	Mass extinctions	Loss of species and higher taxa

[a] First and third tiers are defined temporally as intervals between events of the second and fourth tiers respectively. The changes of the second and fourth tiers can undo any evolutionary changes accumulated at the first and third levels, respectively.

investigates these links in a context which permits detailed historical reconstructions and the dating of events.

A *Climatic forcing of butterfly evolution*

Although butterfly fossils are rare compared to those found for other insects, enough exist to elucidate broad evolutionary patterns (Fig. 4.7). Modern butterflies have ancient roots; all families, many subfamilies and tribes date to the Mesozoic, well before the K–T boundary at 66.5 ma BP. In view of this, pattern diversification among living forms may be regarded as limited. Yet modern genera date only from the mid-Oligocene (c. 30 ma BP) and modern species from the Miocene (c. 10 ma BP), implying a substantial turnover of lineages in the Tertiary (Fig. 4.7c). Despite insufficient data to test correlations between climatic change and butterfly evolutionary events, there are indications that climate has been a prominent abiotic factor in forcing butterfly macroevolution. The impact of mass extinction events, whether extraterrestrial or volcanic in origin, has been mainly through abrupt climatic changes. Together with progressive monotonic planetary cooling, these can have influenced butterfly evolution in several ways. Firstly, climatic changes have radically altered plant life-forms. Angiosperms have undergone several explosive evolutionary episodes, closely associated with mass extinction events (Fig. 4.7d). Butterflies as phytophagous insects are entirely dependent on these changes. The evolution of butterfly families corresponds with the first angiosperm radiation in the Late Albian and Early Cenomanian (100–95 ma BP) and the relatively recent emergence of modern butterfly species with the post-Eocene evolution of modern angiosperm taxa. Secondly, the same changes have resulted in the massive disruption of vegetation 'communities' by substantially altering the frequency and distribution of plant species, e.g., the selection for deciduous life-forms and the emergence and expansion of steppe, taiga and tundra (Tallis, 1991a; Fig. 4.7d). Plants provide much more than nutrition and chemical markers for butterflies. Habitats of hostplants establish an evolutionary context for phytophagous insects, distinguished by specific physical conditions, predators, parasites, competitors and symbionts (Dennis, 1992: 217). The adoption of novel hosts must almost invariably involve adaptation to new evolutionary contexts. Thirdly, the massive disruption of communities, extinction of plant forms, sudden climate changes triggered by mass extinction events, must have had a dramatic impact on phytophage biogeography. For example, the existence of fusain at the K–T boundary and a subsequent 'fern spike' in pollen spectra indicate a successional 're-start' (Moore, 1989; Collinson, 1990). Such large-scale abiotic processes, which enhance extinction, also serve to promote isolation and speciation through fragmentation of biomes and populations (Mayr, 1942). As it is, the evolutionary trajectory of both mass extinction events and gradual planetary cooling has been towards more open, cold tolerant communities favouring dormancy and deciduousness, thus opening new adaptive zones for phytophages (Wolfe, 1987; Collinson, 1990). These sudden massive shifts in conditions, which would have impinged on a large fraction of the fauna, may lead to periods of relatively rapid evolutionary changes under directional selection, especially where circumstances have generated the adoption of and specialization on novel hostplants (Futuyama, 1988).*

* In a local population faced with the abundance of an inferior host and rarity of a normal host, selection may favour attraction to the novel host, even if its defensive chemicals initially reduce survival. Once the behavioural barriers are overcome, selection will favour physiological and morphological adaptations to other properties of the plant and the habitat.

4.7 (overleaf) Evolutionary context for butterflies during the Cenozoic. (a) Age and geological divisions (after Collinson, 1990; *Oxford Dictionary for Scientific Writers and Editors*, 1991). Plio, Pliocene; Q, Quaternary. (b) Oxygen-isotope record for the Cenozoic, determined from Atlantic benthic foraminiferans (adapted from Miller *et al.*, 1987). The record reflects the general trend towards cooler, more glacial conditions throughout the Cenozoic, punctuated by occasional sharp declines (e.g., at 35, 13 and 2.5 ma BP). The vertical line is the $\delta\ ^{18}0$ value considered as the threshold separating generally ice-free and ice sheet conditions, though a positive ice budget is believed to have started earlier (Prentice and Matthews, 1988 in Collinson, 1990). (c) Butterfly taxa (modified from Brown, 1987). The origin of higher taxa has been determined from their distribution in relationship to plate tectonics movements (Shields, 1976, Shields & Dvorak, 1979; see Smith *et al.*, 1981). The meagre fossil record includes some 50 butterfly species dating from the Early Eocene (Zeuner, 1962; Nekrutenko, 1965; Brito & Ribeiro, 1975 in Brown, 1987; Brown, 1976; Durden & Rose, 1978; Tindale, 1980; Jarzembowski, 1980; Whalley, 1986). 1. Earliest butterflies known currently are *Praepapilio colorado* and *P. gracilis* (Papilionidae), *Riodinella nympha* (Lycaenidae) and a satyrine nymphalid (unnamed) from the Green River Shale, Colorado at 48 ma BP. 2. The earliest known British butterflies are *Nymphalites zeuneri* (Nymphalidae) and *Lithopsyche antiqua* (Lycaenidae) from the Bembridge Marls, Isle of Wight at c. 35 ma BP; twelve fossil butterflies from the Florissant shale beds, two of which are placed in extant genera (e.g., *Vanessa amerindica*; Nymphalidae) have a similar date (c. 35 ma BP) (Emmel *et al.*, 1992). 3. Examples of fossils assigned to modern genera are *Vanessa karaganica* (Nymphalidae) and *Cynthia* (*Pyrameis*) *fossilis* from the Karagan horizon of the Miocene, Stavropol, North Caucasus. 4. Examples of fossils assigned to modern species: *Papilio maackii* (Papilionidae), *Belenois cranshayi* (Pieridae) and *Melinaea ethra* (Nymphalidae) (see Brown, 1987). (d) Landforms, climate and plant forms; main events. (Hsu *et al.*, 1977; Suc, 1984; Wolfe, 1987; Bertoldi *et al.*, 1989; Moore, 1989; Collinson, 1990; Crepet & Feldman, 1991; Krassilov, 1991; Tallis, 1991b; Heron, 1992.)

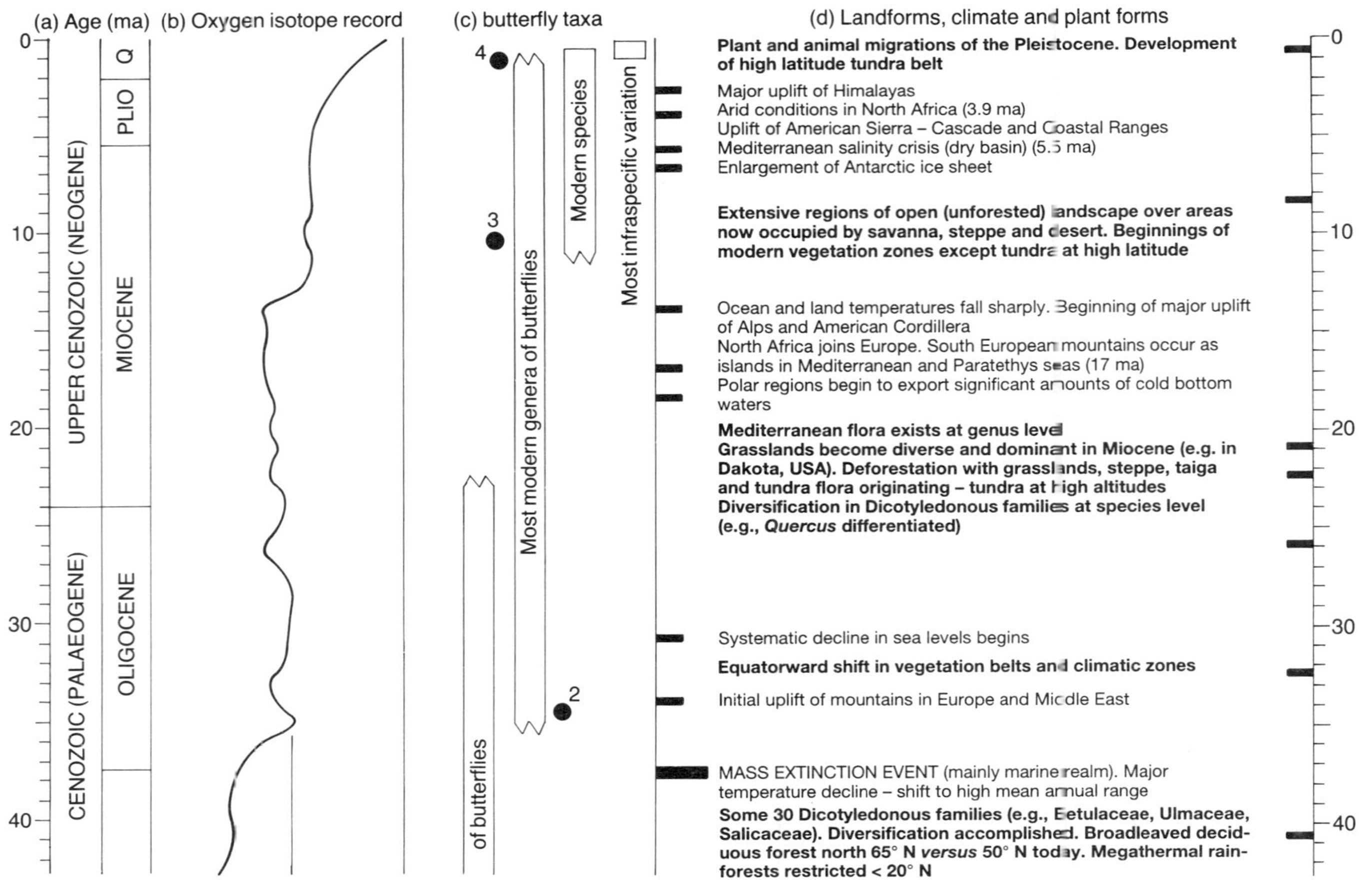
(a) Age (ma)
(b) Oxygen isotope record
(c) butterfly taxa
(d) Landforms, climate and plant forms
0
10
20
30
40
UPPER CENOZOIC (NEOGENE)
CENOZOIC (PALAEOGENE)
Q
PLIO
MIOCENE
OLIGOCENE
4
3
2
Most modern genera of butterflies
Modern species
Most infraspecific variation
of butterflies
Plant and animal migrations of the Pleistocene. Development of high latitude tundra belt
Major uplift of Himalayas
Arid conditions in North Africa (3.9 ma)
Uplift of American Sierra – Cascade and Coastal Ranges
Mediterranean salinity crisis (dry basin) (5.5 ma)
Enlargement of Antarctic ice sheet
Extensive regions of open (unforested) landscape over areas now occupied by savanna, steppe and desert. Beginnings of modern vegetation zones except tundra at high latitude
Ocean and land temperatures fall sharply. Beginning of major uplift of Alps and American Cordillera
North Africa joins Europe. South European mountains occur as islands in Mediterranean and Paratethys seas (17 ma)
Polar regions begin to export significant amounts of cold bottom waters
Mediterranean flora exists at genus level
Grasslands become diverse and dominant in Miocene (e.g. in Dakota, USA). Deforestation with grasslands, steppe, taiga and tundra flora originating – tundra at high altitudes
Diversification in Dicotyledonous families at species level (e.g., Quercus differentiated)
Systematic decline in sea levels begins
Equatorward shift in vegetation belts and climatic zones
Initial uplift of mountains in Europe and Middle East
MASS EXTINCTION EVENT (mainly marine realm). Major temperature decline – shift to high mean annual range
Some 30 Dicotyledonous families (e.g., Betulaceae, Ulmaceae, Salicaceae). Diversification accomplished. Broadleaved deciduous forest north 65° N versus 50° N today. Megathermal rainforests restricted < 20° N

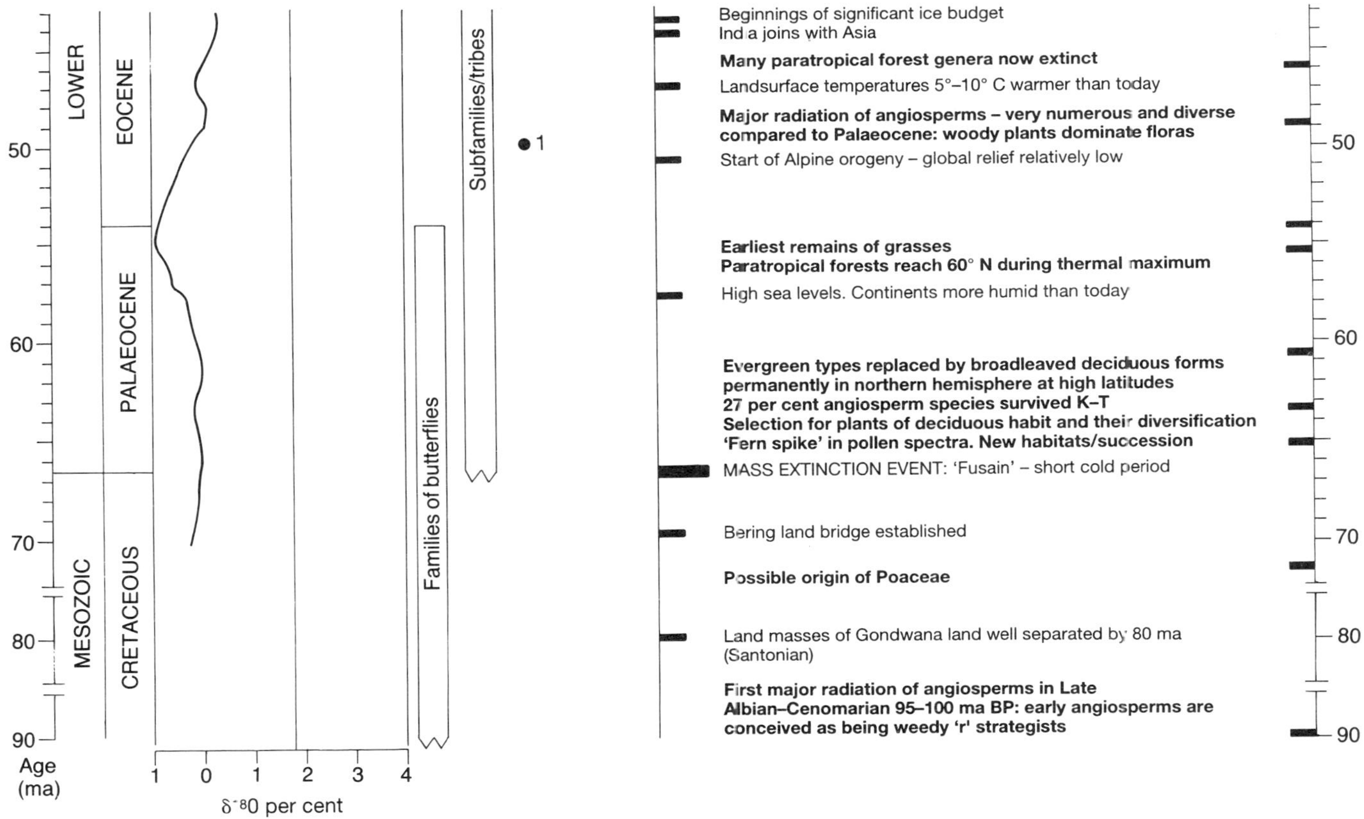
Age (ma)
50
60
70
80
90
LOWER
MESOZOIC
EOCENE
PALAEOCENE
CRETACEOUS
δ¹⁸O per cent
1 0 1 2 3 4
Families of butterflies
Subfamilies/tribes
1
Beginnings of significant ice budget
India joins with Asia
Many paratropical forest genera now extinct
Landsurface temperatures 5°–10° C warmer than today
Major radiation of angiosperms – very numerous and diverse compared to Palaeocene: woody plants dominate floras
Start of Alpine orogeny – global relief relatively low
Earliest remains of grasses
Paratropical forests reach 60° N during thermal maximum
High sea levels. Continents more humid than today
Evergreen types replaced by broadleaved deciduous forms permanently in northern hemisphere at high latitudes
27 per cent angiosperm species survived K–T
Selection for plants of deciduous habit and their diversification
'Fern spike' in pollen spectra. New habitats/succession
MASS EXTINCTION EVENT: 'Fusain' – short cold period
Bering land bridge established
Possible origin of Poaceae
Land masses of Gondwana land well separated by 80 ma (Santonian)
First major radiation of angiosperms in Late Albian–Cenomarian 95–100 ma BP: early angiosperms are conceived as being weedy 'r' strategists

It is suggested that climatic forcing of butterfly macroevolution through the Tertiary has been determined by plant evolution and geography; the sequential adoption of new hostplants and habitats generated a constellation of novel pressures which selected for transformations in behaviour and ecology and subsequently in morphology and physiology. No reciprocal co-evolutionary feedbacks on plants were necessarily involved (Berenbaum, 1990). Climatic cooling has also had a more direct effect on the emergence of a 'cool' fauna. The cold, dark period at the K–T boundary (66.5 ma BP) may have selected against 'warm' faunas (Wolfe, 1991). Contemporary arctic, montane and temperate butterfly faunas probably date from the Eocene–Oligocene abrupt step in planetary cooling when the seasonal range in temperatures also became much wider. The widespread equatorward displacement in vegetation belts occurred, modern vegetation 'communities' developed and dicotyledons diversified at species level. Furthermore, orogenesis during the late Eocene produced high mountain belts such as the Alps and Pyrenees. These and earlier high-latitude montane blocks provided opportunities for the evolution of a cool taxa on a cooling planet. Their steep environmental gradients allowed a rapid response to changing conditions and periodic isolation ensured allopatry (Hsu *et al.*, 1977). The novel seasonal habitats with their cool summers have resulted in a range of adaptations and specializations in behaviour, physiology and morphology. Examples are hibernal diapause, cold tolerance, thermoregulatory behaviour and temperature excesses (T_{ex}), lower developmental zero temperatures (T_0min), lateral resting postures, duller wing phenotypes and microclimate selectivity. The emergence of these specializations and parallel evolution in different lineages (e.g., morphology of Andean *Tatochila* spp., and alpine *Pieris* spp. Shapiro, 1989; Geiger, 1987) are indicative of the severity and monotonicity of directional selection pressures caused by 5°–10°C cooling since the Mid Eocene.

B *Macroevolution during Milankovitch cycles?*

It is generally accepted that the likelihood of evolutionary transformations (speciation) increases with allopatry – in relation to the spatial (inversely) and temporal (directly) dimensions of isolates – and the intensity of selection pressures (Barton, 1988; Coyne, 1992). A traditional view has been that Pleistocene polyglacialism provided just those conditions for evolution among Lepidoptera in Europe (i.e., endemics, sibling species, hybrid zones, Lattin, 1967; Sbordoni & Forestiero, 1984). This view conflicts with that of Coope (1979; 1987), based on fossil carnivorous Carabidae (Coleoptera). For this taxon, little or no evolutionary change is evident throughout the Mid and Upper Pleistocene. Instead, extinction rates have been low and the beetles have been able to track climatic changes over thousands of kilometres. Coope argues that only on Atlantic islands south of the polar front have appropriate conditions, stable environments for small areas, existed for speciation. Similar observations have been made for some other organisms, many modern species dating back at least 1 ma BP (cf. Stanley, 1985).

Despite these compelling findings for the Milankovitch disturbance of evolutionary events, discounting continued macroevolution in the Lepidoptera in areas directly influenced by ice advance and ablation is premature. Firstly, all fossil beetles are defined on unambiguous morphological criteria (morphospecies), whereas many good Lepidopteral biological species exist that cannot be distinguished on consistent physical characters (e.g., *Pontia daplidice* and *P. edusa*, Geiger *et al.*, 1988; Wagener, 1988). Thus, the initial stages of the speciation process may be missed in the fossil record. Furthermore, differences in genitalia may not imply large genetic differences; those that exist in *Tatochila* spp. are either simple Mendelian traits or polygenic and do not prevent gene exchange (Porter & Shapiro, 1990). Genitalia morphology may simply be refractory to evolutionary modification. Secondly, the potential for evolutionary change depends greatly on the capacity for migration. Phytophagous Lepidoptera, because of hostplant and habitat limitations, probably evolve somewhat more and move somewhat less than the Carabidae. It is noteworthy that endemicity in southern Europe is more frequent and widespread among life-forms with lower vagility than butterflies (e.g., nematodes, plants) and less frequent or widespread in organisms with greater migratory capacity than butterflies (e.g., birds) (Topham & Alphey, 1985; Blondel, 1987; Major, 1988). Similar comparisons can be made among taxa within the Lepidoptera (Dennis *et al.*, 1991). Thirdly, microevolutionary rates are known to be fast in Lepidoptera, as much regional variation in morphology, biochemistry, physiology and ecology of genetic origin has developed in areas that can have been colonized only from a single source in the Holocene (Dennis, 1977; Thomson, 1987b). Fourthly, Coope's research has been confined largely to high latitudes. Multivariate analyses of European butterfly distributions reveal faunal regions and faunal elements the component species of which differ widely in their tolerance of climatic conditions and resources and, because of the different regions occupied by them, in environmental history (Fig. 4.8a & b). They also contrast significantly in dimensions and content. Those north of the alpine mountain chain are few in number and comprise extensive meridional units dominated by widespread species which range across the entire Palaearctic or Holarctic; those south of this line are numerous and occupy comparatively small geographical units restricted to mountain blocks, islands and isolated landmasses bordering the Mediterranean Sea. The latter also include a high frequency of endemic species (Dennis *et al.*, 1991; Fig. 4.8c).

These observations suggest that species of strictly northern and southern elements differ significantly in their capacity to migrate and potential to evolve during the Pleistocene (Dennis *et al.*, 1991). Resources have been more extensive and continuous for northern elements than for those equatorward of the northern edge of the alpine chain. Refuges also differ for the two groups. For northern species, they tend to be impermanent and during the shorter interglacials; for southern species, they are more permanent and during longer glacials. The result is that species of southern west palaearctic elements have experienced longer periods of isolation, conducive to independent development, than species for northern elements. Thus, taxa restricted to one of the three blocks in the Mediterranean zone (*viz*., Iberia, Italian peninsula or Balkans) during the current interglacial would have been more isolated during glaciations (see Fig. 4.6). An analogous process, but on a larger scale, would have affected the numerous species belonging to the 'extent' group whose distributions currently envelop much of the west palaearctic (Fig. 4.8b). As for this group each glacial stage heralds the extinction of all populations north of the Alps, 'core' populations in southern refuges continue to be isolated in relatively stable and distinctive environments. There is, then, potential for gradual divergence and ultimately speciation, which is tested when the products of refuge populations come into contact during every interglacial. The discovery of increasing numbers of tension zones and sibling species units within apparently uniform pan-European taxa (Fig. 4.8d) supports the proposition of ongoing evolutionary divergence during Milankovitch cycles in higher latitudes (Barton & Hewitt, 1989; Hewitt, 1990; Dennis, *et al.*, 1991). Although species of arctic and cold temperate environments have less opportunity for cladogenesis, Milankovitch events could trigger two scenarios for isolation and continued divergence: (i) reciprocal colonization between arctic and southern montane environments during deglaciation (e.g., Pyrginae; Jong, 1972) and (ii) temporal isolation of biennial species divided by northern montane blocks or continental ice sheets. Regionally isolated and geographically overlapping population units of closely related arctic and boreal *Erebia* spp. and *Oeneis* spp., as well as moths such as *Xestia* spp. (Noctuidae), have flight periods in odd and even years (Douwes, 1980; Mikkola & Kononenko, 1989).

The differences between European butterfly faunal elements, in structure and environmental history, have wider implications for macroevolutionary dynamics. The same conditions, vanishingly small populations in space and geological time, are conducive to both speciation and extinction. But any regional disparities in the equilibrium between production of species and extinction relate to inbuilt constraints in the speciation process, rather than to the altogether simpler process of gene pool deletion, and the balance of regional conditions which enhance or deter both processes. In conditions appropriate for macroevolution, the longer the speciation process takes, the more probable it is that extinction rates will exceed speciation rates. The coincidence of a number of factors probably accounts for the high endemicity among butterflies in the alpine chain mountains of Europe and

North Africa (Fig. 4.8c). The steep gradients, high elevation and contrasting aspects allow the short-distance tracking of large and sudden climatic changes during the Pleistocene, reducing extinction rates. The highly dissected, three-dimensional, landscape increases the potential for the multifurcation of gene pools during climatic change – isolated refuges on alps and peaks during interglacials and on lower south-facing slopes and nunataks during glacial phases. The degree to which a gene pool multifurcates into smaller geographical units, the more probable is cladogenesis and the less probable is extinction of the taxon. Separate mountain blocks allow for faunal exchange whilst increasing the probability of longer-term isolation. The topography and local climatic contrasts will tend to restrict suitable conditions to small areas, especially for ecological specialists of the subalpine zone. Compact isolated populations have higher evolutionary potential than those spread over broad regions. The gene pool is more integrated,

---

4.8 (overleaf) Biogeographical structures among European butterflies. (a) Summary regionalization of the European butterfly fauna from factor analysis of 85 primary areas (centroids given as dots) on 393 species (number of factors, 9). Faunal regions are labelled I to VIII. The Italian peninsula was not well defined and links up with the east Mediterranean at a marginally lower level than with the west Mediterranean. Loadings for factor 9 were relatively low (< 0.55) and focused on the Alps. The Mediterranean islands were distinguished in analyses which extracted additional factors. Isolines illustrate communalities from the nine factor solution (stippled, low communalities; shaded, high communalities). (From Dennis *et al.*, 1991 by courtesy of Academic Press Ltd.) (b) Faunal elements based on complete linkage analysis of Jaccard coefficients between species. An arbitrary 70 per cent inclusion level is used to determine 'core' regions for faunal elements: thick continuous line, extent element; pecked line, montane element; thin continuous line, southern element; dotted line, northern element. (c) Endemicity in European butterflies. Isolines, the relative frequency (per cent) of 144 species endemic to the mapped zone found within 1 to 10 primary areas (contiguous and non-contiguous areas); few new primary areas were added with the cumulative addition of species occupying > 10 primary areas. Bold line, the distribution of endemics restricted to single primary areas. (d) Tension zones (approximate boundaries) in selected European butterflies: (i) *Mellicta athalia* (north) and *M. celadussa* (south) (Nymphalidae); (ii) *Pyrgus malvae* (north) and *P. malvoides* (south) (Hesperiidae); (iii) *Maniola jurtina* (Satyrinae) east and west genitalia valve forms; (iv) *Pseudophilotes baton* (west) and *P. schiffermuelleri* (east) (Lycaenidae); (v) *Pontia daplidice* (west) and *P. edusa* (east) (Pieridae); (vi) *Pararge aegeria aegeria* (south and west) and *P. aegeria tircis* (north and east) (Satyrinae). (Data from Guillaumin & Descimon, 1976; Higgins & Riley, 1983; Thomson, 1987a; 1990; Geiger *et al.*, 1988; Brakefield & Shreeve, 1992.)

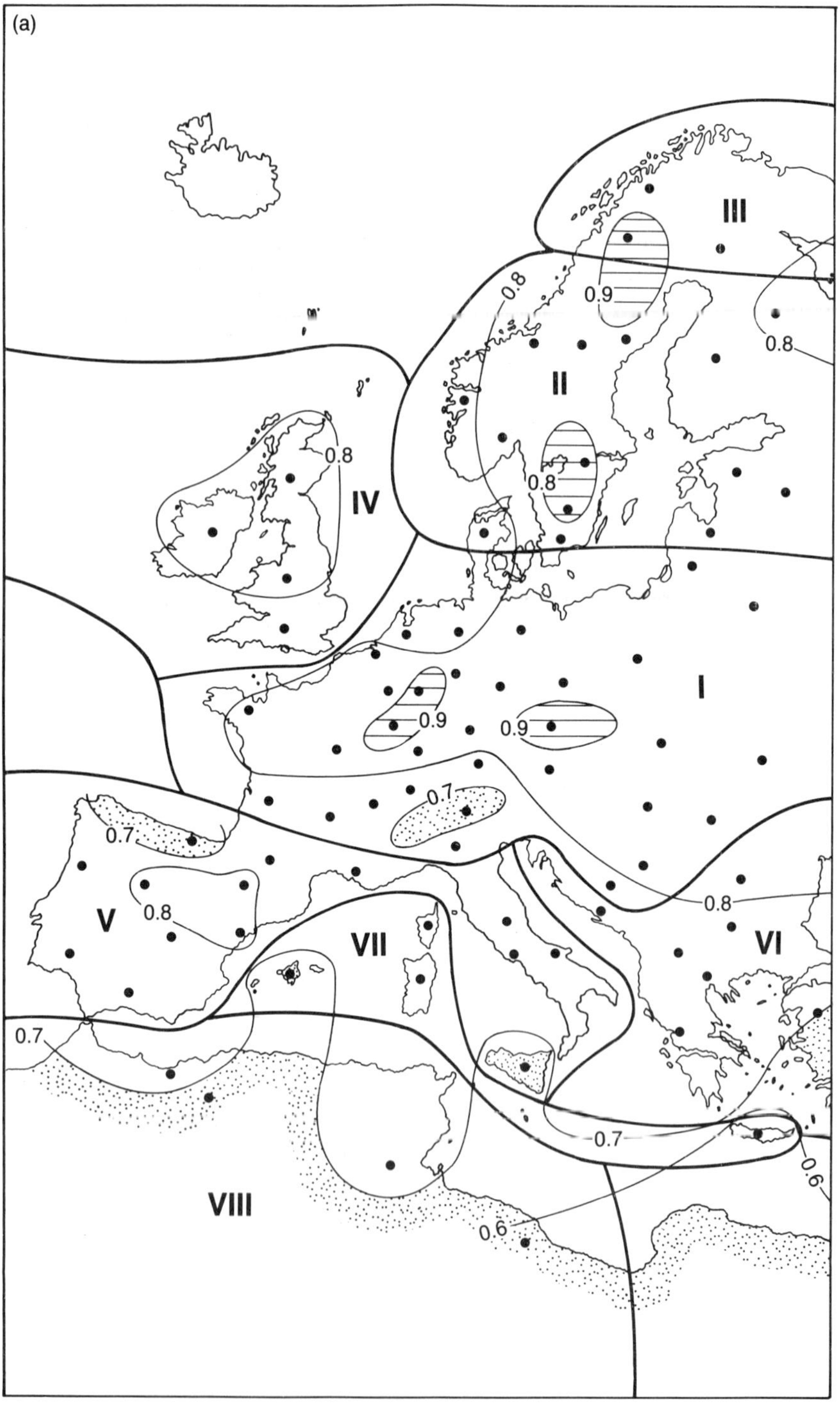
(a)
III
0.9
0.8
0.8
II
0.8
0.8
IV
I
0.9
0.9
0.7
0.7
0.8
0.8
V
VII
VI
0.7
0.7
0.6
0.6
VIII

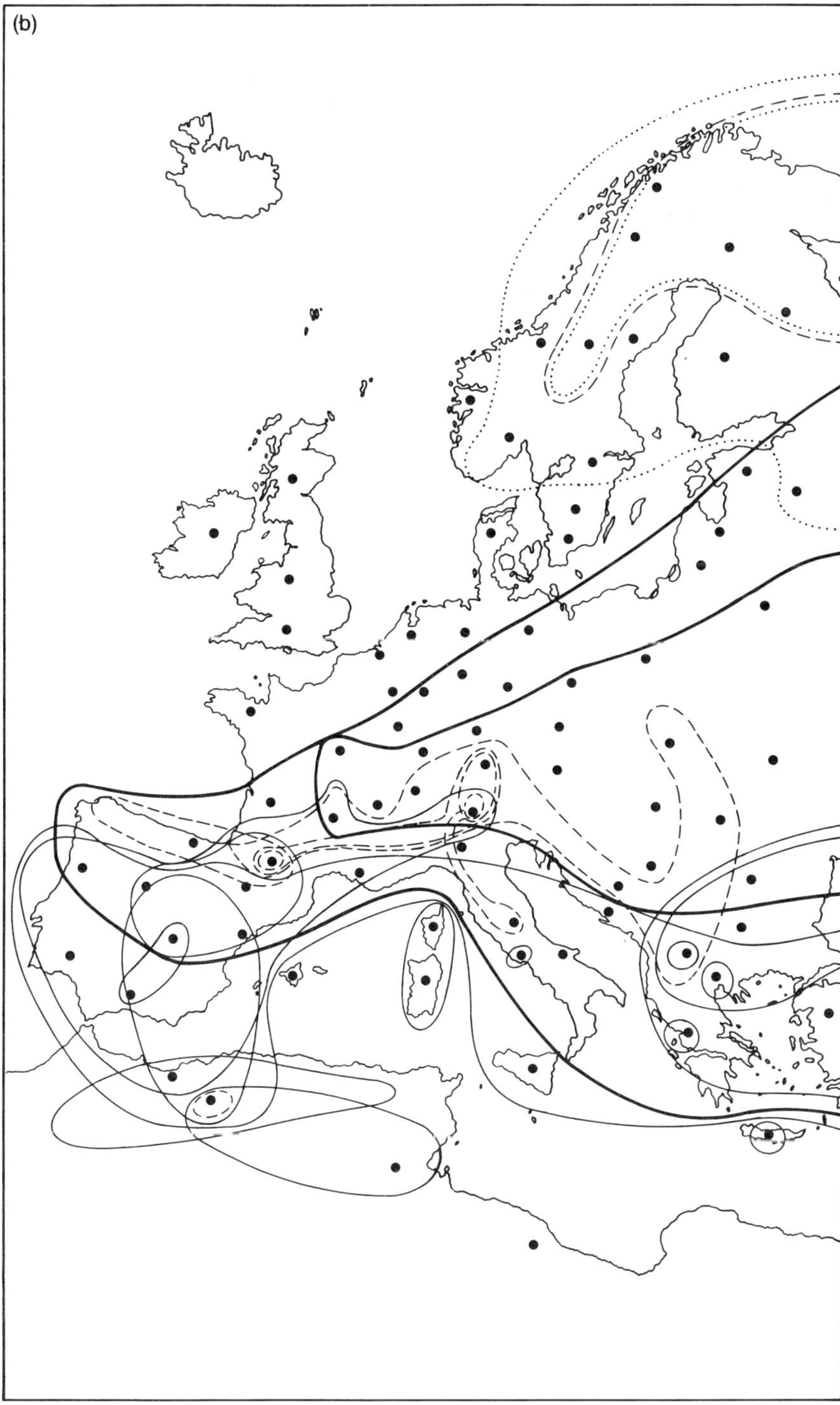
(b)

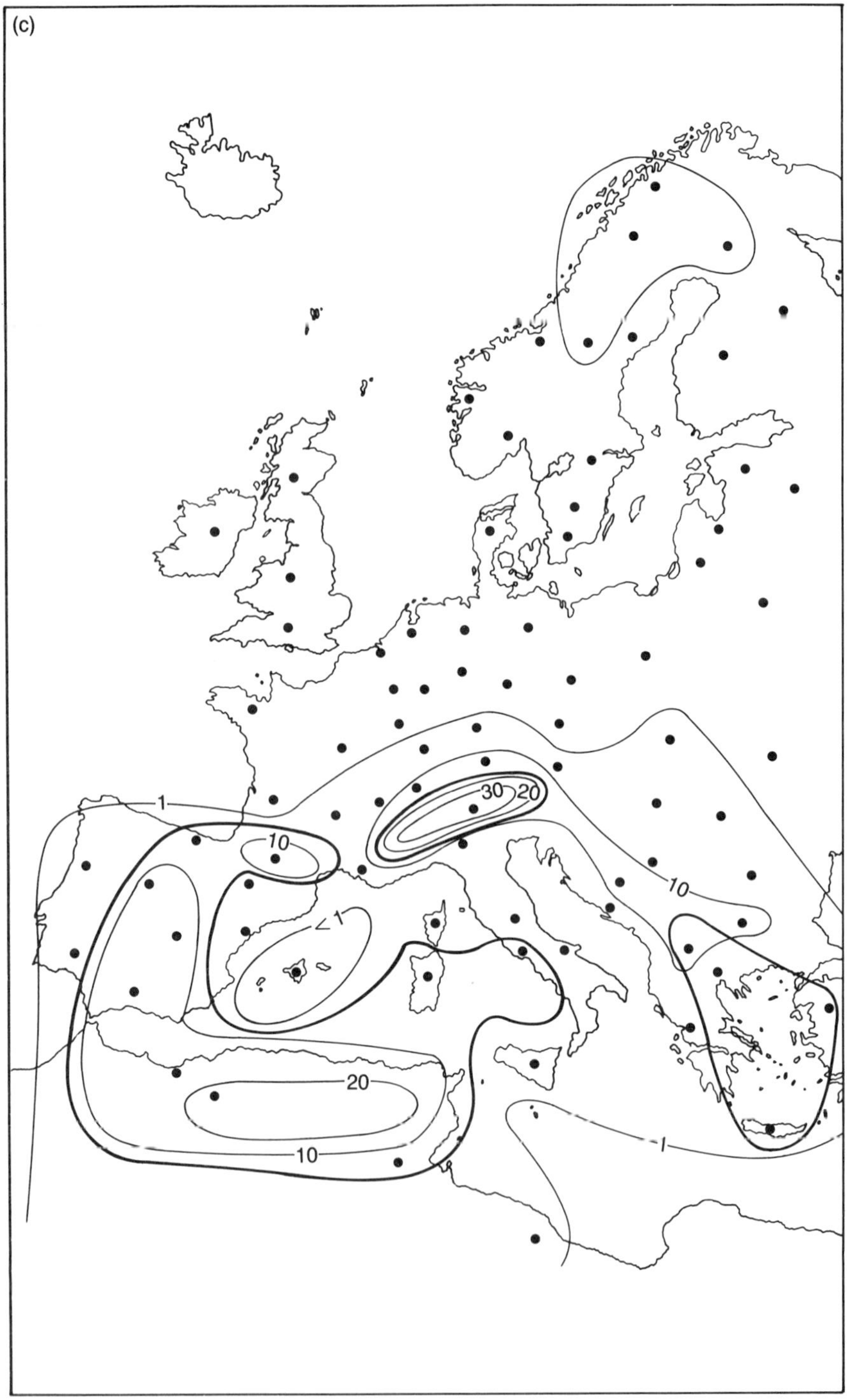
(c)
1
10
30
20
10
<1
10
20
10
1

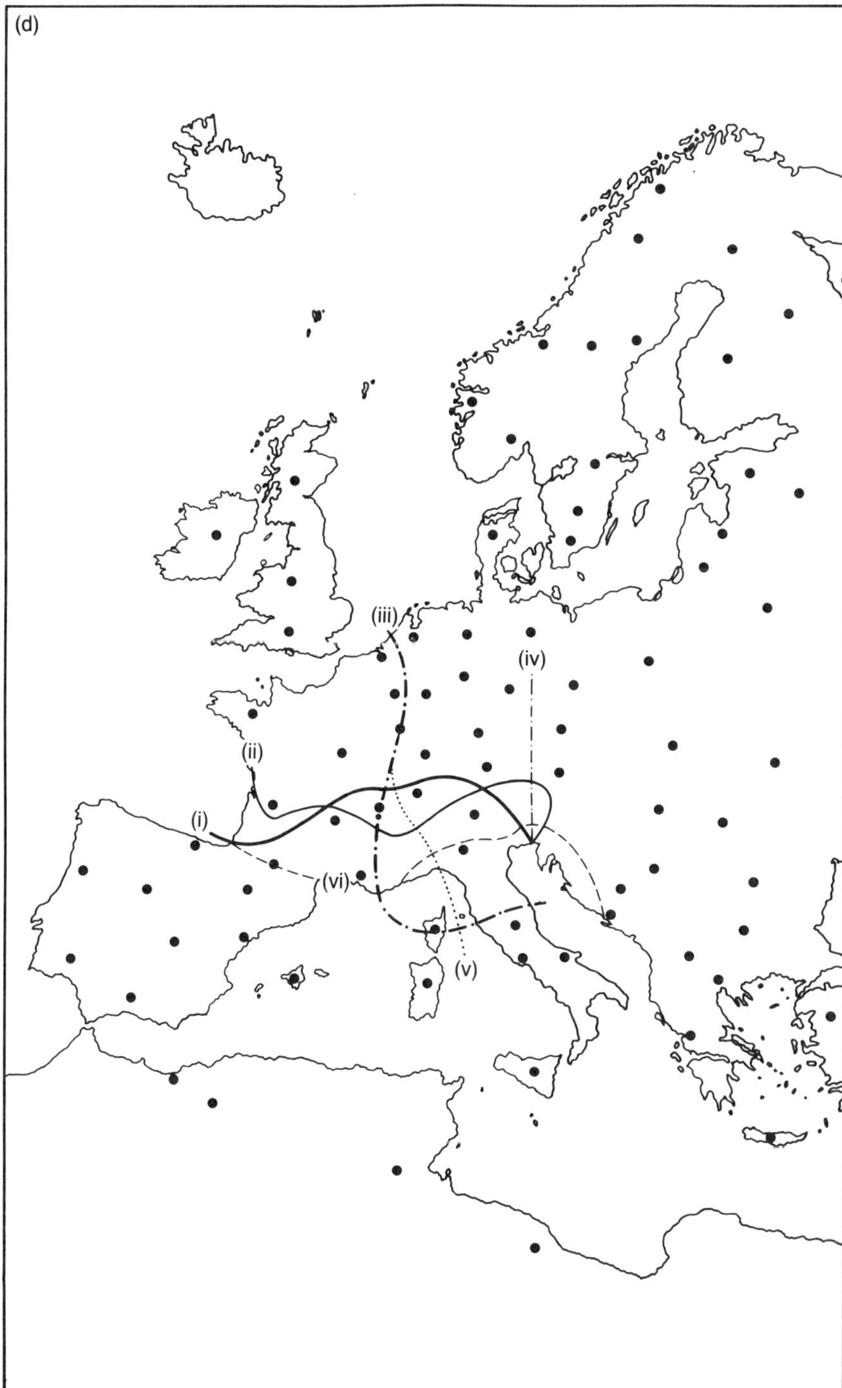
(d)
(i)
(ii)
(iii)
(iv)
(v)
(vi)

individuals will be subjected to a more homogeneous set of selection pressures, and stochastic habitat changes, causing directional shifts in selection, are more likely in marginal but spatially compact areas. The montane context is also one of climatic extremes over both short and long-time scales. Short periods of severe ecological stress, especially those caused by extreme temperatures, are probably highly conducive to rapid morphological evolution, influencing population dynamics and possibly also effecting increased genotypic variability and *de novo* variation (Parsons, 1988; Holt, 1990). This combination of conditions, which may also enhance divergence through genetic drift, contact between isolates and occasional gene flow (Wright, 1982), may well explain the complex pattern of taxonomic links in the *Erebia tyndarus* complex (Lorkovic, 1958; 1990) and between *Pieris napi* and *P. bryoniae* (cf., Guillaumin & Descimon, 1976). All this suggests that montane macroevolution is far more complex than Adams's (1977) evolutionary model for speciation in Andean *Lymanopoda* spp., which assumes isolation to be restricted to short interglacial phases. Some support for the continued turnover in European butterfly species is provided by the wide disparity in morphological differentiation among closely related endemics. As yet, the use of biological clocks applying molecular data is in its infancy and has its pitfalls (Myers & Giller, 1988; Geiger, 1990; Brower & Bryce, 1991), but recent work on the Maniolini and on the *Aricia* complex indicates that macroevolutionary rates in some taxa may be rapid (10^4–10^5 a; Thomson, 1987b; Høegh-Guldberg in Dennis, 1992). As so many different reproductive isolating mechanisms exist, varying in degrees of severity (Lorkovic, 1978), it is as yet unwise to assume that speciation is entirely a long-term process ($> 10^6$ a) (see Coyne, 1992).

C *Microevolution during Holocene climatic changes*

If macroevolutionary transformations are the endproduct of cumulative microevolutionary changes (Barton, 1988), then the processes hypothesized for higher taxonomic differentiation during the Tertiary (section 4.4A) should cause significant microevolution during shorter timescales (10^2–10^4 a) such as the Holocene. As evolutionary hypotheses cannot be tested directly, their validity depends largely on the comparison of alternative scenarios and on internal logic. Minimal requirements are a clear datum for colonization, continuous relevant environmental data for the period and region in question and a detailed bio-profile of any differentiation. Fauna in the British Isles largely fulfil these criteria. The clock for current British species starts at c. 15 ka BP (Fig. 4.3). Owing to population growth, dispersal, panmixia and widespread recombination, colonizing species were probably highly varied. This would have allowed them initially to exploit the developing post-glacial landscape of north Europe, and subsequently to adapt to climatic changes (Dennis *et al.*, 1991). Holocene climatic changes have influenced butterfly microevolution directly and indirectly (i.e., isolating factors and habitat changes), usually conjointly rather than individually. Detailed reconstructions of events surrounding morphological and ecological transformations are discussed elsewhere (Dennis, 1977, 1992). In this section it is possible only to outline briefly the main factors, processes and approximate timing of butterfly microevolution in Holocene Britain.

Infraspecific regional phenetic differentiation within British butterflies mainly postdates 15 ka BP (Dennis, 1977; 1992).* This phenetic differentiation increases to the north and west (Fig. 3.4) and a model has been developed for it in relation to climatic gradients (section 3.3). Although climatic gradients in Britain have persisted throughout evolutionary time, marginal conditions for butterflies in temperature and insolation, hypothesized as the basis for wing phenotype transformations, may date only from 4 ka BP (Fig. 4.3 column i). However, earlier Holocene climatic changes indirectly triggered independent development for a number of regional populations. Temperature increases led to rising sea levels and forest closure and these, in turn, to the isolation and independent development of populations on offshore islands (e.g., *Maniola jurtina cassiteridum*, Isles of Scilly), on mountain tops (e.g., *Erebia epiphron*, Lake Disrict and Grampians) and on ecological islands such as calcareous outcrops (e.g., *Erynnis tages bayensis* and *Hipparchia semele clarensis*, Burren, Eire). Forest closure was complete by 9 ka BP and all British populations had been cut off from the continent by 7.5 ka BP (Fig. 4.3 columns a and e). Climax forest development would have had a profound effect on populations of butterfly species requiring open conditions (section 4.3B). It may explain outstanding issues such as the reversed cline for hindwing marginal ocellation in *Maniola jurtina* in the south-west peninsula (Ford, 1975). At 9 ka BP, populations of *M. jurtina* were probably isolated to cooler, humid acid conditions on siliceous intrusions in the west and to warmer, drier alkaline conditions on calcareous scarplands in the east. Four thousand years of independent development would have been terminated only with Bronze and Iron Age clearances, the consequent deterioration in climate on the western granite moorlands and the spread of the species to lower elevations (Dennis, 1977; Brakefield, 1990; Brakefield & Shreeve, 1992). The cline may subsequently have been maintained by the climatic gradient across south-west England. Other gene pools have been influenced by different factors associated with climatic change. The saturation of all northern populations of *Aricia artaxerxes* by the autosomal recessive gene for forewing white discal spotting points to its emergence before forest closure, as the elimination of the hostplant *Helianthemum*

* There is the prospect that taxonomic revision may identify sibling species pairs within current taxa that occur in Britain and on the continent (cf. voltinism units in *Polyommatus icarus* with *Aricia agestis* and *A. artaxerxes*), thus dating any variation associated with them to the pre-colonization event (Dennis, 1992).

nummularium over an extensive region by vegetation succession and podsolization would have prevented any subsequent spread of the gene. If the gene is restricted to Britain, then it is more probable that it evolved in southern Britain perhaps on the scarplands during the Younger Dryas, when conditions were marginal for the species (Dennis, 1977; 1992). By comparison, the dwarfing and early emergence of adults in two unrelated species, *Hipparchia semele* and *Plebejus argus*, on the west side of the Great Orme's Head in North Wales is associated with forest closure, but perhaps as early as 9.6 ka BP in a moisture – deficient habitat where the effects of locally high insolation levels and high temperatures have been compounded by topography, aspect and permeability of the limestone screes (Dennis, 1977, 1992).

Ecological differentiation in British butterflies may also have resulted from the direct and indirect effects of climatic change. Population differentiation as to habitat associations, hostplant affinities and degree of hostplant specialization can be successfully modelled by population isolation during climax forest development (Dennis, 1977). Populations with obligative specializations (*viz*., monophagy; stenotopy) evolved due to the selective elimination of generalists in locations in which a resource (i.e., hostplant) was solitary or for which survival was significantly higher than for others in the same habitat (Dennis, 1992). Morphological changes often accompany ecological specialization as directional shifts in habitat components result in new systems' equilibria for pressures on phenotypes (Dennis, 1977; 1992). British *Papilio machaon* and *Lycaena dispar* (now extinct) are distinctive in ecology and phenotype. They would have become restricted to the fenland habitat by the climax forest and isolated from their European counterparts by rising sea levels. Monophagous on *Peucedanum palustre* and *Rumex hydrolapathum* respectively, they now differ in wing colour and pattern from continental forms. However, these adaptations are not unique. In Sweden, there are fen populations of *P. machaon* (Wiklund, 1981) and in the Netherlands, there are fenland *L. dispar* feeding on *Rumex hydrolapathum* (Bink, 1986). British races differ primarily in that populations with generalists have been extirpated. Most specializations have been found in species with restricted distributions or in isolated populations of more widely distributed species (e.g., *Erebia aethiops* at Arnside, Cumbria; Dennis, 1992). However, it is possible that specialization may be more widespread among British butterflies. English *Polygonia c-album* discriminates more strongly among potential hostplants than do Swedish stock for plants, permitting faster development (Nylin, 1988), an adaptation that may be related to Britain's cooler, cloudier summers.

The onset of marginal climatic conditions for the current British butterfly fauna in the latter part of the Holocene (Fig. 4.3I) may have resulted in a number of microclimatic associations and adaptations, that is the restriction of populations to: (i) permeable, drier and warmer substrates (e.g.,

Hesperia comma on calcareous outcrops); (ii) southerly aspects (e.g., *Melitaea cinxia*, Isle of Wight); (iii) open habitats exposed to higher insolation (e.g., *Maniola jurtina*, open grasslands) and (iv) earlier seres (e.g., *Maculinea arion*, sward < 4 cm high) (see Table 5.2). Not all these limitations are the direct result of climatic changes. In the case of *Lysandra bellargus* and *L. coridon*, the hostplant *Hippocrepis comosa* would initially have been restricted to open base rich localities during climax forest cover. Neither are all associations necessarily obligative (see section 5.3B). For several British species currently dependent on early seres (e.g., *L. bellargus*; *M. arion* and *H. comma*; Thomas, 1983; 1989; Thomas *et al.*, 1986) humans may have contributed to their survival not only by opening up the forest but, through practices such as firing the vegetation and grazing animals, by maintaining extensive early disclimaxes which record high maximum ground temperatures (Thomas, J. A., pers. comm.). With deforestation and the loss of shelter belts, the high maximum ground temperatures of sparse swards would have compensated for the lowering of ambient temperatures as downlands became exposed, and would have been critical for their continued survival once summer temperatures dropped after 3 ka BP. These species may have been less dependent on short swards during the early climax forest as mean temperatures were then higher (Fig. 4.3B), but their isolation at this time to steep unstable slopes (i.e., cliffs; scarps), dominated by mass-wasting processes, may imply a longer history for this habitat association which could then have been an exaptation to later climatic cooling.

Absolute dating of evolutionary transformations awaits advances in 'biological clocks' based on mitochondrial DNA. None of the evolutionary changes attributed to the Holocene involves large genetic transformations (Dennis, 1977; 1992; Dennis & Shreeve, 1989). For example, the relatively large inter-population morphological differences in British *Coenonympha tullia* (Fig. 3.6) are matched by similar differences in California, but there they equate with low inter : intra population variability in allozymes (Porter & Geiger, 1988). Most changes identified have probably evolved very rapidly (10^2–10^3 a; Dennis, 1977). In directional experiments, populations typically reach response plateaux by 10–20 generations. Plateaux can be transcended because of recombination or mutation; larger populations permit greater final responses to a given intensity of selection as they are more likely to harbour alleles without deleterious pleiotropic effects (Holt, 1990). As such, it is important to realize that spatially small populations can be very large numerically. This is the case for North Wales *Plebejus argus*, in which significant morphological changes have been recorded within 30 years, as it is for many alpine butterfly endemics (Dennis, 1977; pers. obs.).

5

Future atmospheric changes and butterfly populations: predictions and consequences

Global mean temperatures and other climatic statistics show considerable variability on all timescales. This variability is usually described as low (> 10a) or high (< 10a) frequency and the causes classified as external or internal to the Earth–atmosphere system (Wigley & Raper, 1990). Since 1861, global mean surface air temperature has increased by 0.45°C ± 0.15°C (Fig. 5.1; Jones, 1990). This low-frequency trend correlates with the increased concentration of tropospheric infra-red absorbing gases (Houghton *et al.*, 1990). Regardless of any causal link, the potential impact on biological systems by systematic trends either in climate or in atmospheric 'pollutants' is substantial and warrants serious consideration. The first section of the chapter briefly reviews the link between pollutant emissions and climate and predictions of future trends. The basic consequences for butterfly biology of these trends is discussed in the second section, whilst a regional view at two widely different scales is attempted in the third. A brief final section is given over to suggestions for future research.

5.1 Greenhouse gases and environmental changes

Although a high level of confidence may be attached to the global rise in mean temperature and in atmospheric pollutants from emissions resulting from human activities over the past 150 years, there are many uncertainties in attributing the climatic change to the enhancement of greenhouse gases (Kuo *et al.*, 1990; Barnett, 1990). Difficulties arise in determining the efficacy of pollutant sources and sinks and interactions among them are complex and only partly understood (Ramanathan, 1988; Watson *et al.*, 1990). Predictions of change into the future remain speculative.

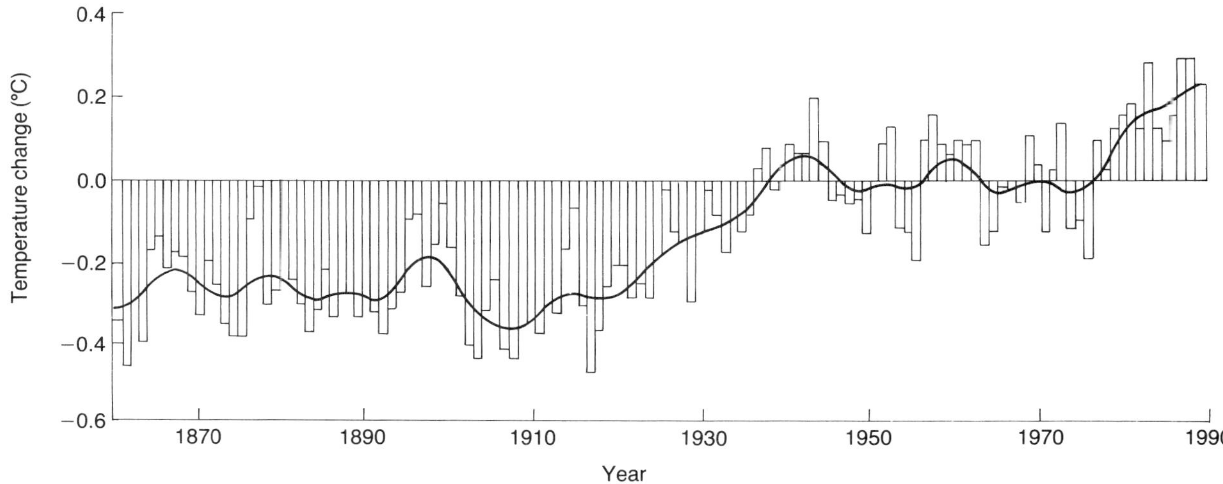

5.1 Global mean combined land–air and sea–surface temperatures, 1861–1989, relative to the average for 1951–80. The smooth curves are filtered values designed to show the longer-term trends more clearly. (From Folland *et al.*, 1990 by courtesy of Cambridge University Press.)

A *Projected transformations in the Earth's atmosphere*
Anthropogenically induced atmospheric pollutants can be divided into two groups: those capable of absorbing substantial amounts of infra-red radiation (e.g., CO_2, CH_4, N_2O, CFCs, HCFCs and HFCs, O_3) and those that have little direct absorbing capacity (e.g., SO_2, CO, NH_3, aerosols). Both groups are capable of interacting to produce further greenhouse gases (e.g., O_3 from photochemical reactions involving NO_x precursors and hydrocarbons) or of modifying the global radiation budget (CFCs on stratospheric ozone; SO_2 on cloud formation). They may also alter ecosystems more directly (e.g., toxification by low pH and NH_3; fertilization by NO_x). Consequently they are all relevant to any enquiry on biosystems. Table 5.1 summarizes statistics on key greenhouse gases during the industrial era and documents anthropogenic sources and major sinks for the gases.

Other greenhouse gases, not in Table 5.1, show evidence of increasing concentration, particularly tropospheric ozone which may now account for as much as 5–10 per cent of radiative forcing. Water vapour is the most significant greenhouse agent in the Earth's atmosphere. Though the total global water budget is unaffected by human activity, a strong positive feedback of increased temperatures on atmospheric water vapour concentrations may greatly enhance radiative forcing (Del Genio *et al.*, 1991). Other gases may contribute indirectly to the greenhouse effect; for example methane (CH_4) is destroyed by reaction with hydroxl (OH) radicals, but carbon monoxide (CO) also reacts with OH, and increasing concentrations of CO could reduce OH availability (Ashmore, 1990).

Predictions of future increases and concentrations in greenhouse gases are based on forecasts of different emissions scenarios. These make assumptions about the growth of economy and population and the efficiency of international controls (e.g., GISS, Hansen *et al.*, 1988; IPCC, Houghton *et al.*, 1990). Even the most restrictive controls, in which carbon dioxide emissions are reduced to 50 per cent of the 1985 levels by 2050 AD, predict a doubling of radiative forcing by 2100 AD, but by a factor of five for a 'business-as-usual' scenario (Fig. 5.2). Carbon dioxide remains the dominant contributor to change because of its extensive reserves, despite its relatively weak radiative effect compared to other greenhouse gases. For example, the cumulative release of CO_2 from fossil fuel use and cement manufacturing from 1850 to 1987 is estimated at 200 ± 20 GtC (Marland, 1989) and from deforestation c. 115 ± 35 GtC (Houghton & Skole, 1990). During the decade 1980–89, c. 48 per cent of anthropogenic emissions have stayed in the atmosphere. Long-term contributions depend on the magnitude of human-made emissions and exchanges in the Earth–atmosphere system. Some predictions exceed the business-as-usual scenario. The aggregation of national projections by the IPCC gives higher emissions (10–20 per cent) of CO_2 and CH_4 by 2025 AD than the business-as-usual scenario, and this may be exacerbated further if rates of respiration exceed

Table 5.1 Summary of key greenhouse gases affected directly by human activities during the industrial era (Modified from Houghton *et al.*, 1990; Ashmore, 1990)

	CO_2	CH_4	N_2O	CFC–11	CFC–12
Pre-industrial concentration (1750–1800) (ppm)	280	0.65	0.28	0	0
Current concentration (ppm)	353	1.72	0.31	0.0003	0.0004
Current rate of increase per annum (per cent)	0.4	0.9	0.25	4	4
Atmospheric residence time (years)	50–230[a]	10	150	65	130
Radiative effect per molecule relative to CO_2	1	30	200	11000	14000
Contribution to radiative forcing (per cent):[b]					
(i) since 1750	61	17	4	12	
(ii) 1980–1990	56	11	6	24	
Major anthropogenic sources	Fossil fuel burning; deforestation; biomass burning	Rice paddies; domestic ruminants biomass burning; natural gas exploitation; landfill	Fertilizer application; changes in land-use; fossil fuel burning	Aerosol propellants; refrigerants; Foam packing	
Major sinks	Photosynthesis; exchange at ocean surface	Atmospheric reaction with OH radicals	Chemical break down in stratosphere	Chemical breakdown in atmosphere	

[a] Absorption of CO_2 by oceans and biosphere is not simple and a single value cannot be given
[b] Residual effects of other gases not accounted. The contribution from topospheric ozone may be significant; some sources estimate its contributions as 5–10 per cent of current warming (see Ashmore, 1990).

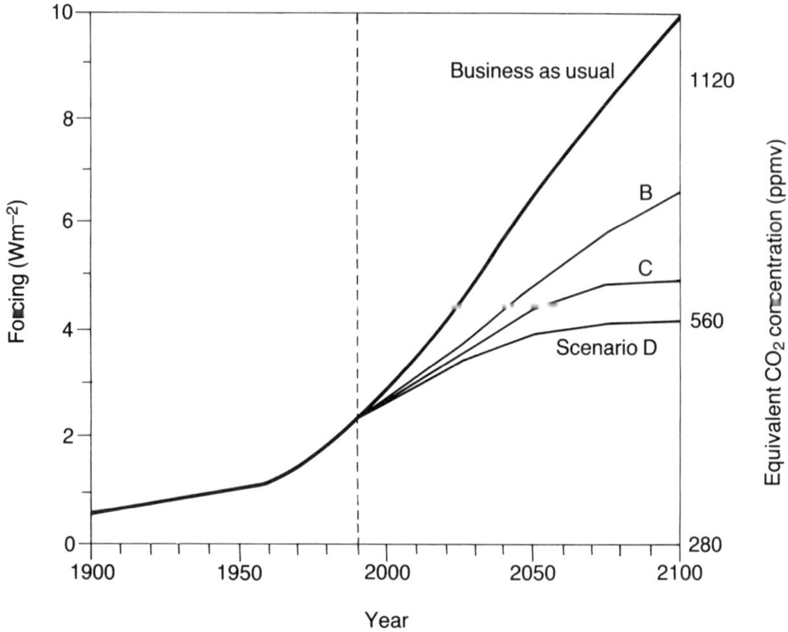

5.2 Increase in radiative forcing by anthropogenically-produced greenhouse gases since the early twentieth century and predicted to result from the four IPCC emissions scenarios. Equivalent carbon dioxide concentrations are also given. Scenarios B, C and D, account for shifts towards increasing efficiency and use of renewables – towards lower carbon fuels (i.e., natural gas), controls on CO emissions and reversal of deforestation in B, and towards renewables and nuclear energy, phasing out of CFCs and radical cuts in CO_2 emissions in the next century, later in C and earlier in D. (From Houghton *et al.*, 1990 by courtesy of Cambridge University Press.)

those of photosynthesis and if global warming has a positive feedback on atmospheric water vapour. Some business-as-usual estimates predict $2 \times CO_2$ *sensu stricto* by 2050 or 2060 AD (Hansen *et al.*, 1988; Dickinson, 1989; Watson *et al.*, 1990) and of equivalent CO_2 doubling by 2020–30 AD (Watson *et al.*, 1990; Warrick & Barrow, 1991). Recently it has been argued that the ability of forests to sequester CO_2 and the increasing amplitude of the seasonal cycle in CO_2 (Woodwell, 1989) point to a growth in net primary production of terrestrial ecosystems and to somewhat lower atmospheric CO_2 concentrations in line with the more limited scenarios (Idso, 1991). However, the calculations on which such contentions have been made are vigorously contested; furthermore, they fail to consider ecosystem limitations, for instance nutrient supply, crowding and competition in plant communities as well as increased respiration rates, which would result from the accumulation of biomass, detritus and soil organic matter (Harvey, 1991).

Concentrations of most pollutants are unevenly spread (e.g., SO_2, NO_x, NH_3, pH levels in precipitation) implying spatial variation in impact on regional or local ecosystems (Woodin, 1989; Smith, 1991). Trends for pollutants also vary nationally and are decreasing in some regions (e.g., SO_2 emissions in UK), but others are increasing dramatically (e.g., European NH_3 emissions increased by 50 per cent between 1950 and 1980; ApSimon *et al.*, 1987). Emission concentration and deposition data are mapped for North America and Europe from a variety of sources, but accuracy of these data is generally rather poor (references in Smith, 1991). Mean annual concentrations and depositions relate closely to emissions sources and changing weather patterns, especially to air flow categories (Lamb, 1977; see Fig. 1.3). For example, in Britain, cyclonic conditions are associated with higher than mean H^+ and a westerly circulation with lower than mean H^+ (Davies *et al.*, 1992). This relationship is influenced by a more recent trend to high stack emissions compared to which fall-out used to be on a more local scale (Battarbee *et al.*, 1985, 1988), although local maxima in pollutant concentrations associated with specific emission sources still remain (e.g., urban motorways).

B *Enhanced greenhouse models for climatic change*

Predictions of climatic changes from increased production of infra-red absorbing gases are based on theoretical general circulation models (GCMs). Use of historical and palaeoenvironmental analogues are unsatisfactory because of the dearth of data and difficulties in establishing replicate conditions (Henderson-Sellers, 1990). Even so, current GCMs have potentially high error loadings because climatic data are of poor quality (Strangeways, 1991), and there is uncertainty about future emissions and about the sensitivity of the Earth–ocean–atmosphere system (Houghton *et al.*, 1990; Mitchell & Zeng, 1991; Perry *et al.*, 1991). Differences of opinion exist regarding the relative efficacy of ocean and terrestrial 'sinks' (Cusbach & Cess, 1990; Gates *et al.*, 1990; Charles & Fairbanks, 1992), particularly the lags in ocean feedbacks (Bethoux *et al.*, 1990; Ellsaesser, 1990; Michaels, 1991), the carbon sequestering ability of forests (Woodwell, 1989; Anderson & Bowler, 1990; Idso, 1991; Harvey, 1991) and the role of clouds (Ramanathan, 1988; Mitchell *et al.*, 1989; Platt, 1989; Del Genio *et al.*, 1991; Rind *et al.*, 1991). Parameterization of fundamental attributes in the models (e.g., tropical convection) has outrun empirical calculations and observations (Idso & Balling, 1991; Mitchell & Zeng, 1991), spatial resolutions are coarse compared to temporal resolutions in the models (Henderson-Sellers, 1990), significant attributes have been omitted (e.g., SO_2 aerosol inhibition of northern hemisphere warming; Wigley, 1991) and computing power is inadequate.

The consensus from GCMs of global mean temperature rise with the doubling in CO_2 atmospheric concentration (ΔT_{2x}) is 1.5° to 4.5°C with a best estimate of 2.5° (Bolin *et al.*, 1986; Woodwell, 1989; Mitchell *et al.*, 1990) and c. 1.8°C by 2030 AD (Mitchell & Zeng, 1991). Much of the range for ΔT_{2x} is attributed to different cloud parameterization (Mitchell in Henderson-Sellers, 1990). Larger estimates for ΔT_{2x} occur (5.5° in Woodwell, 1989; France, 1991) but also some much smaller ones (< 0.4°C) which derive from a number of empirical calculations on historical and contemporary data sets (Idso & Balling, 1991) and evoke a Gaian response to enhanced greenhouse gas emissions (Lovelock, 1988). There are some indications that the increase in global mean temperature of 0.45°C since 1861 (Fig. 5.1) may be causally linked to the enhancement of greenhouse gases. The urbanization component is very small (< 0.05°C; Folland *et al.*, 1990; 0.2°C, Henderson-Sellers, 1990) and there is little contribution from variability in solar irradiance (Kelly & Wigley, 1990). But only a maximum of 0.6°C of the predicted warming of 1°C for the 25 per cent increase in atmospheric CO_2 has occurred. Stratospheric ozone depletion may have masked part of the heating effect of greenhouse gas increases (Lacis *et al.*, 1990). It is possible, too, that the expected rise in temperature of 2.5°C for the current 420 ppmv CO_2 equivalent has not achieved an equilibrium state, but this is doubted (Ellsaesser, 1990; Idso, 1991). Some effects of the rising temperatures have been demonstrated in glacier ablation (Houghton *et al.*, 1990), forest ecosystems (Anderson, 1991) and oceanic deep water (Bethoux *et al.*, 1991), and the 1980s have been the warmest decade since the 1860s (Schneider, 1989). However, even part of the warming that has occurred may be caused by other factors (e.g., stratospheric turbidity; Idso & Balling, 1991) and perhaps part of the difference from expected values may be accounted for in the ability of forests to sequester large amounts of carbon, as much as 2.8 times for a 300 ppmv atmospheric increase in one experiment (Idso, 1991). Although the global trend over the past 135 years is possible to isolate from any oscillatory fractions, it is too small (0.005°C a^{-1}) to allow the signal to be confirmed for another one or two decades (Ghil & Vautard, 1991). In northern hemisphere mid-latitudinal regions, atmospheric cooling may even have occurred over the past 50 years (e.g., USA, Historic Climatic Network data, Karl, T. in Michaels, 1991), a trend that may be related to sulphate pollution and the increase in albedo of low-level clouds (Michaels, 1991; see Wigley, 1991; Lelieveld & Heintzenberg, 1992).

All GCMs show substantial changes in regional climates for ΔT_{2x}, though it is emphasized that there is low confidence for regional predictions (Cannell *et al.*, 1989). Expected rates of temperature increase depend on regional calculations of equilibrium ΔT_{2x} from GCMs. All models demonstrate much greater rises in temperature for high latitudes (>> 8°C; 18°C for the poles in the most depressing scenarios) than equatorial regions (2–4°C) (Schneider, 1989; Mitchell *et al.*, 1990). Under the IPCC business-as-usual emissions of greenhouse gases, a rate of increase of global mean temperature is expected of c. 0.3°C ± 0.2–0.5°C per decade; the winter rate could be 1°C per decade (Woodwell, 1989). The global mean rate may rise to 1.5°C per decade at high latitudes. Although similar rates of temperature increase have occurred over limited areas and for short periods in the recent geological past (e.g., 7°C in 50 years at 10,720a BP in south Greenland; Dansgaard *et al.*, 1989), this predicted increase into the indeterminate future is unprecedented during the Quaternary. Global averages in precipitation are expected to increase (3–15 per cent) directly with increased evaporation and the degree of warming (Bolin *et al.*, 1986). Regionally, precipitation increases are projected for high latitudes throughout the year and for mid latitudes during winters, from increased advection, whereas for low latitudes little overall change may occur, though GCMs predict shifts in rain bands (Mitchell *et al.*, 1990). The trend since 1940 has been for increased precipitation between 35° and 70° north and a decrease between 5° and 35° north (Jones, 1990). Although extreme weather events may accompany greenhouse warming (e.g., tornadoes, drought) it is not possible to predict these directly from current GCMs.

The projected enhancement of greenhouse gases and climatic changes have other environmental consequences: for soil moisture and water balance (Mitchell *et al.*, 1990); for snow and ice cover and sea levels (Warrick & Oerlemans, 1990; Pettier & Tushingham, 1989; Oerlemans & Fortvin, 1992); for natural ecosystems (Melillo *et al.*, 1990; Shugart, 1990; Morison, 1990) and for human systems (Parry *et al.*, 1990; Scott *et al.*, 1990). These are discussed, where relevant, in the following sections.

5.2 **Predicted atmospheric changes and butterfly biology**

With the projected enhancement of greenhouse gases, butterfly biology may be influenced by the increased gas concentrations or by the resulting climatic changes. The atmospheric pollutants may have either a 'toxifying' or a 'fertilizing' effect on butterfly populations, whereas there are at least three aspects of climatic change which are relevant to butterfly biology: (i) the absolute rise in regional temperatures and consequent changes in other climatic attributes particularly precipitation; (ii) the rate at which climate changes; (iii) the frequency and magnitude of extreme weather events, such as droughts, late frosts or storms. These changes will be beneficial or detrimental to butterfly species depending on the local status of habitats and environmental conditions. It is possible that the changes in conditions will be within the tolerance of individuals and populations, in which case adjustments can be made without substantial evolutionary modifications. Where environment changes are more severe the alternatives for organisms are extinction, movement resulting in distribution and range changes, or adaptation *in situ*. As populations within the range of a single taxon may be subject to very different changes and usually experience different initial conditions, these three scenarios are not necessarily mutually exclusive. The first section examines some of the more prominent potential effects of climatic and atmospheric changes on butterfly populations and individuals, whereas the second section focuses on the vulnerability of specific gene pools.

A *Consequences of atmospheric changes for butterfly biology*

Chapter 1 furnishes details of the way in which climate change may directly influence butterfly biology. As butterflies are ectotherms, a rise in temperature may be expected to be largely beneficial. To recapitulate briefly, high ambient temperature reduces the time required to raise body temperatures to flight activity thresholds and would therefore increase flight-dependent activities: mate-location, egglaying, nectaring, predator-evasion and dispersal. Egg production may be increased (e.g., *Colias eurytheme*; Stern & Smith, 1960), and the development of early stages enhanced by reducing the time exposed to predators (Pollard, 1979; Porter, 1983). Faster development rates may lead to additional broods of species with restricted seasonal voltinism as in the case of *Coenonympha pamphilus* (Lees, 1962b), *Aglais urticae* (Pullin, 1986a) and *Pararge aegeria* (Shreeve, 1986b) and also to larger population sizes, as for *Lasiommata megera* in northern Britain (Dennis & Bramley, 1985). There are, however, alternative consequences, especially for species close to their maximum temperature threshold. Excessive temperatures may cause adult heat stress (Kingsolver & Watt, 1983), reducing activity (Dennis & Bramley, 1985) and decreasing longevity and fecundity (Stern & Smith, 1960). Eggs and larvae may also be affected (Taylor, 1988). Moreover, although additional broods of some species may be successful, in others (e.g., *Boloria euphrosyne* in northern Europe) they may fail if insufficient time is available for completion of development to the normal overwintering stage or if larval resources are inadequate in quantity or quality (e.g., *A. urticae*; Dennis, 1984).

A wide range of potential interactions between butterflies, climate and atmospheric gas and pollutant concentrations are also mediated via host-plants and nectar sources. Enhanced rates of photosynthesis in C_3 plants to elevated atmospheric CO_2 levels, compared to the relatively small response in C_4 plants, has been demonstrated in numerous studies (Acock & Allen, 1985; Oechel & Strain, 1985; Cure & Acock, 1986). The effect of CO_2 on photosynthesis increases with temperature (> 18°C) and with nutrient availability (Sage & Sharkey, 1987; Cure, 1985), but can be offset by atmospheric pollutants such as tropospheric ozone and acid fall-out (Cannell *et al.*, 1989; Mooney *et al.*, 1991). These rates are the result of two opposing effects of CO_2, one on stomata and the other on the enzyme ribulose bisphosphate carboxylase/oxygenase (rubisco). A doubling of CO_2 decreases stomatal conductance (e.g., by 36 per cent in 16 herbaceous species; Morison & Gifford, 1984). This has little effect on photosynthesis because gas diffusion is more rapid than the photochemical uptake of CO_2 by plants. With elevated CO_2, internal CO_2 concentration increases, providing an enhanced CO_2 gradient to chloroplasts and enhanced activity of rubisco, thus increased sucrose formation (Mooney *et al.*, 1991). The decreased stomatal conductance leads to an increased assimilation of CO_2 per unit of water transpired, thus water use efficiency (Eamus & Jarvis, 1989). The growth of most non-woody plants increases by c. 33 per cent with a doubling of CO_2 from 300 to 600 ppmv (Kimball, 1983a, b), whereas growth in trees is four times that of non-woody plants (Idso, 1991). Although the proportional decrease of nitrogen to carbon in CO_2-enriched plants would predict deterioration in leaf litter quality and therefore nutrient cycling, there is little evidence for this (Norby *et al.*, 1986), especially as increased temperatures may offset decreasing litter quality (Pastor & Post, 1988). There is also no evidence that this proportional reduction in nitrogen decreases resistence to heat shock in plants (Coleman *et al.*, 1991), but it is possible that CO_2-enriched plants may be less able to produce natural pest-repelling defences (Hanstock, 1991). Response of plants to elevated atmospheric CO_2 is extremely rapid; photosynthesis can respond in seconds, homeostatic adjustments of net ecosystem CO_2 flux begins within a season and complete adjustment to $2 \times CO_2$ can occur within three seasons (Oechel & Riechers, 1986).

Theoretical considerations would suggest that for insect herbivores, the decrease in foliar nitrogen concentrations (a limiting nutrient for Lepidopteral larvae – Mattson, 1980; Lincoln & Couvet, 1989) would be only partly offset by increased plant production. To compensate for the lower foliar nitrogen concentration, larvae would need to assimilate additional elevated CO_2-grown foliage to achieve normal growth rates and take longer to attain full size (Lincoln *et al.*, 1986; Strain, 1987; Taylor, 1988; Fajer *et al.*, 1989). Lepidopteral larvae reared for an entire larval period on enhanced atmospheric CO_2 conditions in the wild may then experience reduction in fitness-related parameters such as reduced pupal mass, and increased mortality due to increased exposure on hostplants and longer development times (Osbrink *et al.*, 1987; Akey & Kimball, 1989). Post-neonate male and female *Junonia coenia*, reared on high CO_2 (double concentration) grown plants of *Plantago lanceolata* grew significantly more slowly, and consequently had greater times to pupation than those reared on low CO_2-grown plants. The mean nitrogen concentrations were about one-ninth lower in plants grown in high CO_2 rather than in low CO_2 environments. Conversely, there was no significant difference in the average mass of larvae reared on artificial diets in either low or high CO_2 environments (Fajer *et al.*, 1991). Such longer development times could lead to fitness reductions resulting from increased exposure to predators and parasites (Price *et al.*, 1980) and/or from a diminished likelihood of completing development on the same host in climatically limited environments (cf., Weiss *et al.*, 1988). Larvae of *Pieris rapae* on low nitrogen plants developed more slowly and devoted more time to feeding than those on high nitrogen collards. When exposed to large (> 2 mm) predators and parasites in a natural setting, the survivorship of larvae on low nitrogen collards was reduced significantly compared to larvae on high nitrogen plants (Loader & Damman, 1991). However, it is possible that longer development times may be offset by higher temperatures that may accompany rising atmospheric CO_2 concentrations. In some butterfly species, egglaying and larval diet is more closely related to the presence of secondary plant metabolites than leaf nitrogen (e.g., *Gonepteryx rhamni*, McKay, 1988). For this, and perhaps other, species which feed on long-lived 'chemically defended' woody hostplants, secondary plant metabolites may be more important than nitrogen in limiting plant growth. The effect of changed climatic conditions on the concentration of such compounds is uncertain; in the experiments on *J. coenia* on *Plantago lanceolata*, Fajer *et al.* (1991) found no significant differences in iridoid glycoside and fibre concentrations nor in foliar water between CO_2 treatments.

Some indirect effects of a temperature increase with enhanced atmospheric CO_2 may be beneficial to butterfly populations. Increased temperatures may result in the increased production of hostplant biomass and longer growing seasons (Campbell, 1989). This is more likely to be the case in the northern hemisphere industrial mid latitudes, where increased cloud cover from aerosol pollution could result in warmer nights and winters but lower diurnal maximum temperatures and thus lower than expected increases in evapotranspiration. Although some GCMs predict increases in precipitation for higher latitudes (Mitchell *et al.*, 1990), this outcome is by no means certain (Schneider, 1989), and soil moisture would decline markedly with the rising evapotranspiration rates associated with temperature rises for high latitudes which are expected to be well above the global mean. Water use, calculated as transpiration rates, of forests in Britain can rise as much as 39 per cent with a 1°C increase in temperature and 2 mb increase in saturation vapour pressure deficit (Cannell *et al.*, 1989: 339). A decrease in soil moisture could have serious indirect effects through hostplant desiccation, leaf abscission and seedling failure which could reduce larval growth and cause starvation. Nectar sources would diminish both by reduction in flowering and through increased nectar viscosity, potentially reducing adult longevity and fecundity in those species which are dependent on nectar for egg production (e.g., *Euphydryas editha, Maniola jurtina, Aphantopus hyperantus*; Murphy *et al.*, 1983; Porter, 1992). Yet in those few species which emerge with a near complete egg-load (e.g., *Eurodryas aurinia*; Porter, 1992) or which can feed on viscous nectar (e.g., *Thymelicus lineola*; Pivnick & McNeil, 1985), such effects may be minimized. Moreover, in a landscape in which hostplants and nectar supplies are increasingly affected by desiccation as the season progresses, additional broods triggered by mild springs and hot summers may fail, causing population declines as is thought to have happened in Britain during 1976 (see section 2.2A). These consequences depend much on whether the effects of drought are offset by increased water use efficiency.

For species at high latitudes which overwinter in the egg stage (e.g., many European Theclinae) or which are deposited as eggs in the early spring (e.g., *Gonepteryx rhamni*) and feed on the leaves of woody plants, the timing of egg-hatch is such that feeding commences before significant accumulation of secondary plant compounds in hostplants. Rising temperatures with enhanced atmospheric CO_2 may affect the timing of budburst of woody species (Murray *et al.*, 1989). Simulation of ΔT_{2x} for trees in Finland induced budburst in winter and subsequently exposed them to greater risk from frost damage, temperatures between –27° and –10°C (Hanninen, 1991). Such a change could increase any tendency for plant/larval asynchrony with possible larval starvation, if hostplants and larvae use different environmental cues for the onset of growth and development. In the long term, high temperatures in the winter months could be detrimental to many trees (Cannell *et al.*, 1989); but increased rates of respiration, inadequate chilling and disruption of phased sequences of development may all not only affect the fitness of arboreal hostplants but have a similar detrimental impact on insect physiology (see section 2.1A). Increased temperatures may also increase the numbers and activity of arthropod predators and parasitoids, but the mortalities they cause are often density-dependent and therefore there may be no net effect (Warren, M.S., pers. comm.). Overall adverse effects of plant-induced reductions of population success may outweigh beneficial effects, but this will depend on the regional provenance of species.

As high increases in temperature equate with large shifts in geographical location (1°C per decade ≈ 100–150 km per decade; Dobson *et al.*, 1989; Woodward, 1989), many organisms may be forced to adopt new habitats, move large distances or adapt to the new conditions. Cooler habitats exist in longer vegetation, under woodland cover and with changes in topography (aspect; slopes shaded from midday insolation) and these may be exploited by the butterfly fauna (see section 5.3B). As regards movement, migration rates may match short distance transfers up mountains for which the lapse rates are c. 1°C for every 160 m, but may not facilitate the tracking of climatic changes with latitude. The maximal post-glacial migration rate of plant species, such as spruce *Picea abies*, was some 2 km a^{-1}; a rate of 6.4 km a^{-1} would be required to accommodate a rise of 2°C by 2040 AD (Huntley, 1988; Dobson *et al.*, 1989). On this basis and assuming an adult butterfly lifespan of approximately five days, typical for many temperate species (Warren, 1992a), migration would need to occur at a rate of 213 mh^{-1} for 6 hd^{-1} in order for populations to track their current climates. A number of temperate species are more than capable of making such movements (e.g., Nymphalids, *Aglais urticae*; Pierids, *Pieris rapae*; Baker, 1978). Yet most butterflies surveyed in high latitude environments, in which larger than mean temperature changes are predicted to occur, have recorded much smaller daily movements than required to track thermally identical habitats (e.g., *Plebejus argus*; < 50 md^{-1}; Thomas, 1985). However, data on the mobility and colonizing ability of most species are rare, and the most commonly used research methods (i.e., mark–release–recapture techniques) underestimate migration and dispersal. Anecdotal observations demonstrate much larger movements than those obtained from limited area studies (Shreeve & Dennis, 1991). Dispersal and changes in distribution of organisms may also be enhanced by, or be dependent upon, random, long-distance movements (Hengeveld, 1988). For butterflies, the chance of this occurring may be promoted by warmer conditions which stimulate adult activity and also by the deterioration of existing habitats (Warren, 1987), concomitant on enhanced greenhouse conditions.

Short-lived organisms have the greatest capacity to adapt to the projected changes of enhanced greenhouse gases. Butterfly physiology and morphology have clearly been able to adapt to the rapid environmental and climatic changes of the Late Glacial and Holocene (sections 3.3, 4.3 and 4.4); both the degree of polygenic inheritance demonstrated for such traits, the amount of genetic variance in populations (Majerus, 1979; Brakefield, 1984; Pullin, 1986c) and the wide range of climates occupied by most species (Dennis *et al.*, 1991) suggest that evolutionary adjustments may be possible.

B *Vulnerability of butterfly populations to environmental perturbations*

Despite the various indications for the effects of increases in atmospheric infra-red absorbing gases and climatic warming on butterfly biology, predictions are not easily made. Responses to environmental perturbations will be species-specific; as such, the factors involved are best examined in some detail for a case group, the British fauna. Survival of environmental perturbations depends on external (environmental) and internal (biological) factors. External factors differ in type (e.g., climatic, terrestrial), in causal basis (natural, anthropogenic), in magnitude and sign. They may also vary in frequency, an important distinction being drawn between random unpredictable oscillations that generally occur at high frequency and significant monotonic changes in conditions usually at lower frequency. As all kinds of perturbations inflict stress on populations, and as other environmental changes may dominate or reinforce atmospheric influences on populations, the following discussion examines the vulnerability of butterflies to sudden massive environmental changes unrestricted to type. Internal factors fall into two groups: (i) the capacity of organisms to withstand change, and (ii) their flexibility of response to changes.

The capacity of organisms to withstand environmental changes is defined by the spatial extent and variety of resources. The larger the number of alternative resources available to an organism (outlets of escape), the more probable is it that it will survive environmental changes. The most successful organisms are likely to be those which can use the greatest variety of resources (i.e., habitats, hostplants, nectar sources, roost sites, mate-location sites etc.), and have the most abundant resources in space (i.e., ubiquity) and time (i.e., frequency of production; lifespan) (Table 5.2). Butterflies which have access to a variety of hostplants and habitats can switch between these resources during environmental perturbations that affect the synchronization between the insect and its food supply or which selectively eliminate habitat components. In the same way, butterflies which have the widest geographical range or the most dense distribution probably experience the widest range in regional conditions, local climates and microclimates which similarly provides the greatest buffer against environmental perturbations. However, frequency of resource production and resource lifespan, as measures of temporal resource distribution, provide conflicting opportunities. Climax-communities have the greatest lifespan but are not necessarily any more robust than early seral associations and develop over very long time-spans. Thus, the lag time to colonization by insects for conditions which benefit specific habitats is greatest for forest climax components. Because forest communities are undergoing monotonic decline in the mid and tropical latitudes, species dependent on them are more vulnerable than those founded on early seres. Early successional stages are the least vulnerable since their taxonomic components tend to have life-histories and population dynamics that facilitate rapid colonization of temporary areas. This rapid turnover also makes possible genetic adaptation to perturbations more readily than for long-lived trees. On the basis of capacity measures alone, the butterfly species least vulnerable to environmental perturbations are those polyphagous on ubiquitous herbs occupying extensive ranges, continuous distributions and using varied and extensive habitats. The most vulnerable are monophagous on scarce arboreal substrate-dependent climax components having narrow ranges, punctiform distributions and restricted to a single habitat.

Flexibility of response (Table 5.2) to environmental perturbations is more difficult to determine: (i) there are numerous attributes in life-history strategies that are not well known for many species (e.g., realized fecundity); (ii) these attributes combine in complex ways for different organisms, and (iii) their response to environmental perturbations is likely to depend more on the type of perturbation. One obvious prediction is that 'r-strategists', species adapted to early successional 'communities', might respond more flexibly to environmental change than 'K-strategists', organisms thriving in mature species-saturated communities (Pianka, 1972; see p. 43). An analysis of life history traits for 145 species from north-west Europe by Bink (1989) revealed no simple partition into r-types and K-types nor any obvious

Table 5.2 Factors influencing the vulnerability of resident British butterflies to environmental perturbations

	Capacity factors (CF)						
	A	B	C	D	E	F	Σ_A^F
Carterocephalus palaemon	1	1	3	3	3	2	13
Thymelicus sylvestris	3	4	3	3	3	4	20
Thymelicus lineola	2	2	3	4	3	3	17
Thymelicus acteon	1	1	1	3	3	1	10
Hesperia comma	1	1	1	4	4	1	12
Ochlodes venata	3	3	3	4	3	4	20
Erynnis tages	4	2	3	4	4	2	19
Pyrgus malvae	2	2	3	4	4	2	17
Papilio machaon	2	1	1	1	1	1	7
Leptidea sinapis	2	1	3	3	3	2	14
Gonepteryx rhamni	3	3	2	3	2	2	15
Pieris brassicae	4	4	3	4	4	1	20
Pieris rapae	4	4	3	4	4	2	21
Pieris napi	4	4	3	4	4	3	22
Anthocharis cardamines	4	3	3	4	3	3	20
Callophrys rubi	4	2	4	3	3	3	19
Thecla betulae	2	1	1	4	2	2	12
Quercusia quercus	3	2	2	3	1	1	12
Satyrium w-album	3	2	2	2	1	1	11
Satyrium pruni	1	1	1	4	2	1	10

Flexibility factors (FF)					
G	H	I	J	Σ_G^J	Index
2	1	2	2	7	1.96
2	1	2	2	7	2.54
2	1	2	1	6	2.17
1	1	2	2	6	1.59
2	1	2	1	6	1.75
2	1	2	2	7	2.54
2	2	2	2	8	2.59
2	2	2	3	9	2.54
2	2	2	3	9	1.71
2	2	2	3	9	2.29
4	1	4	4	13	2.88
4	3	3	3	13	3.29
4	3	3	3	13	3.38
4	3	3	3	13	3.46
3	2	3	3	11	3.04
2	1	2	3	8	2.59
2	1	2	1	6	1.75
2	1	2	1	6	1.75
2	1	2	1	6	1.67
1	1	1	1	4	1.33

Table 5.2 (cont.)

	Capacity factors (CF) A	B	C	D	E	F	Σ_A^F
Lycaena phlaeas	4	3	2	4	4	4	21
Lycaena dispar	2	1	1	1	1	1	7
Cupido minimus	4	1	1	2	4	2	14
Plebejus argus	2	1	4	4	4	2	17
Aricia agestis	2	2	4	4	4	3	19
Aricia artaxerxes	3	1	3	3	4	1	15
Polyommatus icarus	4	4	3	4	4	4	23
Lysandra coridon	2	1	1	1	4	1	10
Lysandra bellargus	1	1	1	1	4	1	9
Celastrina argiolus	3	3	3	3	3	2	17
Maculinea arion	1	1	1	2	4	1	10
Hamearis lucina	3	1	2	2	3	2	13
Ladoga camilla	2	2	1	3	3	1	12
Apatura iris	1	1	2	3	1	1	9
Aglais urticae	4	4	1	4	3	4	20
Nymphalis polychloros	2	1	3–4	2	1	1	11
Inachis io	4	3	1	4	3	3	18
Polygonia c-album	3	4	3	4	3	2	19
Boloria selene	4	2	2	3	4	3	18
Boloria euphrosyne	4	1	2	3	4	2	16

Flexibility factors (FF)					
G	H	I	J	Σ_G^J	Index
3	4	3	2	12	3.25
2	1	1	2	6	1.34
1	2	2	2	7	2.04
1	1	3	1	6	2.17
1	2	2	2	7	2.46
1	1	2	2	6	2.00
2	4	3	2	11	3.29
2	1	2	1	6	1.59
1	2	2	2	7	1.63
3	4	2	3	12	2.92
1	1	1	2	5	1.46
2	1	2	3	8	2.09
2	1	2	2	7	1.88
2	1	2	2	7	1.63
4	3	3	4	14	3.42
3	1	2	4	10	2.17
4	2	3	4	13	3.13
4	3	3	4	14	3.34
2	2	2	2	8	2.50
2	2	2	2	8	2.34

Table 5.2 (cont.)

	Capacity factors (CF)						
	A	B	C	D	E	F	Σ_A^F
Argynnis adippe	3	1	2	3	4	2	15
Argynnis aglaja	4	2	2	3	4	3	18
Argynnis paphia	2	2	2	3	2	2	13
Eurodryas aurinia	4	1	2	3	3	2	15
Melitaea cinxia	1	1	1	3	4	1	11
Mellicta athalia	1	1	2	3	3	2	12
Pararge aegeria	4	3	3	3	2	2	17
Lasiommata megera	3	4	3	4	4	4	22
Erebia epiphron	2	1	1	3	1	1	9
Erebia aethiops	2	1	3	3	3	2	14
Melanargia galathea	3	2	3	4	3	3	18
Hipparchia semele	4	1	3	4	4	2	18
Pyronia tithonus	3	4	3	4	3	4	21
Maniola jurtina	4	4	3	4	3	4	22
Aphantopus hyperantus	4	3	3	4	3	3	20
Coenonympha pamphilus	4	4	3	4	4	4	23
Coenonympha tullia	3	3	3	3	3	1	16

Capacity (resources) to withstand changes

A Range, based on the latitudinal extent of species on the British mainland: 1, < 25; 2, < 50; 3, < 75; 4, < 100 (per cent).

B Distribution, based on the proportion of 10 km squares occupied within the range of the species: 1, < 25; 2 < 49; 3, < 75; 4, < 100 (per cent).

C Hostplant type: 1, monophagous; 2, oligophagous – 1 species per habitat; 3, oligophagous – > 1 species per habitat; 4, polyphagous.

		Flexibility factors (FF)			
G	H	I	J	Σ_G^J	Index
3	1	2	1	7	2.13
3	1	3	2	9	2.63
2	1	2	2	7	1.96
1	1	2	2	6	2.00
2	2	2	2	8	1.92
2	1	2	2	7	1.88
2	4	3	2–3	12	2.92
2	4	2	2	10	3.09
2	1	1	2	6	1.50
2	1	2	2	7	2.04
2	1	2	2	7	2.38
3	1	3	2	9	2.63
2	1	2	2	7	2.63
2	1	4	2	9	2.96
2	1	2	2	7	2.54
2	4	3	2	11	3.29
2	1	2	2	7	2.21

D Hostplant abundance: 1, substrate dependent; 2, patchy within habitats; 3, ubiquitous within habitat types; 4, ubiquitous and cosmopolitan.

E Vulnerability of major habitat seral stage occupied: 1, climax woodland or plagioclimactic bog; 2, pre-climax forest; 3, shrubs, tall forbs and grasses; 4, bare ground, short forbs and grasses.

F Range of semi-natural habitat types occupied: 1, < 5; 2, < 9; 3, < 14; 4, < 18 (maximum number).

Table 5.2 (cont.)

Flexibility to withstand changes

G Dispersal ability: 1, closed populations with little evidence of movement outside colonies; 2, colonial species with evidence of dispersal outside colonies; 3, open population structures with evidence of frequent movements between habitat units; 4, migrants and species which are vagrant and known to engage in long-distance movements.

H voltinism: 1, biennial or univoltine; 2, univoltine but with occasional partial second broods; 3, bivoltine; 4, multivoltine.

I Length of the flight period for the longest brood in the year (egglaying females): 1, < one month; 2, 1–2 months; 3, 2–3 months; 4, > 3 months.

J Overwintering stage: 1, egg; 2, larva; 3, pupa; 4, adult.

Coding of attribute states: 1–4, most to least susceptible, respectively, to environmental perturbation. Index = $(\Sigma^{F}_{A}$ CF/6 + Σ^{J}_{G} FF/4) 0.5. Correlations: $CF_{(A-F)}$ with $FF_{(G-J)}$, $r_p = 0.57$, $p < 0.001$; $CF_{(A-F)}$ with $FF_{(G-I)}$, $r_p = 0.63$, $p < 0.001$. Note: equal weight is given to all variables, but this may be unrealistic since critical variables are not identifiable for all species with the current state of knowledge. Ant associations in some lycaenidae may interact with factors C, D, E and F, but the precise nature of such interactions is not comprehensively known and is probably accommodated in factor F. (Data from Thomas, 1986; Emmet & Heath, 1989; Dennis, 1992; see Dennis & Shreeve, 1991).

ecological 'periodic table' based on those traits (Southwood, 1977). However, it is manifest that a number of life-history traits are closely related to habitat components (*viz*., pre-oviposition, larval growth rate, voltinism, overwintering stage and dispersal), therefore, specifically to sequential aspects of environmental stress. Bink's study revealed the decline of stress tolerators and 'fixtures' (i.e., sedentary species restricted to stable environments), mainly due to habitat loss or decline of the stress factor (e.g., due to soil fertilization or microclimatic warming associated with forest growth). The 34 per cent of species still thriving are all species which disperse well or migrate and have a high annual increase (high fecundity or several generations per year), that is, are typically ruderal or 'r-type'. These observations have a good logical basis. Species which can disperse well or migrate can move to new 'resource' outlets rapidly, and those which have a high intrinsic level of increase can respond rapidly to environmental pulses that enhance population growth and establish large enough numbers to withstand environmental downturns. This group of species also tends to overwinter as adults or pupae, occasionally as half-grown larvae. The significance of this was not discussed by Bink (1989), but probably relates to the vulnerability of different stages to direct habitat damage by agricultural or land-use practices (cf. Grime, 1989). The length of the flight season over which ovipositing females eclose may also influence vulnerability to perturbations, inasmuch as those which emerge over a longer period have a greater probability of surviving periodic catastrophes such as frost, drought, or intense episodes of predation, collecting, grazing or mowing.

The two groups of components describing vulnerability to environmental perturbations are to some extent related (i.e., species with greater dispersal capabilities will tend to have wider geographical ranges and a denser distribution pattern), but as the maximum correlation between combinations of variables in the two groups is modest ($r_p = 0.63$), a large part of the variance (60 per cent) is specific to the capacity and flexibility measures. An index of vulnerability based on the sum of the two groups of variables (Table 5.2) corresponds approximately with the current spatial dynamics of British butterflies. Those species with low values (Index < 2) are generally experiencing contractions in range and distribution, whereas the status of those with high values (Index > 3) remains unchanged or they are undergoing range and distribution expansions (see Table 5.3). However, significant exceptions exist; among those with intermediate values (Index 2 to 3), the fritillaries are in sharp decline, whereas the satyrines are maintaining their status. Such disparities highlight significant difficulties with vulnerability measures. Firstly, it is unlikely that all variables can be adequately accounted. Secondly, some attributes, such as dispersal potential, may not be scored accurately. Thirdly, the ranked scores for attributes reflect simple ranges. This may not accurately score any bias in conditions held by a species for a resource nor regional differences in resource use; for example, the dependence of fritillaries in eastern England on the groundlayer violets of woodlands places them in the most vulnerable categories for seral stage and range of habitats for that region (Table 5.2E, F). Finally, systematic changes in the landscape (i.e., deforestation, decline of coppicing, conifer afforestation, hedgerow removal) tend to have a particular significance for specific biological attributes, only some of which are relevant in a vulnerability index. A monotonic change to a warming climate has less predictable consequences for life-history traits than direct terrestrial land abuse. For example, the effects of drought or 'late' frosts, resulting from climatic warming, depends much on their variability in timing and duration as of their severity; thus, an increasing trend to the occurrence of late summer drought may disadvantage multivoltine compared to univoltine species as in 1976, despite their capacity for rapid recovery if normal conditions of soil water balance are re-established. The specific effects of potential warming on the British fauna are examined further in section 5.3B.

5.3 The regional impact of enhanced greenhouse conditions on butterfly biogeography

Although current GCMs lack the sophistication to make accurate predictions of the regional effects of increased concentrations in atmospheric infra-red absorbing gases, disparities in consequences for regional fauna will be caused by significant variations in environmental impact. Not only is regional variation expected in climatic parameters such as temperatures and precipitation, but substantial spatial variation in terrestrial conditions induces greater sensitivity in some ecosystems than others. The two sections that follow examine the ramifications of such regional variation at a broad global scale and over a smaller region diverse in climate and topography.

A *Global aspects*

At a global scale, butterflies can be classified spatially by two sets of criteria: (i) into geographical provenances or biogeographical regions reflecting on the location and timing of emplacement of major physical barriers (i.e., oceans, deserts, mountains) associated with plate tectonics, and (ii) into biomes, that is, regional ecosystems (e.g., tundra, taiga, tropical rain forest), associated with quasi-uniform climates and vegetation (Sbordoni & Forestiero, 1984). Of the two, biomes, which vary markedly in productivity, may be particularly sensitive to changes in climate. Projected changes in both temperature and precipitation for a doubling of infra-red absorbing gases have far-reaching consequences for the length of the growing season and thus productivity. Temperate plant formations on which butterfly fauna depend correlate more highly with water balance than with either temperature or precipitation, probably because it incorporates lags from interactions between these two variables (Stephenson, 1990). ΔT_{2x} implies a large shift northwards (1°C for 100–150 km, Woodwell, 1989; 1°C for 150–200 km., Parry *et al.*, 1990) and to high elevations (1°C for 150–170 m; Sedjo & Solomon, 1989) of natural zones. Increases in potential evapotranspiration at higher latitudes are likely to be largely offset by predicted increases in precipitation and water use efficiency, except in northern mid-latitudinal continents during summer (Mitchell *et al.*, 1990). No such increases in water availability are predicted for lower latitudes. The direct impact on regional ecosystems of elevated CO_2 is more difficult to determine. The absolute response to increases in atmospheric CO_2 is usually greater with elevated nutrient availability and in drought-stressed warmer climates (Mooney *et al.*, 1991). Collectively, these projections disadvantage cold ecosystems compared to warm ecosystems and savanna and semi-arid ones compared to low-latitude forests, but do not consider repercussions from disease, pests, fire or human activity (Clark, 1988).

Analysis of these changes suggests repercussions for forests (Pickett & White, 1985; Overpeck *et al.*, 1990), grasslands and tundra (Emmanuel *et al.*, 1985) and arid communities (Neilson, 1986, 1987). The greatest concern is for the tundra and alpine communities (Peters & Darling, 1985), which are predicted to shrink at the expense of the taiga. Continental interior ecosystems, such as temperate grasslands and arid communities, may increase in area, and because of the shape of continents arid environments will spread northwards (Woodwell, 1989). The UK Meteorological Office GCM projection of the climate under a 2 × CO_2 scenario for the Mediterranean region reveals a large reduction in the growing season (months with mean monthly temperature > 5°C and rainfall > potential evapotranspiration) (Brouwer, 1989) and advancing aridity. However, it is essential to reiterate that many uncertainties exist in the predictions, for instance as regards: the triggers and thresholds for monsoonal regions and thus for water balances in tropical continental interiors (Hanstock, 1991); the relative balance of photosynthesis and respiration, poleward extension *versus* lower latitudinal and arid margin decay of temperate forests (Rosenberg, 1981; Woodwell, 1989; Idso, 1991); the migration rates of trees (Davis, 1988) and human responses, which are expected to be most deleterious to wildlife where water balances are adversely affected and land is lost from rises in sea level (Hekstra, 1989; Scott *et al.*, 1990).

Climatic changes for a 2 × CO_2 scenario have important implications for butterfly biogeography, for ranges, species richness and the distribution of faunal elements. Generally, as the Earth warms and climatic optima for species shift northwards, range extensions will occur and species richness will increase in those areas not affected by serious drought. As temperature changes are predicted to be greater at high latitudes, these changes will be more pronounced at higher than lower latitudes. However, the resulting community developing at any latitude, due to the temperature rise, may become increasingly composed of species whose overall average range is narrower (France, 1991). This conclusion follows from Rapoport's rule (Stevens, 1989), but depends largely on whether increases in temperature are accompanied by decreases in seasonal extremes; because species living at higher latitudes are exposed to fluctuating annual climatic conditions, natural selection has favoured wide tolerances, and these organisms are less restricted in habitat use and are characterized by broader geographical ranges. Increased temperatures at higher latitudes relieve physiological constraints. Thus, because winter temperatures are expected to increase more than summer temperatures, range extensions northwards may occur among those species unable to diapause over winter, though this movement will be biased to maritime regions and will depend on extensions in the range or growth of hostplants. A candidate for such range extension is *Colias evagore* which has maintained a foothold on the coastal strip of southern Spain since 1983. It is unable to diapause and requires 162.5°d for larval development above a T_o min of 12°C. However, it is also restricted to *Capparis spinosa* which is unavailable throughout most of Spain in winter (Jordano *et al.*, 1991).

As distinctive faunal elements among butterflies tend to correspond with biomes (Larsen, 1984b), their status under a 2 × CO_2 scenario is closely linked to that of the vegetation zones they occupy. Taiga and arid zones have relatively impoverished butterfly faunas compared to contiguous biomes; the expansion of these biomes heralds a decrease in species richness. Nevertheless, it opens up larger areas to species affiliated to these regional ecosystems (e.g., *Megathymus* spp. and *Agathymus* spp. on agave and yucca in arid environments). The reduction in tundra and alpine ecosystems could result in the loss of distinctive faunas over wide areas, though resource outlets on slopes with different aspects and shading or marshland over discontinuous permafrost amid taiga could facilitate survival. The decline in deciduous forests, should this occur, may not result in a large reduction in species richness, since most species of this biome utilize herbaceous hostplants, but it could well lead to contracting distributions of species dependent on arboreal hostplants and habitat components. There is little point in speculating on the biogeographical changes of butterflies within the tropical rain forest biome from climatic changes under a 2 × CO_2 scenario, as such changes are insignificant compared to the wholesale destruction of this ecosystem which shows no signs of abating. Suffice it to say that, unlike temperate forests, attempts to reafforest areas that have been cleared lack the knowledge, resources and time to re-establish the complex communities that are required by rain forest butterflies and other organisms.

The effect of climatic change on faunal elements depends to a large extent on the classification of their component species for vulnerability measures (see Table 5.2). In Europe, the faunal unit most likely to benefit from climatic change under atmospheric CO_2 doubling is the extent group which includes c. 40 per cent of the entire fauna (Fig. 4.8b). Many of the component species of this group have the facility to spread north and west towards Scandinavia and Britain. Conversely, faunal units least likely to benefit from projected climatic changes comprise species which score highly for vulnerability measures. These include sub-alpine and Mediterranean zone endemics (Fig. 4.8c) for which temperature increases may cause a further diminution and isolation of habitats, in the former due to elevational shifts in vegetation zones and in the latter due to drought (see Thomsen, 1990). Many Mediterranean butterflies may depend on the maintenance of riparian forests which characteristically have higher species diversity than adjacent open areas (Viejo *et al.*, 1992). A number of these species could become extinct if the 'worst scenarios' for atmospheric 2 × CO_2 are realized. Nevertheless, the effects of climatic change may run counter to vulnerability measures for some groups. Thus, arctic and northern elements with a broad resource base could very easily contract, particularly as these areas become more suitable for cultivation and afforestation with conifers. Similarly, many pan-Mediterranean species intolerant of cold, or whose hostplants are intolerant of frost, may continue to be restricted by the marked physical and climatic barrier of the southern European mountain chain; valleys, gaps and cols regularly experience severe inversion conditions and cold northern winds such as the Mistral and Boro. Moreover, the distribution of some elements with a narrow resource base in Europe could greatly expand. Twenty-four butterfly species of tropical provenance could invade a large part of the Mediterranean region, particularly as nineteen of these species are migrants (e.g., *Danaus chrysippus*), and a number of their larvae feed on plants having a wide distribution (e.g., *Lampides boeticus* on Leguminosae), or grown as garden and ornamental plants (e.g., *Catopsilla florella* on *Cassia* spp.), or grown as crops (e.g., *Dendorix livia* on dates and pomegranates) (Larsen, 1986; Masó, 1992).

B *Britain: the highland–lowland divide*

Regional contrasts in climate arising from increases in atmospheric infrared absorbing gases are influenced by two sets of factors: (i) the sensitivity of climatic systems in the different regions to any changes in climatic attributes, and (ii) regional disparities for changes in climatic attributes that may be predicted by GCMs. In Britain, substantial contrasts in climate exist between the north-west and the south-east (Fig. 5.3a). These two regions, traditionally labelled highland and lowland Britain respectively and crudely divided by the Tees–Exe line, differ in relief, in meridional and zonal influences and thus in heating and moisture budgets. Compared to lowland Britain, the highland region is cooler, cloudier, wetter, more humid and windier (see Fig. 2.8). The effects of these climatic contrasts in Britain on butterfly biology have been reviewed earlier (sections 2.3B, 3.3B). These regional climatic differences also translate into contrasts for ecosystem components such as soil conditions and plant growth which index the combined effects of different climatic attributes. Thus, the growing season (months with mean temperatures > 6°C) decreases markedly with increasing altitude in highland Britain, and soil moisture deficit increases systematically towards the south-east (Fig. 5.3b; Gregory, 1954; Green, 1964). The steep lapse rates of climatic parameters with altitude makes highland environments more robust and lowland environments more sensitive to any climatic change uniformly imposed over Britain. However, as organisms respond to specific thresholds in external conditions (e.g., T_0min for larval development), the regional impact of climatic change will depend largely on the prior value of climatic parameters in different regions.

Climatic change scenarios for the United Kingdom and equivalent 2 × CO_2 conditions indicate uniformity for some parameters over the British Isles and regional disparities for others. Warrick and Barrow (1991), utilizing the same model for business-as-usual emissions applied by the IPCC scientific assessment (Houghton *et al.*, 1990) suggest equivalent 2 × CO_2 to occur by about 2028 AD with a global mean temperature + 1.4°C warmer than today (Fig. 5.3b). A summary of climatic conditions for this mid-term scenario of 2028 AD for the UK notes the following. (i) A regionally uniform change of 1.4° for the summer season (June to August), comparable

(a)

(b)

5.3 Environmental contrasts in highland and lowland Britain. (a) Upland areas: stippled, > 100 m; black, > 400 m. Many butterflies in Britain are restricted to below 400 m but may have access to higher elevations up to 600 m with ΔT_{2x} equivalent of 1.4˚C assuming a lapse rate of 0.6˚C for every 100 m. The northern slopes of downland (areas > 100 m) in southern Britain could provide essential cooler and damper habitats for many species with climatic warming. Continuous thick line, boundary separating highland from lowland Britain, but drawn south of the North Yorkshire Moors rather than to the Tees estuary. Pecked line, land below 5 m above OD in lowland Britain in danger of flooding with rises in sea level (Tooley, 1989). (b) Annual mean soil moisture deficit (cm) and projected mean surface air temperature changes. Soil moisture deficit from Green, 1964; shaded, low deficits; stippled, high deficits. Projected temperature changes (˚C) based on a business-as-usual enhanced greenhouse gas scenario, adapted from Warrick & Barrow, 1991 by courtesy of the Institute of British Geographers.

Zone	*2010*	*2030*	*2050*	
A	1.2	2.5	3.7	Winter (˚C)
B	1.1	2.1	3.2	
C	0.9	1.8	2.6	
A, B, C	0.7	1.4	2.1	Summer (˚C)

to the global mean value. (ii) A gradient of enhanced warning with latitude in the winter season of + 1.8°C for southern Britain increasing to + 2.5°C for northern Scotland. Thus, the warming gradient implies a trend towards more uniform average winter temperatures over the UK (Fig. 5.3b). (iii) Precipitation is most likely to increase in winter a best estimate of + 5 per cent on average. (iv) For summer precipitation changes, there is a large uncertainly. Independent climate model results differ in direction and amount. A best estimate is probably no change with an uncertainly of 11 per cent by 2028. The model-average results indicate that, although uncertainties are large, the chances of drier summers in the future be slightly higher in the south of the UK than in the north, that is compared to current regional means. These changes, if they occur, will have a significant impact on regional environments. The 'growing' season (> T_0min 10°C) will increase dramatically over Britain by as much as 550 degree days; the largest increases will be in the south-west but the greatest percentage increases will occur in northern Scotland (Fig. 5.4; Warrick & Barrow, 1991). Higher temperatures imply greater evapotranspiration, especially in the south of the UK; this has serious consequences for reduced soil moisture, for runoff and for natural vegetation (Mitchell *et al.*, 1990). Climatic change heralds the greatly increased likelihood of extreme events. The probability of hot summers such as that of 1976 rises from 0.1 per cent in 1990 AD to 10 per cent in 2030 AD and 36 per cent by 2050 AD, and of warm winters such as that of 1988/89 from c. 2 per cent in 1990 AD to 22 per cent in 2030 AD and 44 per cent by 2050 AD (Warrick and Barrow, 1991). Thus, despite the uncertainties about projections for future precipitation, there is the prospect of more prolonged and severe droughts in lowland Britain.

Regional climatic analogues for highland and lowland Britain for an equivalent atmospheric 2 × CO_2 scenario based on a global mean temperature change of + 1.4°C would approximate to conditions in southern England and north-west France respectively. The projected central England summer (16.7°C) and winter (5.6°C) means at the boundary between the two regions correspond with the summer conditions near to Strasbourg and the winter conditions near to Bordeaux (Warrick & Barrow, 1991). Although 'geographical analogues' provide useful models for comparison, there are strict limitations to their usefulness. Variability in climatic conditions, particularly extreme events, depend on weather systems (see Fig. 1.3) determined to some extent by unique geographical influences (i.e., land and sea configuration, relief, proximity to air mass sources, sun angle). Climatic change could induce extreme weather patterns over a region unlike those found in any alleged analogues. Even so, there is no evidence to link unusual weather patterns that have been experienced in Britain, as over the past decade (e.g., mild and windy winters, Hulme & Jones, 1991; heatwaves, Brugge, 1991; storms in southern Britain, Cannell & Coutts, 1988), to enhanced greenhouse conditions.

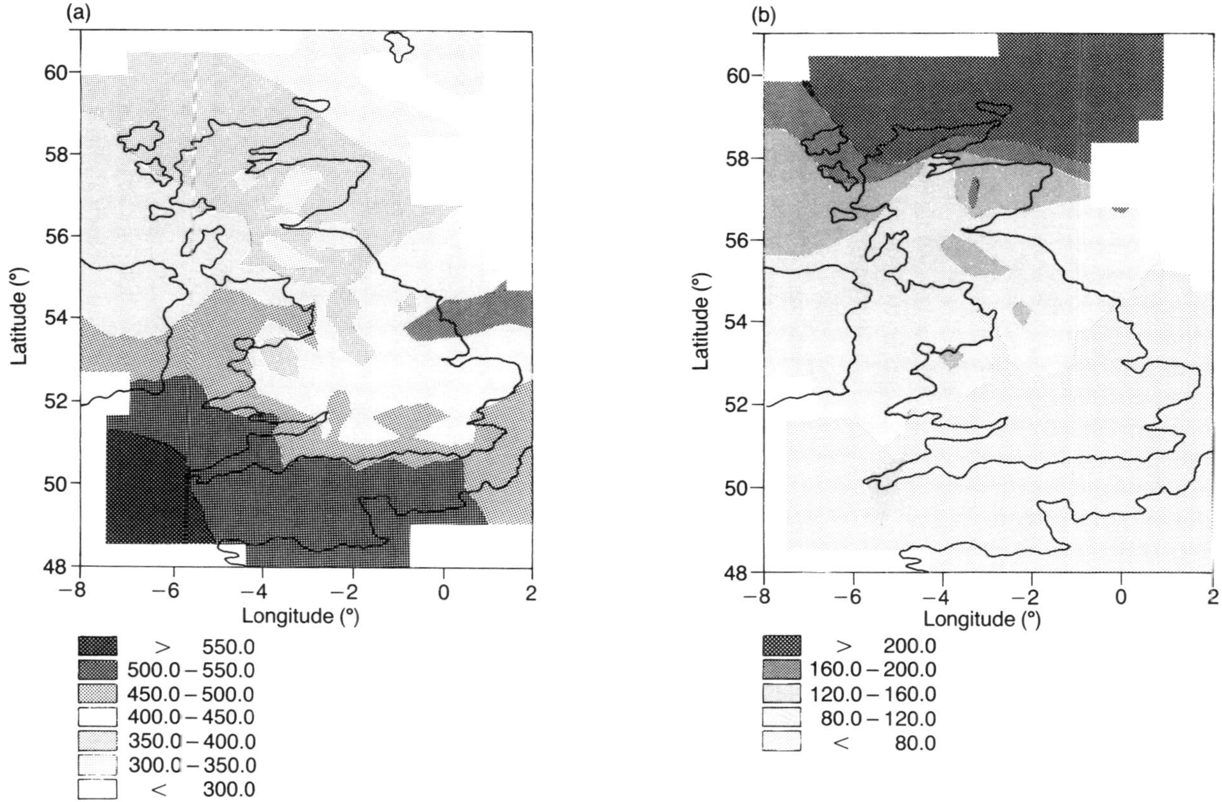

5.4 Projected increases in the number of degree days above a threshold of 10°C for 2050. (a) Absolute increases. (b) Per cent increases. (From Warrick & Barrow, 1991 by courtesy of the Institute of British Geographers.)

All but one of the current resident butterfly species are characteristic indicators of the largest group of European butterflies, the 'extent' group, which is distributed over the whole continent (Fig. 4.8; Dennis *et al.*, 1991). The exception is *Erebia epiphron*, which belongs to the widespread subgroup of the European mountain element. These species have been capable of colonizing a vast expanse of territory during the Holocene and of tolerating wide extremes in conditions (section 4.3). All but five of these species occur within the Mediterranean zone and are therefore tolerant of much higher temperatures and extreme summer drought. Mean winter and summer temperatures in the Mediterranean zone are over 6°C warmer than in lowland Britain. Neither the Mediterranean mean temperatures nor the extreme seasonal rainfall pattern approximate to those conditions projected for ΔT_{2x}; it is extremely unlikely that the UK would be dominated by the subtropical high pressure cell in summer to cause the typical seasonal winter–wet/summer–dry Mediterranean seasonal climate (Warrick & Barrow, 1991).

A more conservative comparison with regional analogues for equivalent atmospheric 2 × CO_2 at 2028 AD suggests that the majority of British butterfly species will benefit from climatic change (Table 5.3A, B). Such analogue comparisons assume the widespread availability of suitable hostplants and habitats and the ability of organisms to migrate. This, clearly, will not be the case for some species, such as for example the Fenland races (*Papilio machaon* and *Lycaena dispar*) and butterflies dependent on calcicole hostplants (*Lysandra bellargus* and *L. coridon*). As it is, for 11 species, summer temperatures may be marginal in central England, especially if the lower (+ 0.7°C) rather than best estimate of temperature rise is experienced by 2028 AD (Table 5.3A); these butterflies also tend to have strict habitat requirements. The analogue comparisons imply distinct problems for four butterflies currently restricted to northern Britain (*Aricia artaxerxes, Erebia aethiops, E. epiphron* and *Coenonympha tullia*), but there is no evidence to suggest that these species or their habitats will not benefit from a rise in temperature; each is found where mean summer temperatures attain 16°C, so too are their hostplants, and montane *E. epiphron* survived higher mean summer temperatures (+ 2°C) in the Holocene. Although hostplants for these two species are vulnerable to drought (*Erebia aethiops* and *Coenonympha tullia*), droughts in highland Britain may be partly offset by higher rainfall. These brief observations point to the limitations of regional analogues. On the one hand they may not adequately address climatic tolerances; on the other, profound ecological differences may persist between a region and its supposed analogue. Specific habitat components could be resistant to migration (e.g., trees) or to development (e.g., edaphic conditions and specific plant associations) detracting from any comparison. Furthermore, fundamental differences may be perpetuated in land-use during climatic change for a variety of reasons, the most obvious of which are human population density, differences in land management and governmental directives and incentives to agriculture. Comparisons with alleged analogues also ignore differences in genotypes and in phenotypic plasticity between regionally distinct populations of the same species.

Butterfies which belong to the extent group, as do all but one species in Britain, by sharing widespread distributions over Europe do so in spite of widely divergent life-history patterns (Bink & Siepel, 1986). These different 'strategies' have evident not handicapped their Holocene colonization of north-west Europe, but will almost certainly influence their relative status in the event of any sudden climatic warming. It is clear that the impact of changing climate will vary between species and also between regions. This is dependent not just on the tolerances and flexibility of species to change (Table 5.2), and on the magnitude of climatic extremes and regional differences during the processes of warming, but on the vulnerability of species to particular events that may be linked to climatic warming, for instance to drought, fire, windthrow, nutrient enrichment and sea level rise (Table 5.3).

Among projected extreme events are a succession of years of drought. Drought will clearly have an adverse effect on butterflies of damp habitats (e.g., *Aphantopus hyperantus*), on those species whose hostplants attain the appropriate growth form in damp habitats (e.g., *Eurodryas aurinia* on *Succisa pratensis; Anthocharis cardamines* on *Cardamine pratensis*), but particularly on those species whose hostplants are restricted to mires and wetlands (e.g., *Coenonympha tullia*; *Lycaena dispar* and *Papilio machaon britannicus*) (Table 5.3C). Also vulnerable are species which have second broods on permeable substrates (i.e., chalk, limestone, sands and gravels) for which there is evidence of decline in the dry periods of the 1970s and 1980s (e.g., *Lysandra bellargus, Lycaena phlaeas*; Thomas & Merrett, 1980; Pollard *et al.*, 1986; Pollard, 1988), especially those that are monophagous and lack alternative habitats.

As soil moisture deficit increases from north to south in Britain (Fig. 5.3b), drought heralds more serious consequences for butterfly populations in southern rather than northern Britain. Butterfly species that are particularly vulnerable to droughts, severe enough to impinge on all hostplants, are those dependent on mature long-lived components (trees) subject to maximal periods of regeneration (see Cannell *et al.*, 1989) (e.g., *Quercusia quercus* on *Quercus spp*.). Enhanced growth in oaks with increased cyclonic activity and precipitation (Kelly *et al.*, 1989) indicates potential for drought stress among climax components in southern England. Forests stressed by drought may be more susceptible to pest infestation, disease and eventually fire (Clark, 1988). The 'drying factor' for forest fuels increases with temperature at any given saturation deficit (Cannell *et al.*, 1989). Thus, fires may become more common, especially in southern Britain, as

Table 5.3 Ecological data on British butterflies, their hostplants and habitats, pertinent to climatic warming

	A	B	C	D	E	F
Hesperiidae *Carterocephalus palaemon*	+	+	−1, 2	+1, 2	+	+1, 2?
Thymelicus sylvestris	+	+	–	–?	+	+
Thymelicus lineola	(+)	+	+2? −1?	+2	+	+2 −1
Thymelicus acteon	(+)	+	+	+	+	+
Hesperia comma	+	+	+	–	+	?
Ochlodes venata	+	+	+2? −1, 3	+1, 2, 3	+	+1, 2, 3?
Erynnis tages	+	+	−1, 2, 3	+1, 2?	+	+
Pyrgus malvae	+	+	−1, 3	+3?	+1, 2	+
Papilionidae *Papilio machaon*	+•	+	–	–?	+	+
Pieridae *Leptidea sinapis*	+	+	−2, 3, 4, 5	+6?	+1–5 –6	+
Gonepteryx rhamni	+	+	−1, 2?	–?	+1, 2 –3	+1 –2
Pieris brassicae	+	+	−1, 2	?	+	+
Pieris rapae	+	+	−1, 2, 3, 5	?	+	+
Pieris napi	+	+	+5, 7 −1, 2, 3, 4, 6	−1, 2, 6	+	+

G	H	I
−1, 2L	–	[1] *Brachypodium sylvaticum*; [2] *Molinia caerulea* (920 m); [3] *Calamagrostis canescens* (?)
+	−/+	[1] *Holcus lanatus* (610 m)
?	0	[1] *Holcus mollis* (580 m); [2] *Dactylis glomerata*
+U −L	0	[1] *Brachypodium pinnatum* (380 m, but probably < 200 m)
–	–	[1]*Festuca ovina** (1310 m)
−1, 3L	0	[1] *B. sylvaticum* (380 m); [2] *D. glomerata*; [3] *M. caerulea*
–	0	[1] *Lotus corniculatus** (915 m); [2] *L. uliginosus**; [3] *Hippocrepis comosa**
+1U, 2 −1L, 3, 4	0	[1] *Potentilla erecta**; [2] *P. reptans**; [3] *Fragaria vesca* (640 m); [4] *Agrimonia eupatoria**
–	–	[1] *Peucedanum palustre* (probably < 50 m)
–	–	[1] *Lathyrus pratensis** (460 m); [2] *L, montanus*; [3] *L. uliginosus**; [4] *L, corniculatus**; [5] *Vicia cracca*; [6] *Shaded habitats*
?	0	[1] *Rhamnus catharticus*; [2] *Frangula alnus*; [3] woodland hibernation sites
I	0	[1] *Brassica oleracea**; [2] *Brassica spp.**; [3] *Tropaeolum majus*; [4] *Reseda lutea*
+	0	[1] *B. oleracea**; [2] *Alliaria petiolata* (350 m); [3] *Sisymbrium officinale**; [4] *R. lutea*; [5] *T. majus*
+3, 4, 5, 6 −1, 2, 7	0	[1] *Cardamine pratensis** (1000 m); [2] *C. amara**; [3] *C. hirsuta**; [4] *S. officinale**; [5] *A. petiolata*; [6] *Rorippa n–aquaticum**; [7] *Sinapis arvensis**

Table 5.3 (cont.)

	A	B	C	D	E	F
Anthocharis cardamines	+	+	+5, 7 −1, 2, 3, 4	+7 −1, 2	+	+
Lycaenidae *Callophrys rubi*	+	+	+2, 3, 4 −1, 5	+2, 3	+	+1, 5, 7 −2, 3, 4
Thecla betulae	+	+	+1 −2?	+2? −1?	+	+
Quercusia quercus	+	+	+3? −1, 2, 4?	+?	−	+2 −1
Satyrium w-album	+	+	+3 −1, 2, 4	+?	−	+
Satyrium pruni	(+)•	+	+1 −2	+2? −1?	+1 −2	+
Lycaena phlaeas	+	+	+2 −1?	+2	+	+1 −2
Lycaena dispar	+•	+	−	−?	+	+
Cupido minimus	+•	+	−	?	+	+
Plebejus argus	+	+	+1, 5, 7 −2, 3, 6, 8	+1, 5, 7, 8?	+	+2, 3, 4 −5, 6, 7
Aricia agestis	+	+	+2 −1, 4	+4	+	+
Aricia artaxerxes	−	−	+2 −1, 3	+3	+	+
Polyommatus icarus	+	+	+4 −1, 2, 3	+1?	+	+
Lysandra coridon	(+)•	+	−	+2	+	+

G	H	I
+3, 4 −1, 2, 5, 7	−/+	[1] *C. pratensis** (1000 m); [2] *C. amara**; [3] *A. petiolata*; [4] *S. officinale**; [5] *S. arvensis*; [6] *Barbarea vulgaris**; [7] *Arabis hirsuta**; [8] *Hesperis matronalis*
+2U, 3 −1, 2L, 5	0	[1] *Helianthenum nummularium**; [2] *Vaccinium myrtillus* (1300 m); [3] *Ulex europaeus;* [4] *Sarothamnus scoparius;* [5] *L. corniculatus**; [6] *Genista tinctoria;* [7] *Thelycrania sanguinea*
?	−	[1] *Prunus spinosa* (415 m); [2] Tree crown territories
?	0	[1] *Quercus petraea* (460 m); [2] *Q. robur* (< 300 m but 460 m on Dartmoor); [3] *Ant symbiont;* [4] Tree crown territories
−	−	[1] *Ulmus procera;* [2] *U. glabra* (535 m, seedlings not > 300 m); [3] Dutch elm disease vector; [4] Tree crown territories
?	0	[1] *Prunus spinosa* (415 m); [2] *mature shrubs of hostplant and tree crown territories*
+2 −1	0	[1] *Rumex acetosa** (1240 m); [2] *R. acetosella** (1050 m)
−	E, I	[1] *Rumex hydrolapathum**
−	−	[1] *Anthyllis vulneraria**
−2–7 +1	−	[1] *Ulex spp.* (640m); [2] *L. corniculatus**; [3] *H. nummularium**; [4] *H. canum**; [5] *Calluna vulgaris** (1100 m); [6] *Erica tetralix**; [7] *E. cinerea**; [8] Ant symbiont habitat
−1	0	[1] *H. nummularium** (660 m); [2] *Geranium molle**; [3] *Erodium cicutarium**; [4] Ant symbiont habitat
−1	0	[1] *H. nummularium** (660m); [2] *Geranium sanguineum;* [3] Ant symbiont habitat
+3, 4 −1, 2	0	[1] *L. corniculatus** (915 m); [2] *L. uliginosus**; [3] *Trifolium repens**; [4] *T. dubium**
−?	0	[1] *H. comosa**; [2] Ant symbiont habitat

Table 5.3 (cont.)

	A	B	C	D	E	F
Lysandra bellargus	(+)•	+	–	+2	+	+
Celastrina argiolus	+	+	+	+3 –1, 2?	+	+
Maculinea arion	(+)•	+	+	+?	+	+
Hamearis lucina	+	+	–	?	+	+
Nymphalidae *Ladoga camilla*	(+)	+	–1?	+1? –2	+1 –2	–?
Apatura iris	(+)	+	–	+?	+1, 2, 3 –4	+1–3
Aglais urticae	+	+	–1?	+1? 2	+1 –2	+
Nymphalis polychloros	(+)	+	–	+?	–	+
Inachis io	+	+	–1?	+1? 2	+1 –2	+
Polygonia c-album	+	+	–2, 3, 5	+1, 2, 3 –7?	+1, 4, 5, 6 –7	+1, 2, 3, 5
Boloria selene	+	+	–2	+1? 2	+2	+
Boloria euphrosyne	+	+	–2	+1? 2	+2	+
Argynnis adippe	+	+	?	+1?	+?	+
Argynnis aglaja	+	+	–1, 2? 3, 4	+2? 3 –1	+1, 3	+
Argynnis paphia	+	+	?	+1, 2, 3	+1, 2? –3	+

G	H	I
–?	–	[1] *H. comosa**; [2] Ant symbiont habitat
+3 –1?	0/+	[1] *Ilex aquifolium* (550 m); [2] *Hedera helix* (610 m); [3] *U. europaeus*
–	E, I	[1] *Thymus praecox** (1200 m); [2] *Myrmica ants,* especially *M. sabuleti*
–	–	[1] *Primula veris* (750 m); [2] *P. vulgaris* (760 m)
?	–/+	[1] *Lonicera periclymenum* (610 m); [2] Shaded habitats for early stadia
+ ?4	–	[1] *Salix caprea;* [2] *S. cinerea;* [3] *S. fragilis;* [4] Tree crown territories; *Salix* spp. up to 880 m)
+	0	[1] *Urtica dioica* (840 m); [2] woodland hibernation sites
–	E?	[1] *U. procera;* [2] *U. glabra* (535 m); [3] woodland hibernation sites
+	–/+	[1] *U. dioica* (840 m); [2] woodland hibernation sites
+1, 5 –2, 3, 4	–/+	[1] *U. dioica* (840 m); [2] *U. procera;* [3] *U. glabra;* [4] *Humulus lupulus;* [5] *Salix* spp.; [6] *Ribes spp.;* [7] woodland hibernation sites
–	–	[1] *Viola riviniana* (1020 m); [2] *V. palustris**
–	–	[1] *V. riviniana* (1020 m); [2] *V. palustris** (rarely used)
–	–	[1] *V riviniana* (1020 m); [2] *V. odorata*
–	–	[1] *V. hirta* (600 m); [2] *V. riviniana;* [3] *V. palustris**; [4] *V. canina*
–	–	[1] *V. riviniana* (1020 m); [2] *V. odorata;* [3] tree trunk egglaying sites

Table 5.3 (cont.)

	A	B	C	D	E	F
Eurodryas aurinia	+	+	–	–?	+	–(pH > 7.5)
Melitaea cinxia	(+)	+	+1	?	+	+
Mellicta athalia	+	+	+2 –1, 3?	?	+	+
Pararge aegeria	+	+	+5, 6, 7 –2? 3	+2, 5? 7	+	+
Lasiommata megera	+	+	+7 –1, 3, 5	+1, 5, 7	+	+1, 2, 3, 5 –4, 6
Erebia epiphron	–	–	?	+	+	–
Erebia aethiops	–	–	+2? –1	+1	+	+
Melanargia galathea	+	+	+	–2	+	+1
Hipparchia semele	+	+	+	+4 –2	+	+3
Pyronia tithonus	+	+	+1? 2, 3	–2	+	+3 –1
Maniola jurtina	+	+	+1	+3	+	+1, 3, 4 –2
Aphantopus hyperantus	+	+	+	+	+	+
Coenonympha pamphilus	+	+	+	–2	+	+1, 3
Coenonympha tullia	–	–	–	+?	+	–

G	H	I
–	–	[1] *Succisa pratensis** (970 m)
+1 –2	0	[1] *Plantago lanceolata** (790 m); [2] *P. coronopus**
+2 –1, 3	–	[1] *Melampyrum pratensis* (915 m); [2] *P. lanceolata**; [3] *Veronica chamaedrys*
+3, 5 –2, 6	–/+	[1] *D. glomerata*; [2] *B. sylvaticum* (380 m); [3] *H. lanatus*; [4] *Bromus ramosus*; [5] *Agrostis stolonifera**; [6] *Festuca rubra**; [7] *shaded habitat and wood edge territories*
+3, 5U –1, 5L	–/+	[1] *B. sylvaticum* (380 m); [2] *D. glomerata*; [3] *H. lanatus* (610 m); [4] *Agrostis tenuis*; [5] *B. pinnatum*; [6] *Deschampsia flexuosa*; [7] bare ground territories
+	0	[1] *Nardus stricta** (1010 m)
–1L	0	[1] *M. caerulea* (920 m); [2] *Sesleria caerulea*
–	–	[1] *F. rubra** (1050 m); [2] *F. ovina** (1310 m)
–2, 3	0	[1] *Agrostis curtisii*; [2] *F. ovina** (1310 m); [3] *F. rubra**; [4] Bare ground territories
–2, 3	–/+	[1] *A. capillaris* (1040 m); [2] *F. ovina**; [3] *F. rubra**
–1, 3	0	[1] *Poa pratensis** (950 m); [2] *A. tenuis*; [3] *B. sylvaticum*; [4] *Helichotrichon pubescens*
+	0	[1] *A. stolonifera* (750 m)
–	0	[1] *P. pratensis** (950 m); [2] *F. ovina**; [3] *F. rubra**
+U –L	–	[1] *Eriophorum vaginatum** (940 m)

Table 5.3 (cont.)

A Analogue indicator of climatic change ($\Delta T_{2x} CO_2$) for highland Britain; analogue region, lowland Britain.
B Analogue indicator of climatic change ($\Delta T_{2x} CO_2$) for lowland Britain; analogue region, north-west France, north of a line from Bordeaux to Stuttgart in Germany.
+, increases and –, decreases in population size, distributions and geographical ranges; (+) marginal conditions based on lower estimate of climatic warming of 0.7°C by 2028 AD; •, major obstacle to expansion presented by hostplant geography.
C Drought. Resilience of hostplants to drought conditions: –, susceptible; +, tolerant.
D Fire. Resilience of hostplants to firing of the vegetation: –, vulnerable; +, resilient or benefiting.

C and D identify hostplant and habitat components known to be vulnerable and tolerant of a succession of seasonal droughts and of low temperature surface fires. Persistent long–term severe drought conditions and intense fires (which would rarely occur in deciduous woodland habitats) systematically drive plant communities to bare gorund conditions, suitable for only a few British butterflies (e.g., *Lasiommata megera*). Plants of pioneer seres may not benefit immediately from the exposure of bare ground although they usually readily seed in such zones (e.g., *Lotus corniculatus* can establish itself on bare ground but is intolerant of soil moisture deficiency). Successional restarts may have longer–term benefits for butterfly populations with the resumption of higher soil water levels and growth. Climax deciduous woodland is not easily burnt and thus only severe fire or persistent long–term drought damage to woodlands may have serious consequences for high woodland specialists. Low temperature surface fires may be beneficial to most woodland butterflies by opening spaces required by the majority of species dependent on herbaceous hostplants and by re–establishing lighter shade for others (e.g., *L. sinapis*), whilst having little if any effect on the high canopy. The timing of droughts and fires in insect life–cycles is also crucial. It is conceivable that surface fires may benefit the hostplant but have a deleterious effect on the butterfly. Each species requires precise life–history modelling for potential 'interference' effects (e.g., the habitat of *Maculinea arion* and its ant symbiont may benefit from winter fires when they are underground, but summer fires could exacerbate drought effects on butterfly and ant food supply and butterfly oviposition sites) (J.A. Thomas, pers. comm.).

E Windthrow and windsnap. Opportunites presented by tree fall and canopy opening: –, loss of arboreal hostplants or essential climax habitat components; +, open spaces for herbs and shrubs; *, against species in hostplant list indicates species that are not typical components of ancient woods (*sensu* Peterken & Game, 1984), but which also benefit from scrub clearance on sites.

F pH increase. Potential effects of rising pH levels on hostplants: –, detrimental and +, beneficial to plant. Note: plant species are often affected by specific pH thresholds.

G Hostplant trend. Current status in Britain: –, decreasing; +, increasing; L, lowland Britain; U, Highland Britain.

H Butterfly species trend. Summary of current status: –, decreasing and +, increasing distributions and ranges; –/+, past reduction in range subsequently followed by more recent increase (the ranges and distribution areas of a number of species have increased only recently in response to the sequence of warm years, see section 4.1A); 0, status of range much unchanged, but populations probably decreasing since 1945; E, extinct; I, introduced.

I Hostplants and essential habitat components. Full generic terms given only at first mention. The upper altitudinal limit for the main hostplant (or substitute, if available) is given where known in metres above ordnance datum. *, species which are not typical components of ancient woods.

Numbers in the body of the table refer to species and habitat components in the hostplant list: Note, the data on drought, fire and pH are largely deficient and require surveys over a wide geographical area, especially as hostplant species may benefit in one ecological context but not in another. Information in the table for these variables must be interpreted cautiously as indicators of what, on balance, may happen, not what will necessarily occur. (Data from Thomas & Merrett, 1980; Peterken & Game, 1984; Grime *et al.*, 1988; Clapham *et al.*, 1989; Kelly *et al.*, 1989; Warrick & Barrow, 1991; Dennis, 1992; J.A. Thomas, pers. comm.; M.S. Warren, pers. comm.)

---

during the summers of 1976 and 1990 (Brugge, 1991). The high human population density in southern Britain increases this risk. Fortunately, deciduous trees do not burn easily (Rackham, 1986) and fires within conifer forests will largely benefit butterfly populations. However, the ΔT_{2x} introduces the possible interaction of pathogens with pollution, fires and climatic stress (Dohmen, 1988); waldsterben may be caused by greenhouse gases (e.g., tropospheric ozone), by those gases not major constituents of enhanced warming (e.g., SO_2), as well as by drought, and stressed trees are vulnerable to disease (Woodin, 1989). The effect of disease will not be restricted to southern Britain. Climatic warming may result in the northward spread of insect parasitoids and hostplant pests. There is the prospect that Dutch elm disease will counter any extension in the range of *Satyrium w-album*, especially as the more northern *Ulmus glabra*, unlike *Ulmus procera*, does not produce sucker regenerative growth (Grime *et al.*, 1988).

As over much of northern Britain annual precipitation exceeds annual evapotranspiration, drier conditions may benefit rather than limit many northern butterfly populations. Moreover, there are species whose hostplant and habitat components are resilient to drought (e.g., *Hesperia comma* on *Festuca ovina*) and fire (e.g., *Callophrys rubi* on *Ulex europaeus* and *Vaccinium myrtillus*), as long as these agents are not overly prolonged, too widespread or too severe (Table 5.3 C, D). Some butterfly species may actually benefit from the consequences of drought, fire and disease of habitat components, for example by the opening-up of the woodland canopy (e.g., *Gonepteryx rhamni* on *Frangula alnus*). Low temperature surface fires effectively encourage plants of early seres, benefiting the majority of British butterfly species which are dependent on herbs and shrubs. Because annual soil field capacity is often attained and exceeded in north and west Britain, prolonged and severe drought is less likely to be injurous to the majority of butterfly populations in the north-west than in the south-east. This is particularly so in upland margins, where surface moisture currently suppresses maximum temperatures in ground microhabitats (Fig. 4.4; see below), thereby adversely affecting the development and survival of early stages.

Very little can be said about the regional distribution of other climatic variables such as windiness, storminess, precipitation intensity and cloudiness. Currently, there is no reason to doubt (i) that the geographical pattern in these variables will be largely maintained, and (ii) that storm systems (i.e., depressions, convectional supercell storms and tornadoes) will continue to interrupt more stable conditions and avert the kind of long-term stability characteristic of the Mediterranean zone. However, it is possible that the frequency of some of these events (i.e., supercell storms, tornadoes; Elsom & Meaden, 1985) may increase with climatic warming. Windthrow and canopy windsnap from storms (Table 5.3E) have much the same positive and negative effects on butterfly habitats as drought, disease and fire. Although windthrow affects a relatively small portion of the total land area (c. 8 per cent equally divided among conifer and deciduous woodland), forest biotopes, or at least woodland clearings, represent an extremely important component of semi-natural habitats available to butterflies (Fuller & Warren, 1991; Warren & Key, 1991). The intense rainfall associated with depressions and supercell storms may contribute mainly by offsetting the effects of drought, but can have negative effects by washing eggs off hostplants and drowning larvae and adults.

One indirect effect of climatic warming with potentially serious consequences for British environments is rising sea levels. Global mean sea levels have risen by 2.4 ± 0.9 mm a^{-1} in the last 50 years (Meier, 1990). The IPCC best estimates for a business-as-usual scenario are + 18 ± 10 cm at 2030 AD and + 44 ± 23 cm at 2070 AD (Warrick & Oerlemans, 1990). Northern Britain is still responding to glacio-isostatic uplift (+ 1.92 mm a^{-1} in central Scotland; Tooley, 1991). However, southern Britain is subject to tectonic downwarping (– 1.0 to – 1.5 mm a^{-1}; Tooley, 1991). Moreover, drainage of coastal areas in lowland Britain has led to sediment consolidation and peat oxidation, both of which have resulted in the lowering of ground altitudes. In part of the western Fenlands, ground altitudes have been lowered from + 1.6 m to – 2.3 m OD between 1848 and 1978 (30.5 mm a^{-1} on average) (Hutchinson in Tooley, 1991). Thus, any sea level rises could significantly influence the ecology of the coastal fringe in the south, especially the Fenland, and accelerate coastal erosion and marine transgressions (see Fig. 5.3a). Few butteffly habitats are vulnerable to sea level rise in southern Britain and the Fenland has a low butterfly diversity (Fig. 2.8). However, a relative sea level rise could threaten the hostplant habitat of one British butterfly restricted to coastal wetlands on the Norfolk Broads, *Papilio machaon britannicus* which is monophagous on *Peucedanum palustre*. There is the prospect that with increased sea level, salinization of ground and surface water could be aggravated by decreases in precipitation, by increases in evapotranspiration and by water extraction from aquifers; therefore from reduced soil water throughflow and runoff (Scott *et al.*, 1990).

The response of butterflies to the effects of climatic warming depends on their adaptability to changing selection pressures, their capacity to adopt new habitats and on their ability to colonize them. Highland and lowland Britain present regionally distinct opportunities for novel habitats, adaptations and dispersal. In the context of habitat changes, the ability of individuals to locate or 'track' appropriate habitats may become essential for survival. At least two factors enhance migration and colonization. Firstly, warmer conditions stimulate adult activity and may promote increasingly long-distance movements; deterioration of existing habitats may have the same effect (Baker, 1984). A number of nymphalids and pierids have the capacity to cover large distances, but observations also exist of individuals of species with 'closed populations' moving > 1 km from the nearest colonies in warm weather (Horton, 1977; Warren, 1984a; Shreeve & Dennis, 1991). Secondly, the potential for rapid spread is influenced by the widespread distribution of hostplants for British butterflies; all but four (for five butterfly species) extend to northern Scotland and most occur at high altitudes (Fig. 5.5; Table 5.3). Countering these points is the observation that butterflies often require resources additional to hostplants; hostplant quality, symbionts, microclimates and habitat structure are often limiting. Within the British butterfly fauna there is a tendency for a number of species to have become restricted to fewer habitat types or seral stages during the Holocene forest development and the subsequent alleged climatic deterioration (Dennis, 1977; 1992; Fig. 4.3). As the most stenoecious species are found in the south (e.g., *Lysandra bellargus; Maculinea arion; Satyrium pruni*; see Table 5.2), and habitat fragmentation by intensive human exploitation is also more severe in the south, this could result in differences in colonization potential between highland and lowland Britain. Furthermore, in highland Britain, the relatively high species richness in the vicinity of uplands facilitates short-distance migrations upslope as habitats change with climatic warming.

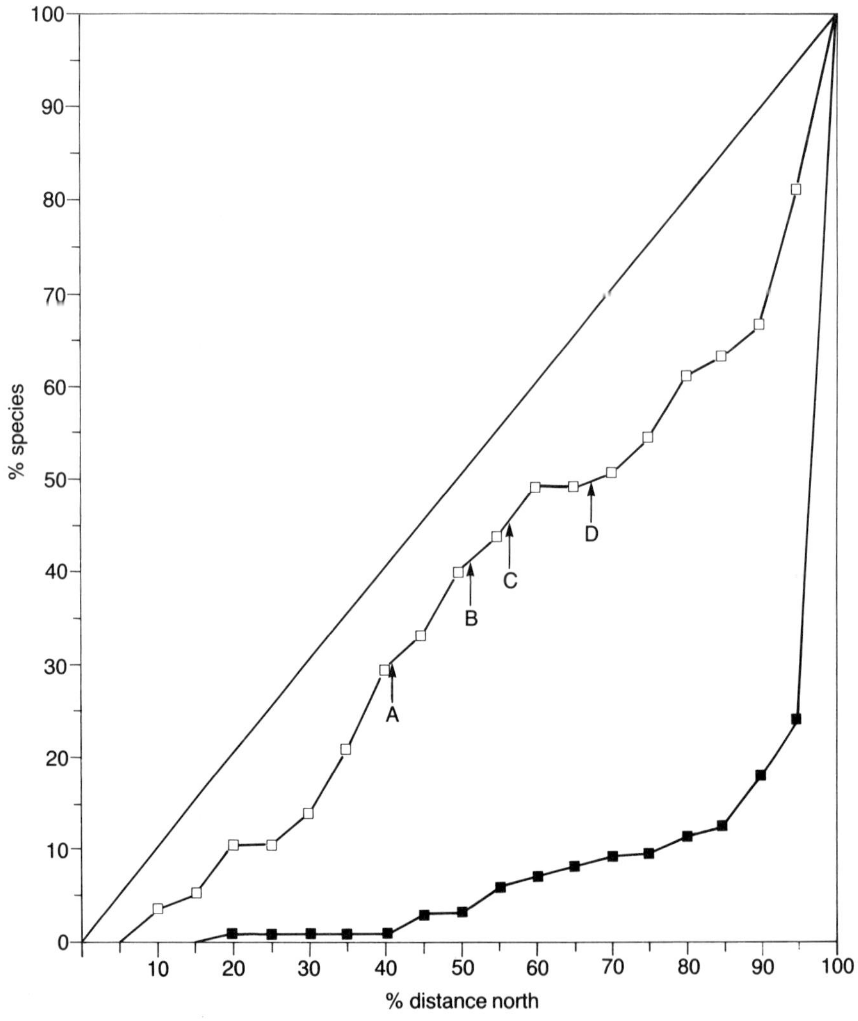

5.5 Lorenz (cumulative frequency) curves comparing the range limit of butterfly and hostplant species in Britain. Open square, butterflies, N = 57; black square, hostplants, N = 95. The diagonal line describes an even decline in the range of species. Butterfly species decline more evenly than their hostplants, most of which extend to the Shetland Isles. Only (A) *Peucedanum palustre* (*Papilio machaon*), (B) *Hippocrepis comosa* (*Lysandra bellargus; L. coridon*), (C) *Frangula alnus* (*Gonepteryx rhamni*) and (D) *Brachypodium pinnatum* (*Thymelicus acteon*) would limit the range extensions of butterflies dependent on them into northern Scotland. Other plant species that have southern limits to their ranges are subsidiary hostplants. Note: hostplant quality is not considered (see text). (From Dennis & Shreeve, 1991 (by courtesy of *Biological Conservation* and Elsevier Science Publications Ltd.)

Adoption of novel habitats during climatic warming may reduce the need for long-distance migration. Because of the projected regional distinctions in evapotranspiration : precipitation ratio, the type of habitats that may become suitable could differ substantially between northern and southern Britain. Survival in lowland Britain may require adjustments to the use of cooler, damper habitat types – the adoption of shaded open woodlands (e.g., *Maniola jurtina*), of taller sward (e.g., *Lysandra bellargus*) and of habitats with northern aspects (e.g., *Lysandra coridon*) (Fig. 5.3a; Dennis & Shreeve, 1991). As *Hamearis lucina* already prefers north-facing slopes, fewer options of escape are open to it. These changes may be imposed on species by the effect of drought on hostplants either directly or through competition with more drought resistant plants, but may also be induced by the spread of hostplants and symbionts to cooler habitats and the ability of different butterfly stadia to function in cooler habitats as mean temperatures increase. There is the possibility of changes in hostplant use (i.e., different hostplant species, broadening of hostplant use and of hostplant quality) as butterfly species occupy habitats with different plant associations and larval development times become less limiting on 'novel' hostplants (e.g., *Celastrina argiolus*). This scenario would reverse adaptations to limited hostplant use and warmer microclimates which are suggested to have evolved with Holocene forest climax development and the advent of modern climatic absolutes (Fig. 4.3). These changes would be of great significance in expanding the range of butterflies currently restricted to steep, south-facing slopes in southern Britain (e.g., *Melitaea cinxia*) and to short turf habitats with high maximum temperatures (e.g., *Hesperia comma*) (Thomas & Simcox, 1982; Thomas *et al.*, 1986).

In contrast to southern Britain, much of northern Britain is upland > 400 m, and currently too cold, wet and windy for most butterfly species. A rise in mean temperatures is likely to be beneficial not only to butterflies directly but also to their hostplants. Warmer conditions lead to drier aerobic soil horizons, a larger soil organic community, faster organic decomposition rates, more rapid nutrient cycling and higher soil pH levels. The rate of plant litter breakdown may be increased by a third if temperature is raised from 14° to 17°C. When soils dry out, their surface temperature can become much higher than air temperature and the rate of organic breakdown exceeds that from an equivalent ΔT_{2x} (Hendricksen, 1985). The soil surface conditions may become increasingly suitable for those species currently restricted to the coastal lowlands which require dry substrates for early stages (e.g., *Hipparchia semele*). Local climates and microclimates may be enhanced further with the invasion of *Calluna* moorland and ombrogenous mires by deciduous trees and scrub such as birch, in the reversal of podsolization and palludification and the formation of windbreaks and sheltered habitats. These transformations depend largely on grazing pressure, but, where they occur, would also encourage the spread of herbaceous hostplants, creating opportunities at higher elevations for a number of butterflies which are found on lower slopes and valley sides (e.g., *Boloria selene* on *Viola riviniana; Callophrys rubi on Vaccinium myrtillus*) (see Miles, 1981; Hester *et al.*, 1991a, b). Not all butterfly species may necessarily benefit from the indirect effects on hostplants in northern Britain. Hostplant populations of those butterfly species restricted to mires and acid soils may diminish. In particular, *Coenonympha tullia*, which is dependent in Britain on *Eriophorum vaginatum* and oligotrophic mires (raised and blanket bogs) is particularly vulnerable to a decrease in the evapotranspiration : precipitation ratio.

Despite changes in absolutes, climatic gradients in Britain would persist. In such circumstances, species whose pattern of voltinism varies clinally in relation to current climatic conditions would face adjustments, though, as photoperiods remain constant, responses to rising temperatures may be rapid and within the capacity of species (Lees, 1962b; Lees & Archer, 1980; Pullin, 1986b). It is also possible that current wing pattern phenotypes of regionally variable species would be maladaptive (see section 3.3B), but how rapidly populations would be able to change is a matter of conjecture. The increased morphological variation among populations of satyrids in north-western Scotland (e.g., *Maniola jurtina, Coenonympha tullia*; Dennis, 1977; Dennis *et al.*, 1986), the degree of plasticity and polygenic inheritance demonstrated for both physiological and morphological attributes and the amount of genetic variance in populations (Majerus, 1979; Lees & Archer, 1980; Brakefield, 1984; Pullin, 1986a) indicate that very rapid adaptive responses within a few generations may be possible and in keeping with climatic change (see Dennis, 1977: 173). Britain is also diverse in geology and local climate; this may provide a geographical mosaic of 'buffers' to change for different organisms in the face of a significant shift in climate.

A summary of projected contrasts between highland and lowland Britain for different aspects of butterfly biology during climatic warming is given in Table 5.4. It is necessary to emphasize at this point that these are expected average changes over the long term under the full influence of equivalent ΔT_{2x}, and that variation in annual population sizes and migration will remain considerable. These changes also depend fundamentally on other factors, namely: (i) the pattern and intensity of land-use changes (i.e., cultivation practices; forestry; grazing pressures; urbanization), and (ii) the stochastics of migrules for plants and butterflies, which respectively affect plant associations and butterfly colonization. Projections for groups of organisms linked to climatic warming scenarios cannot ignore the current status of those organisms and their hostplants. The ranges of 35 (20/57) per cent of British resident species continue to decline and it is clear that the populations of many more have become less dense and increasingly sparse since 1945 (Warren, 1992b; Table 5.3H). Sparser populations may become a prominent feature in southern Britain as droughts become more

Table 5.4 Summary of changes in butterfly biology projected for the British fauna with ΔT_{2x} equiv CO_2

Butterfly biology	*Highland Britain*	*Lowland Britain*
Migration	Assisted by higher T_a	Assisted by higher T_a, but potential for desiccation stress. Enforced by drought, nectar failure and hostplant dieback.
	Short distance – uphill	Short distance – to cool habitats in longer swards, woods and northern slopes
	Long distance – habitat stepping stones emerging	Long distance linear flight paths across country with reduction in quality of semi-natural habitats (depending on changes in agriculture)
Colonization	Extensive on new, dry warmer habitats	Patchy on fragments of semi-natural habitats exposed with reduction in cultivated acreage, and of cooler habitats
	New colonists from lowland Britain	New colonists from continent on southern shores and man-made introductions
Populations	Larger, denser more stable metapopulations	Smaller, less dense and volatile meta-populations
Extinctions	Loss of populations of species dependent on hostplants in mires (e.g., *Coenonympha tullia* on ombrogenous peat)	Potential for extensive losses of populations on sandy and calcareous short turf habitats, especially on south-facing slopes, and of species dependent on damp habitats. Possible (continuing) extinctions of species within Britain without appropriate intervention

Table 5.4 (cont.)

Butterfly biology	*Highland Britain*	*Lowland Britain*
Habitat changes of resident species	Successional habitats – those becoming increasingly drier and nutrient enriched, especially on south-facing slopes and with sheltered aspects (dependent on agricultural practices) Extension to higher altitudes and advances on to northern slopes	Shaded, longer swards, northerly aspects and river basins with cooler damper conditions adopted
	Minor adjustments in voltinism. Directional selection for more broods	
Adaptations	Morphological changes in wing patterns – increased variability in phenotypes with oscillation of annual conditions. See section 3.3B for modifications expected	Possible shifts to increased polyphagy, larger size and slower development times with survival of slower developing individuals with increased length in the development season, moisture availability permitting.
Species diversity	Increasing markedly with expansions in ranges of species in southern Britain and spread through upland areas of northern species (e.g., *Eurodryas aurinia*)	Potential for increases with colonists from the continent and permanent residence of some seasonal migrants. Possible losses among English residents with failure to adjust to combination of climate and landuse changes (e.g., *Eurodryas aurinia*)

frequent and intense (Dennis & Shreeve, 1991). It is significant that hostplants of many declining butterflies are also decreasing (Table 5.3G). However, the prospects are not entirely bleak; eight (14 per cent) of the resident fauna have expanded their ranges in recent years and this seems to be related, in part at least, to a period of warmer summers (section 4.1). With rising temperatures there are also increased opportunities for continental butterflies to colonize southern Britain (Bretherton, 1968; Dennis, 1992). Occasional immigrants (e.g., *Papilio machaon gorganus*), deliberate introductions (e.g., *Araschnia levana, Clossiana dia*) and adventives (e.g., *Lasiommata maera*) (Cribb, 1986; Emmet & Heath, 1989, with ubiquitous hostplants, could establish permanent populations. Common summer immigrants (e.g., *Vanessa atalanta, Cynthia cardui* and *Colias croceus*) may become permanent residents as winter conditions become warmer in southern Britain, and infrequent immigrants (e.g., *Lampides boeticus; Issoria Lathonia*) may become more regular visitors.

5.4 **Topics for future research**

Enhanced greenhouse conditions provide at least two stimuli for future research into the effects of climatic change on butterfly biology. Firstly, as outlined in the foregoing sections, such climatic changes could have dire consequences for many butterfly populations, as well as for wildlife generally. Research programmes are at least required to determine conservation priorities and policies. Secondly, the projected unprecedented shifts in climatic absolutes and rates of change present a unique experiment for research institutions with the foresight to plan for these changes. Not all the research need be long-term; as the ramifications of the predicted climatic changes for butterfly biology are extensive, numerous shorter-term research projects are required to solve specific problems. It would be particularly satisfying to see short-term research projects being properly integrated into regional synoptic programmes and for issues at different scales to be fully addressed in future planning. Limitations of space preclude a detailed inventory of potential research topics, but some important objectives are briefly discussed below.

The conservation of organisms is entirely frustrated by the lack of basic geographical and taxonomic data. There are large areas of the globe, including regions where it might be considered that good data exist (e.g., Europe), where mapping, taxonomic and systematic programmes are desperately needed. Mapping programmes need to address formats for integrating data, ultimately, on species at a global scale with the burgeoning environmental data base (i.e., geographical information systems).

The Butterfly Monitoring Scheme in Britain has proved to be an outstanding experiment (see Pollard, for references; Pollard, E. & Yates, T. J., *Monitoring Butterflies for Ecology and Conservation*. Chapman & Hall; in press). In its eighteenth year (1993 season), it is the only long-term wide-ranging synoptic programme for monitoring changes in butterfly populations. It has been responsible for an increasing number of extremely valuable insights into butterfly population dynamics as the data have accumulated. Substantial advantages would accrue with the extension of this programme over wider regions, across major ecological gradients (i.e., with latitude, longitude and altitude). Analyses of data across state boundaries could at least be co-ordinated for species with extensive ranges, such as the 'extent' element in Europe (see Fig. 4.8).

The projected climatic changes place a high priority on autecological research and more expansive programmes on 'community' structures. Conservation policies require a balanced approach to the association of species within particular sites. Some excellent work has been done on individual butterfly species (see Dennis, 1992 for references to J.A. Thomas and M.S. Warren in Britain). However, resources have been inadequate for solving issues in competing population dynamics models and for determining the adaptability of populations to the kind of serious environmental perturbations and systematic changes threatened by enhanced greenhouse warming. Holistic assessments for entire associations of species within habitats are simply unavailable. Much greater pooling of resources and co-operation is required than hitherto, if variability in response and over suitably long time periods is to be addressed. The opportunities for research are numerous; for instance, in regions which may be substantially influenced by climatic warming (e.g., mountain gradients) and in areas which may be regarded as analogues. Certainly, far more research is required on the tolerances of vulnerable elements (e.g., subalpine fauna), and on the effects of external agents, such as drought, fire and nutrient enrichment on butterfly 'communities' and their hostplants and symbionts.

Very little useful information currently exists on the adaptability of individual species (e.g., thermal tolerances, developmental limits and rates; hostplant and habitat use) to different conditions of weather and climate (chapters 1 and 2), fundamentally on the genetic variability and phenotypic plasticity in populations (e.g., for phenology, life-history traits and variation). This information not only is essential to predict the future impact of climatic change, but will provide the means to reconstruct better evolutionary scenarios for individual traits (chapter 3) and for whole faunas (chapter 4).

Glossary of symbols

AD — approximate digestibility: (IG – F × 100/IG).

A_d — wind direction.

Adj R^2 — an R^2 statistic adjusted for the number of independent variables in the multiple regression equation and the number of cases:

$$\text{Adj } R^2 = R^2 - (q - 1/N - q)\,(1 - R^2)$$

where q is the number of independent variables and N is the number of cases.

A_v — wind velocity ($m.s^{-1}$).

b — unstandardized regression coefficient.

B — biomass gained (assimilated food used for growth): (B = (IG – F) – M).

Beta — standardized regression coefficient.

BP — ^{14}C years before present 1950.

CR — consumption rate: ($IG.d^{-1}$).

DF — degrees of freedom.

ECD — efficiency of conversion of digested food: (B × 100/IG – F).

ECI — efficiency of conversion of ingested food: (B × 100/I).

F — undigested food + excretory products (feces).

FAR — T_b for voluntary flight activity range.

GR — growth rate: ($B.d^{-1}$).

h^2 — heritability; $h^2 = V_A/V_P$ where V_A is the genetic variance resulting from additive effects and V_P is the phenotypic variance.

H_r — relative humidity.

I — summation of mean weekly counts for the Butterfly Monitoring Scheme (BMS) transect (see Pollard *et al.*, 1986).

IG — food ingested.

K — carrying capacity determined by resources (see p. 43).

M	assimilated food metabolized.
NAR	relative nitrogen accumulation rate: (mg N_2 gained.d^{-1}.g^{-1} average larval biomass).
NCR	relative nitrogen consumption rate: (mg N_2 ingested d^{-1}.g^{-1} average larval biomass).
NUE	nitrogen utilization efficiency: (mg N_2 gained/mg N_2 ingested × 100).
OAR	T_b for optimal flight acivity range.
p	probability level.
r	intrinsic rate of increase in populations (see p. 43).
R	solar radiation.
R^2	multiple coefficient of determination.
r_a	yearly rate of increase in populations.
RGR	relative growth rate: (B.d^{-1}.mg^{-1} average larval biomass).
r_p	correlation coefficient (Pearson) for interval data.
r_s	correlation coefficient (Spearman) for ordinal data.
SEb	standard error for the unstandardized regression coefficient.
SE_y	standard error of estimate.
T_a	ambient air temperature (°C), often shade temperature at 1.5 m.
T_b	body temperature (°C) usually taken in thorax by thermistor probe or thermocouple, by 'grab & stab' or by tethering.
T_{bask}	T_b during maximal basking posture.
T_{bg}	black globe temperature (°C) – using a hollow, 0.15 m diameter copper sphere, painted flat black, with a mercury thermometer at its centre (see Pivnick & McNeil, 1987).
T_c	Supercooling point (°C).
T_{ex}	$T_b - T_a$.
T_0max	maximum developmental threshold temperature (°C).
T_0min	minimum developmental threshold temperature (°C).
T_s	temperature of substrate (°C), usually perch site of insect.
T_{sc}	Stevenson screen temperature (°C).
W_b	body water (hemolymph osmolality mOsm).

SI units are used throughout where possible. For the computation form of statistical parameters, see Nie *et al.*, 1975). All indices of larval growth are in dry weights (see Scriber & Slansky, 1981).

Bibliography

Abramsky, Z. & Rosenzweig, M.L., 1984. Tilman's predicted productivity-diversity relationship shown by desert rodents. *Nature*, 309: 150–151.

Ackery, P.R., 1984. Systematic and Faunistic studies on butterflies. In R.I. Vane-Wright & P.R. Ackery (Eds), *The Biology of Butterflies*. Royal Entomological Symposium No. 11. pp. 9–21. London: Academic Press.

Ackery, P.R., 1988. Hostplants and classification: nymphalid butterflies. *Biological Journal of the Linnean Society*, 33: 95–203.

Ackery, P.R., 1991. Hostplant utilization by African and Australian butterflies. *Biological Journal of the Linnean Society*, 44: 335–351.

Ackery, P.R. & Vane-Wright, R.I., 1984. *Milkweed Butterflies. Their Cladistics and Biology*. London: British Museum (Natural History).

Acock, B. & Allen, L.H. Jr., 1985. Crop responses to elevated carbon dioxide concentrations. In B.R. Strain & J.D. Cure (Eds), *Direct Effects of Increasing Carbon Dioxide Concentration on Vegetation*, 53–97. Washington DC: US Department of Energy.

Adam, D.P., 1988. *Palynology of two Upper Quaternary cores from Clear Lake, Lake County, California*. US Geological Survey Prof. Paper 1363.

Adams, J.M., 1989. Species diversity and productivity of trees. *Plants Today*, 2: 183–187.

Adams, J.M. & Woodward, F.I., 1989. Patterns in tree species richness as a test of the glacial hypothesis. *Nature*, 339: 699–701.

Adams, M.J., 1977. Trapped in a Columbian Sierra. *Geographical Magazine*, 49: 250–254.

Adams, P.A. & Heath, J.E. 1964. Temperature regulation in the sphinx moth, Celerio lineata. *Nature*, 201: 20–22.

Adler, P.H. & Pearson, D.L., 1982. Why do male butterflies visit mud puddles? *Canadian Journal of Zoology*, 60: 322–325.

Akey, D.H. & Kimball, B.A., 1989. Growth and development of the beet armyworm on cotton grown in an enriched carbon dioxide atmosphere. *Southwestern Entomologist*, 14: 255–260.

Alvarez, W., Alvarez, L.W., Asaro, F. & Michel, H.V., 1982. Current status of the impact theory for the terminal Cretaceous extinction. *Geological Society American Special Papers*, 190: 305–315.

Anderson, I., 1991. Global warming rings true. *New Scientist*, 21 September: 23.

Anderson, I. & Bowler, S., 1990. Oceans spring surprise on climate modellers. *New Scientist*, 10 March: 24.

Andrewartha, H.G. & Birch, L.C., 1954. *The Distribution and Abundance of Animals*. Chicago: University Press.

Antonova, E.M., 1980–84. Atlas of butterflies. In K.B. Gorodkova (Ed.), *Insect Faunal Areas of the European Part of the USSR*. (In Russian.) Leningrad: Zoological Academy of Sciences.

ApSimon, H.M., Kruse, M. & Bell, J.N.B., 1987. Ammonia emissions and their role in acid deposition. *Atmospheric Environment*, 21: 1939–1946.

Arms, K., Feeny, P. & Lederhouse, R.C., 1974. Sodium stimulus for puddling behaviour by tiger swallowtail butterflies, *Papilio glaucus*. *Science*, 185: 372–374.

Ashmore, M., 1990. The greenhouse gases. *Trends in Ecology and Evolution*, 5: 296–297.

Aspinall, R.J., 1987. The geography of the small copper butterfly *Lycaena phlaeas* (L.) in Northumberland. *Entomologist's Gazette*, 38: 159–165.

Aspinall, R.J., 1988. The changing geography of the common blue butterfly (*Polyommatus icarus*) in Northumberland. *Entomologist's Record and Journal of Variation*, 100: 1–6.

Atkins, A., 1989. The drinking and perching habits of skippers. *Antenna*, 13: 103–104.

Atkinson, T.C., Briffa, K.R. & Coope, G.R., 1987. Seasonal temperatures in Britain during the past 22,000 years, reconstructed using beetle remains. *Nature*, 325: 587–592.

Auerbach, M. & Shmida, A., 1987. Spatial scale and the determinants of plant species richness. *Trends in Ecology & Evolution*, 2: 238–242.

Baker, H.G. & Baker, I., 1973. Amino-acids in nectar and their evolutionary significance. *Nature*, 241: 543–545.

Baker, H.G. & Baker, I., 1975. Studies in nectar constitution and pollinator – plant coevolution. In L.E. Gilbert & P.H. Raven (Eds), *Coevolution of Animals and Plants*. Symposium V, First International Congress of Systematic and Evolutionary Biology. pp. 11–140. Austin, Texas: University of Texas Press.

Baker, R.R., 1978. *The Evolutionary Ecology of Animal Migration*. London: Hodder & Stoughton.

Baker, R.R., 1984. The dilemma: when and how to go or stay. In R.I. Vane-Wright & P.R. Ackery (Eds), *The Biology of Butterflies*. Royal Entomological Society Symposium, No. 11. pp. 279–296. London: Academic Press.

Bakken, G.S. & Gates, D.M., 1975. Heat-transfer analysis of animals: some implications for field ecology, physiology and evolution. In D.M. Gates & R.B. Schmerl (Eds), *Perspectives of Biophysical Ecology*. pp. 255–290. Berlin: Springer.

Bale, J.S., 1987. Insect cold hardiness: freezing and supercooling – an ecophysiological perspective. *Journal of Insect Physiology*, 33: 899–908.

Barbour, D., 1986. Expansion of range of the speckled wood *Pararge aegeria* L., in north-east Scotland. *Entomologist's Record and Journal of Variation*, 98: 98–105.

Barnett, T.P., 1990. Beware greenhouse confusion. *Nature*, 343: 696–697.

Barnola, J.M., Raynaud, D., Korotkevich, Y.S. & Lorius, C. 1987. Vostok ice core provides 160,000-year record of atmospheric CO_2. *Nature*, 329: 408–414.

Barnosky, C.W., 1989. A Late-Glacial and Post-Glacial pollen record from the Dingle Peninsula, County Kerry. *Proceedings of the Royal Irish Academy*, 88B: 23–37.

Barron, E.J., 1985. Explanations of the Tertiary Global cooling trend, *Palaeo*, 50: 45–61.

Barry, R.G. & Chorley, R.J., 1982. Atmosphere, *Weather and Climate*. London: Methuen.

Bartlein, P.J. & Prentice, I.C., 1989. Orbital variations, climate and paleoecology, *Trends in Ecology and Evolution*, 4: 195–199.

Barton, N.H., 1988. Speciation. In A.A. Myers & P.S. Giller (Eds), *Analytical Biogeography*. pp. 185–218. London: Chapman & Hall.

Barton, N.H. & Hewitt, G.M., 1989. Adaption, speciation and hybrid zones. *Nature*, 341: 497–502.

Battarbee, R.W., Flower, R.J., Stevenson, A.C. & Rippey, B., 1985. Lake acidification in Galloway: a palaeoecological test of competing hypotheses. *Nature*, 314: 350–352.

Battarbee, R.W., Anderson, N.J., Appleby, P.G., Flower, R.J., Fritz, S.C., Haworth, E.Y., Higgitt, S., Kreiser, A., Munro, M.A.R., Natkanski, J., Oldfield, F., Patrick, S.T., Raven, P.J., Richardson, N., Rippey, B. & Stevenson, A.C., 1988. *Lake Acidification in the United Kingdom*. Palaeoecology Research Unit. University College London. ENSIS Publishing, London.

Baust, J.G., 1982. Environmental triggers to cold-hardiness. *Comparative Biochemistry and Physiology*, 73A: 563–570.

Beaufoy, E.M., Beaufoy, S., Dowdeswell, H.W. & McWhirter, K.G., 1970. Evolutionary studies on *Maniola jurtina*: the southern English stabilization, 1961–68. *Heredity*, 25: 105–112.

Beck, S.D., 1980. *Insect Photoperiodism*. New York: Academic Press.

Becker, B., Kromer, B. & Trimhorn, P., 1991. A stable-isotope tree-ring time scale of the Late Glacial/Holocene boundary. *Nature*, 353: 647–649.

Begon, M., Harper, J.L. & Townsend, C.R., 1986. *Ecology: Individuals, Populations and Communities*. Oxford: Blackwell Scientific Publications.

Begon, M. & Mortimer, M., 1981. *Population Ecology. A Unified Study of Animals and Plants*. Oxford: Blackwell Scientific Publications.

Behrens, W., Hoffman, K.H., Kempa, S., Gassler, S. & Merkel-Wallner, G., 1983. Effects of diurnal thermoperiods and quickly oscillating temperatures on development and reproduction of crickets, *Gryllus bimaculatus*. *Oecologia*, 59: 279–287.

Beirne, B.P., 1943. The distribution and origin of the British Lepidoptera. *Proceedings of the Royal Irish Academy*, (B), 49: 27–59.

Beirne, B.P., 1947. The origin and history of the British Macrolepidoptera. *Transactions of the Royal Entomological Society*, 98: 275–372.

Beirne, B.P., 1955. Natural fluctuations in the abundance of British Lepidoptera. *Entomologist's Gazette*, 6: 21–52.

Bengtson, S.A., 1981. Does bird predation influence the spot-number variation in *Maniola jurtina*? *Biological Journal of the Linnean Society*, 15: 23–27.

Bennett, A.F., Dao, K.M. & Lenski, R.E., 1990. Rapid evolution in response to high-temperature selection. *Nature*, 346: 79–81.

Bennett, K.D., 1989. A provisional map of forest types for the British Isles 5000 years ago. *Journal of Quaternary Science*, 4: 141–144.

Bennett, K.D., 1990. Milankovitch cycles and their effects on species in ecological and evolutionary time. *Paleobiology*, 16: 11–21.

Bennett, K.D., Tzedakis, P.C. & Willis, K.J., 1991. Quaternary refugia of north European trees. *Journal of Biogeography*, 18: 103–115.

Benson, W.W., 1978. Resource partitioning in passion vine butterflies. *Evolution*, 32: 493–518.

Berenbaum, M.R., 1990. Coevolution between herbivorous insects and plants: tempo and orchestration. In F. Gilbert (Ed.), *Insect Life Cycles. Genetics, Evolution and Co-ordination*. pp. 87–99. London: Springer-Verlag.

Berger, A.L., 1978. Long term variations of caloric insolation resulting from the Earth's orbital elements. *Quaternary Research*, 9: 139–167.

Berger, A.L., 1984. Accuracy and frequency of the Earth's orbital elements during the Quaternary. In A. Berger, J. Imbrie, J. Hays, G. Kukla & B. Saltzman (Eds), *Milankovitch and Climate*. pp. 3–39. Dordrecht: Reidel.

Berner, R.A., Lasaga, A.C. & Garrels, R.M., 1983. The carbonate–silicate geochemical cycle and its effects on atmospheric carbon dioxide over the past 100 million years. *American Journal of Science*, 283: 641–683.

Berry, L. & Cloudsey-Thompson, J.L., 1960. Autumn temperatures in the Red Sea hills. *Nature*, 188: 843.

Berry, R.J., 1979. The Outer Hebrides: where genes and geography meet. *Proceedings of the Royal Society of Edinburgh*, 77B: 21–43.

Bertoldi, R., Rio, D. & Thunell, R., 1989. Pliocene–Pleistocene vegetational and climatic evolution of the south – central Mediterranean. *Palaeo*, 72: 263–275.

Bethoux, J.P., Gentili, B., Raunet, J. & Tailliez, D., 1990. Warming trend in the western Mediterranean deep water. *Nature*, 347: 660–662.

Biever, K.D. & Wilkinson, J.D., 1978. A stress-induced granulosis virus of *Pieris rapae*. *Environmental Entomology*, 7: 572–573.

Bink, F.A., 1972. Het onderzoek naar de grote vuurvlinder (*Lycaena dispar batava* Oberthur) in Nederland (Lep., Lycaenidae). *Entomologische Berichten*, 12: 225–239.

Bink, F.A., 1986. Acid stress in *Rumex hydralopathum* (Polygonaceae) and its influence on the phytophage *Lycaena dispar* (Lep., Lycaenidae). *Oecologia*, 70: 447–451.

Bink, F.A., 1989. *The Butterflies of the Future, Their Strategy*. Private publication; Box 46, 3956 ZR Leersum, The Netherlands. 7 pages.

Bink, F.A. & Siepel, H., 1986. Life History tactics and strategies in butterflies. *Proceedings of the 3rd European Congress of Entomology*, Amsterdam. pp. 409–412.

Birks, H.J.B., 1980. British trees and insects: a test of the time hypothesis over the last 13,000 years. *American Naturalist*, 115: 600–605.

Blackford, J.J. & Chambers, F.M., 1991. Proxy records of climate from blanket mires: evidence for a Dark Age (1400 BP) climatic deterioration in the British Isles. *The Holocene*, 1: 63–67.

Blau, W.S., 1981. Latitudinal variation in the life-histories of insects occupying disturbed habitats: a case study. In R. Denno & H. Dingle (Eds), *Insect Life History Patterns*. pp. 75–96. New York: Springer-Verlag.

Block, W., 1982. Cold hardiness in invertebrate poikilotherms. *Comparative Biochemistry & Physiology*, 73A: 581–593.
Block, W., 1990. Cold tolerance of insects and other arthropods. *Philosophical Transactions of the Royal Society of London*, B. 326: 613–633.
Blondel, J., 1987. From Biogeography to life history theory: a multithematic approach illustrated by the biogeography of vertebrates (birds). *Journal of Biogeography*, 14: 405–422.
Boggs, C.L., 1981. Nutritional and life-history determinants of resource allocation in holometabolous insects. *American Naturalist*, 117: 692–709.
Boggs, C.L., 1986. Reproductive strategies of female butterflies: variation in and constraints on fecundity. *Ecological Entomology*, 11: 7–15.
Boggs, C.L., Smiley, J.T. & Gilbert, L.E., 1981. Patterns of pollen exploitation by *Heliconius* butterflies. *Oecologia*, 48: 284–289.
Boggs, C.L. & Watt, W.B., 1981. Populations structure of pierid butterflies. IV. Genetic and physiological investment in offspring by male *Colias*. *Oecologia*, 50: 320–324.
Bolin, B., Doos, B.R., Jagaer, J. & Warrick, R.A., 1986. *The Greenhouse Effect, Climatic Change and Ecosystems*. SCOPE, 29. Chichester: John Wiley & Sons.
Boppré, M., 1984. Chemically mediated interactions between butterflies. In R.I. Vane-Wright & P.R. Ackery (Eds), *The Biology of Butterflies*. Royal Entomological Symposium No. 11. pp. 259–275. London: Academic Press.
Bowden, S.R., 1979. Subspecific variation in butterflies: adaptation and dissected polymorphism in *Pieris* (*Artogeia*) (Pieridae). *Journal of the Lepidopterists' Society*, 33: 77–111.
Bowden, S.R., 1985. Taxonomy for a variable butterfly, *Entomologist's Gazette*, 36: 85–90.
Bowers, M.D., Brown, I. & Wheye, D., 1985. Bird predation as a selective agent in a butterfly population. *Evolution*, 39: 93–103.
Bowers, M.D. & Wiernasz, D.C., 1979. Avian predation of the palatable butterfly, *Cercyonis pegala. Ecological Entomology*, 4: 205–209.
Bradley, R.S., 1985. *Quaternary Paleoclimatology: Methods of Paleoclimatic Reconstruction*. Boston: Allen & Unwin.
Brakefield, P.M., 1984. The ecological genetics of quantitative characters of *Maniola jurtina* and other butterflies. In R.I. Vane-Wright & P.R. Ackery (Eds), *The Biology of Butterflies*. Royal Entomological Society Symposium No. 11. pp. 167–190. London: Academic Press.
Brakefield, P.M., 1987a. Geographical variability in, and temperature effects on, the phenology of *Maniola jurtina and Pyronia tithonus* (Lep: Satyrinae) in England and Wales. *Ecological Entomology*, 12: 139–148.
Brakefield, P.M., 1987b. Tropical dry and wet season polyphenism in the butterfly *Melanitis leda* (Satyrinae): phenotypic plasticity and climatic correlates. *Biological Journal of the Linnean Society*, 31: 175–191.
Brakefield, P.M., 1989. The variance in genetic diversity among sub-populations is more sensitive to founder effects and bottlenecks than is the mean: a case study. In A. Fontdevila (Ed.), *Evolutionary Biology of Transient Unstable Populations*. pp. 145–161. Berlin: Springer-Verlag.
Brakefield, P.M., 1990. Case studies in ecological genetics. In O. Kudrna (Ed.),

Butterflies of Europe. Volume 2. Introduction to Lepidopterology. pp. 307–331. Weisbaden: Aula-Verlag.

Brakefield, P.M. & Larsen, T.B., 1984. The evolutionary significance of dry and wet season forms in some tropical butterflies. *Biological Journal of the Linnean Society*, 22: 1–12.

Brakefield, P.M. & Reitsma, N., 1991. Phenotypic plasticity, seasonal climate and the population biology of *Bicyclus* butterflies (Satyrinae) in Malawi. *Ecological Entomology*, 16: 291–303.

Brakefield, P.M. & Shreeve, T.G., 1992. Case studies in evolution. In R.L.H. Dennis (Ed.), *The Ecology of Butterflies in Britain*. pp. 197–216. Oxford: Oxford University Press.

Bretherton, R.F., 1968. The butterflies of Britain in relation to those in adjacent parts of the continent. *Proceedings of the British Entomological Natural History Society*, 1: 7–15.

Bridge, M.C., Haggart, D.A. & Lowe, J.J., 1990. The history and palaeoclimatic significance of subfossil remains of *Pinus sylvestris* in blanket peats from Scotland. *Journal of Ecology*, 78: 77–99.

Brockie, R.E., 1972. Evolutionary studies on *Maniola jurtina* in Sicily. *Heredity*, 28: 337–345.

Brooke, M. de L., Lees, A.D., & Lawman J.M., 1985. Spot distribution in the meadow brown L. (Lep., Satyridae): South Welsh populations. *Biological Journal of the Linnean Society*, 24: 337–348.

Brouwer, F., 1989. Determination of broad-scale landuse changes by climate and soils. *Journal of Environmental Management*, 29(1): 1–15.

Brower, A.V.Z. & Bryce, T.M., 1991. Mitochondrial DNA variation in monarch butterflies. *Evolution*, 45: 1281–1286.

Brower, L.P., 1984. Chemical defence in butterflies. In R.I. Vane-Wright & P.R. Ackery (Eds), *The Biology of Butterflies*. Royal Entomological Symposium No. 11. pp. 109–134. London: Academic Press.

Brower, L.P. & Calvert, W.H., 1985. Foraging dynamics of bird predators on overwintering monarch butterflies in Mexico. *Evolution*, 39: 852–868.

Brown, F.M., 1976. *Oligodonta florrisantensis* gen. n., sp. n. (Lepid., Pieridae). *Bulletin of the Allyn Museum*, Sarasota, 37: 1–4.

Brown, J.H., 1981. Two decades of homage to Santa Rosalia: toward a general theory of diversity. *American Zoolologist*, 21: 877–888.

Brown, J.H., 1984. On the relationship between abundance and distribution of species. *American Naturalist*, 124: 255–279.

Brown, J.H., 1988. Species diversity. In A.A. Myers & P.S. Giller (Eds), *Analytical Biogeography*. pp. 57–89. London: Chapman & Hall.

Brown, K.S., 1987. Biogeography and evolution of Neotropical butterflies. In T.C. Whitmore & G.T. Prance (Eds), *Biogeography and Quaternary History in Tropical America*. pp. 66–104. Oxford: Clarendon Press.

Brugge, R., 1991. The record-breaking heatwave of 1–4 August 1990 over England and Wales. *Weather*, 46: 2–10.

Brugge, R., 1992. Three years of warm weather over the British Isles, 1988–90. *Weather*, 7: 230–237.

Burga, C.A., 1988. Swiss vegetation history during the last 18,000 years. *New Phytologist*, 110: 581–602.

Burns, J.M., 1964. Evolution in skipper butterflies of the genus *Erynnis*. *University of California Publication in Entomology*, 37: 1–214.
Bursell, E., 1974. Environmental aspects – temperature. In M. Rockstein (Ed.), *The Physiology of Insecta*. New York: Academic Press.
Calow, P., 1976. *Biological Machines. A Cybernetic Approach to Life*. London: Edward Arnold.
Calow, P., 1983. *Evolutionary Principles*. London: Blackie.
Calow, P., 1989. Proximate and ultimate responses to stress in biological systems. *Biological Journal of the Linnean Society*, 37: 173–181.
Calvert, W.H. & Brower, L.P., 1986. The location of monarch butterfly (*Danaus plexippus* L.) overwintering colonies in Mexico in relation to topography and climate. *Journal of the Lepidopterists' Society*, 40: 164–187.
Calvert, W.H., Brower, L.P. & Lawton, R.O., 1992. Mass flight response of overwintering monarch butterflies (Nymphalidae) to cloud-induced changes in solar radiation intensity in Mexico. *Journal of the Lepidopterists' Society*, 46: 97–105.
Calvert, W.H. & Cohen, J.A., 1983. The adaptive significance of crawling up onto foliage for the survival of grounded overwintering monarch butterflies (*Danaus plexippus*) in Mexico. *Ecological Entomology*, 8: 471–474.
Campbell, G.S., 1977. *An Introduction to Environmental Biophysics*. New York: Springer-Verlag.
Campbell, I.M., 1989. Does climate affect hostplant quality? Annual variation in the quality of balsam fir as food for spruce budworm. *Oecologia, Berlin*, 81: 341–344.
Cannell, M. & Coutts, M., 1988. Growing in the wind. *New Scientist*, 21January: 42–46.
Cannell, M.G.R., Grace, J. & Booth, A., 1989. Possible impacts of climatic warming on trees and forests in the United Kingdom. A review. *Forestry*, 62: 337–364.
Cantello, W.W., 1974. Diapause in a tropical strain of the tobacco hornworm. *Annals of the Entomological Society of America*, 67: 828–830.
Capinera, J.L., Wiener, L.F. & Anamosa, P.R., 1980. Behavioural thermoregulation by late-instar range caterpillar larvae, *Hemileuca oliviae* Cockrell (Lep: Saturnidae). *Journal of the Kansas Entomology Society*, 53: 631–636.
Cappuccino, N. & Kareiva, P., 1985. Coping with a capricious environment: a population study of the rare woodland butterfly *Pieris virginiensis*. *Ecology*, 66: 152–161.
Carcasson, R.H., 1964. A preliminary survey of the zoogeography of African butterflies. *East African Wildlife Journal*, 2: 122–157.
Carpenter, G.D.H., 1939. Birds as enemies of butterflies, with special reference to mimicry. *VII International Congress of Entomology* (1938), 2: 1061–1074.
Casey, T.M., 1976. Activity patterns, body temperature, and thermal ecology in two desert caterpillars. *Ecology*, 57: 485–497.
Casey, T.M. & Hegel, J.R., 1981. Caterpillar setae: insulation for an ectotherm. *Science*, 214: 1131–1133.
Casey, T.M., Joos, B., Fitzgerald, T.D., Yurlina, M.E. & Young, P.A., 1988. Synchronized group foraging, thermoregulation and growth of eastern tent caterpillars in relation to microclimate. *Physiological Zoology*, 61: 372–377.
Chai, P., 1986. Field observations and feeding experiments on the response of

rufous-tailed jacamars (*Galbula ruficauda*) to free-flying butterflies in a tropical rain forest. *Biological Journal of the Linnean Society*, 29: 166–189.

Chai, P. & Srygley, R.B., 1990. Predation and the flight, morphology, and temperature of Neotropical rain-forest butterflies. *American Naturalist*, 135: 748–765.

Chandler, T.J., 1965. *The Climate of London*. London: Hutchinson.

Chaplin, S.B. & Wells, P.H., 1982. Energy reserves and metabolic expenditure of monarch butterflies overwintering in southern California. *Ecological Entomology*, 7: 249–256.

Chappellaz, J., Barnola, J.M., Raynaud, D., Korotkevich, Y.S. & Lorius, C., 1990. Ice-core record of atmospheric methane over the past 160,000 years. *Nature*, 345: 127–131.

Charles, C.D. & Fairbanks, R.G., 1992. Evidence from Southern Ocean sediments for the effect of North Atlantic deep-water flux on climate. *Nature*, 355: 416–419.

Chen, C.P., Denlinger, D.L. & Lee, R.E., 1987. Cold-shock injury and rapid cold-hardening in the flesh fly *Sarcophaga crassipalpis*. *Physiological Zoology*, 60: 297–304.

Chew, F.S., 1975. Coevolution of Pierid butterflies and their cruciferous foodplants. I. The relative quality of of available resources. *Oecologia*, 20: 117–127.

Chew, F.S. & Robbins, R.K., 1984. Egglaying in butterflies. In R.I. Vane-Wright & P.R. Ackery (Eds), *The Biology of Butterflies*. Royal Entomological Society Symposium No. 11. pp. 65–79. London: Academic Press.

Chinery, M., 1989. *Butterflies and Day-Flying Moths of Britain and Europe*. London: Collins.

Clapham, A.R., Tutin, T.G. & Moore, D.M., 1989. *Flora of the British Isles*. Third Edition. Cambridge: Cambridge University Press.

Claridge, M.F., 1987. Insect assemblages – diversity, organization and evolution. In J.A.R. Gee & P.S. Giller (Eds), *Organization of Communities: Past and Present*. Oxford: Blackwell Scientific Publications.

Clark, J.S., 1988. Effect of climate change on fire regimes in Northwestern Minnesota. *Nature*, 334: 233–235.

Clarke, C.A. & Sheppard, P.M., 1962a. The evolution of dominance under disruptive selection. *Heredity*, 14: 73–87.

Clarke, C.A. & Sheppard, P.M., 1962b. The evolution of mimicry in the butterfly *Papilio dardanus*. *Heredity*, 14: 163–173.

Clarke, C.A. & Sheppard, P.M., 1962c. Disruptive selection and its effect on a metrical character in the butterfly *Papilio dardanus*. *Evolution*, 16: 214–226.

Clarke, C.A. & Sheppard, P.M., 1972. Genetic and environmental factors influencing pupal colour in the swallowtail butterflies, *Battus philenor and Papilio polytes*. *Journal of Entomology*, A, 46: 123–133.

Clench, H.K., 1966. Behavioural thermoregulation in butterflies. *Ecology*, 47: 1021–1034.

Clifford, H.F. & Boerger, H., 1974. Fecundity of mayflies (Ephemeroptera), with special reference to mayflies of a brown-water stream of Alberta, Canada. *Canadian Entomologist*, 106: 1111–1119.

Clube, S.V.M. (Ed.), 1989. *Catastrophes and Evolution. Astromonical Foundations*. Cambridge: University Press.

COHMAP, 1988. Climatic changes of the last 18,000 years: observations and model simulations. *Science*, 241: 1043–1052.

Coleman, J.S., Rochefort, L., Bazzaz, F.A. & Woodward, F.I., 1991. Atmospheric CO_2, plant nitrogen status and the susceptibility of plants to an acute increase in temperature. *Plant, Cell and Environment*, 14: 667–674.

Colinvaux, P.A., 1973. *Introduction to Ecology*. New York: Wiley.

Collins, N.M. & Morris, M.G., 1985. *Threatened Swallowtail Butterflies of the World*. The IUCN Red Data Book. Cambridge: IUCN.

Collinson, M.E., 1990. Plant evolution and Ecology during the early Cainozoic diversification. *Advances in Botanical Research*, 17: 1–98.

Connell, J.H., 1978. Diversity in tropical rainforest and coral reefs. *Science*, 199: 1302–1310.

Connell, J.H. & Orias, E., 1964. The ecological regulation of species diversity. *American Naturalist*, 98: 399–441.

Cook, L.M., 1961. Influence of larval environment on adult size and fecundity in the moth, *Panaxia dominula* L. *Nature*, 192: 282.

Coope, G.R., 1979. Late Cenozoic fossil Coleoptera: Evolution, biogeography and ecology. *Annual Review of Ecology & Systematics*, 10: 247–267.

Coope, G.R., 1987. The response of Late Quaternary insect communities to sudden climatic changes. In J.H.R. Gee & P.S. Giller (Eds), *Organization of Communities. Past and Present*. pp. 421–438. Oxford: Blackwell Scientific Publications.

Coope, G.R., 1991. The study of the 'nearly-fossil'. *Antenna*, 15: 158–163.

Courtillot, V., Feraud, G., Maluski, H., Vandamme, D., Moreau, M.G. & Besse, J., 1988. Deccan flood basalts and the Cretaceous/Tertiary boundary. *Nature*, 333: 843–846.

Courtney, S.P., 1980. *Studies on the Biology of Butterflies Anthocharis cardamines (L.) and Pieris napi (L.) in relation to Speciation in the Pierinae*. Ph.D. thesis, University of Durham, UK.

Courtney, S.P., 1981. Coevolution of pierid butterflies and their cruciferous foodplants. III. *Anthocharis cardamines* (L.) survival, development and oviposition on different hostplants. *Oecologia*, 51: 91–96.

Courtney, S.P., 1982a. Coevolution of pierid butterflies and their cruciferous foodplants. IV. Crucifer apparency and *Anthocharis* (L.) oviposition. *Oecologia*, 52: 258–265.

Courtney, S.P., 1982b. Coevolution of pierid butterflies and their cruciferous foodplants. V. Habitat selection, community structure and speciation. *Oecologia*, 54: 101–107.

Courtney, S.P., 1984. The evolution of batch oviposition by Lepidoptera and other insects. *American Naturalist*, 123: 276–281.

Courtney, S.P., 1986. The ecology of pierid butterflies: dynamics and interactions. *Advances in Ecological Research*, 15: 51–131.

Courtney, S.P. & Duggan, A.E., 1983. The population biology of the orange tip butterfly, *Anthocharis cardamines* in Britain. *Ecological Entomology*, 7: 131–137.

Cox, C.B. & Moore, P.D., 1985. *Biogeography: An Ecological and Evolutionary Approach*. Oxford: Blackwell. 4th Edition.

Coyne, J.A., 1992. Genetics and speciation. *Nature*, 355: 511–515.

Creed, E.R., Dowdeswell, W.H., Ford, E.B. & McWhirter, K.G., 1959. Evolutionary studies on *Maniola jurtina*: the English mainland 1956–57. *Heredity*, 13: 363–391.

Crepet, W.L. & Feldman, G.D., 1991. The earliest remains of grasses in the fossil record. *American Journal of Botany*, 78: 1010–1014.

Cribb, P.W., 1986. Weavers fritillary taken in Surrey. *Bulletin of the Amateur Entomological Society*, 45: 193.

Cronin, T.M. & Schneider, C.E., 1990. Climatic influences on species: evidence from the fossil record. *Trends in Ecology and Evolution*, 5: 275–279.

Crowe, P.R., 1971. *Concepts in Climatology*. Harlow: Longman.

Cubasch, U. & Cess, R.D., 1990. Processes and modelling. In J.T. Houghton, G.J. Jenkins & J.J. Ephraum (Eds), *Climate Change. The IPCC Scientific Assessment*. pp. 69–92. Cambridge: University Press.

Cure, J.D., 1985. Carbon dioxide doubling responses: a crop survey. In B.R. Strain and J.D. Cure (Eds), *Direct Effects of Increasing Carbon Dioxide Concentration on Vegetation*. pp. 99–116. Washington DC: US Department of Energy.

Cure, J.D. & Acock, B., 1986. Crop responses to CO2 doubling: a literature survey. *Agricultural & Forestry Meteorology*, 38: 127–145.

Currie, D.J. & Paquin, V., 1987. Large-scale biogeographical patterns of species richness of trees. *Nature*, 329: 326–328.

D'Abrera, B.L., 1980. *Butterflies of the Afrotropical region*. Melbourne: Lansdowne.

D'Abrera, B.L., 1981. *Butterflies of the Neotropical Region. Part I Papilionidae & Pieridae*. Faringdon: Classey & Lansdowne.

D'Abrera, B.L., 1984. *Butterflies of the Neotropical Region. Part II Danaidae, Ithomiidae, Heliconiidae & Morphidae*. Victoria: Hill House.

D'Abrera, B.L., 1987. *Butterflies of the Neotropical Region. Part III Brassolidae, Acraeidae & Nymphalidae (Partim)*. Victoria: Hill House.

Danilevskii, A.S., 1965. *Photoperiodism and Seasonal Development of Insects*. Edinburgh: Oliver & Boyd.

Danks, H.V., 1978. Modes of seasonal adaptation in insects. I. Winter survival. *Canadian Entomologist*, 110: 1168–1205.

Dansgaard, W., White, J.W.C. & Johnsen, S.J., 1989. The abrupt termination of the Younger Dryas climate event. *Nature*, 339: 532–534.

Darwin, C., 1874 (2nd Edition). *The Descent of Man and Selection in Relation to Sex*. London.

Davies, T.D., Pierce, C.E., Robinson, H.J. & Darling, S.R., 1992. Towards an assessment of the influence of climate on wet acidic deposition in Europe. *Environmental Pollution*, 75: 111–120.

Davis, M.B., 1988. Ecological systems and dynamics. In H.A. Mooney (Ed.), *Toward an Understanding of Global Change*. pp. 69–106. Washington DC: National Academy Press.

Davis, M.B., 1990. Biology and paleobiology of global climatic change: introduction. *Trends in Ecology and Evolution*, 5: 269–270.

De Beaulieu, J.L. & Reille, M., 1984. A long upper Pleistocene pollen record from Les Echets, near Lyon, France. *Boreas*, 13: 111–132.

Del Genio, A.D., Lacis, A.A. & Ruedy, R.A., 1991. Simulations of the effect of a warmer climate on atmospheric humidity. *Nature*, 351: 382–385.

Dempster, J.P., 1967. The control of *Pieris rapae* with DDT. I. The natural mortality of the young stages of *Pieris. Journal of Applied Ecology*, 4: 485–500.

Dempster, J.P., 1983. The natural control of populations of butterflies and moths. *Biological Review*, 58: 461–481.

Dempster, J.P., 1989. Insect introductions: natural dispersal and population persistence in insects. *The Entomologist*, 108: 5–13.

Dempster, J.P., King, M.L. & Lakhani, K.H., 1976. The status of the swallowtail butterfly in Britain. *Ecological Entomology*, 1: 71–84.

Dempster, J.P. & Pollard, E., 1986. Spatial heterogeneity, stochasticity and the detection of density-dependence in animal populations. *Oikos*, 46: 413–416.

Denlinger, D.L., 1986. Dormancy in tropical insects. *Annual Review of Entomology*, 31: 239–264.

Denlinger, D.L., Tanaka, S., Downes, W.L., Wolda, H. & Guardia, M., 1991. Does lack of diapause result in less insect seasonality? *Oecologia, Berlin*, 87: 152–154.

Dennis, R.L.H., 1970. Note on the relation of *Aglais urticae* L. to direct and shade temperatures. *Entomologist's Record and Journal of Variation*, 82: 302.

Dennis, R.L.H., 1977. *The British Butterflies. Their Origin and Establishment.* Faringdon, Oxon.: Classey.

Dennis, R.L.H., 1982. Observations on habitats and dispersal made from ovipositional markers in north Cheshire *Anthocharis cardamines* (L.) (Lep: Pieridae). *Entomologist's Gazette*, 33: 151–159.

Dennis, R.L.H., 1982–83. Mate locating strategies in the wall brown butterfly, *Lasiommata megera* (L.) (Lep: Satyridae): wait or seek? *Entomologist's Record and Journal of Variation*, 94: 209–214; 95: 7–10.

Dennis, R.L.H., 1983a. Hierarchy and spatial pattern in the responses of ovipositing *Anthocharis cardamines* (Lep; Pieridae). *Vasculum*, 68: 27–43.

Dennis, R.L.H., 1983b. Egg-laying cues in the wall brown butterfly, *Lasiommata megera* (L.) (Lep: Satyridae). *Entomologist's Gazette*, 34: 89–95.

Dennis, R.L.H., 1984a. The 'edge effect' in butterfly oviposition: batch siting in *Aglais urticae* (L.) (Lep: Nymphalidae). *Entomologist's Gazette*, 35: 157–173.

Dennis, R.L.H., 1984b. Egg-laying sites in the common blue butterfly, *Polyommatus icarus* Rott. (Lep. Lycaenidae): the edge effect and beyond the edge. *Entomologist's Gazette*, 35: 85–93.

Dennis, R.L.H., 1985. Voltinism in British *Aglais urticae* (L.) (Lep., Nymphalidae): variation in space and time. *Proceedings of the British Entomological and Natural History Society*, 18: 51–61.

Dennis, R.L.H., 1986a. Selection of roost sites by *Lasiommata megera* (Lep: Satyridae) on fencing at Brereton Heath country park, Cheshire. *Nota Lepidopterologica*, 9: 39–46.

Dennis, R.L.H., 1986b. Motorways and cross-movements. An insect's 'mental map' of the M56 in Cheshire. *Bulletin of the Amateur Entomologists' Society, UK*. 45: 228–243.

Dennis, R.L.H. (Ed.) 1992. *The Ecology of Butterflies in Britain*. Oxford: Oxford University Press.

Dennis, R.L.H. & Bramley, M.J. 1985. The influence of man and climate on dispersion patterns within a population of adult *Lasiommata megera* (L.) (Satyridae). *Nota Lepidopterologica*, 8: 309–324.

Dennis, R.L.H., Porter, K. & Williams, W.R., 1984. Ocellation in *Coenonympha tullia*. (Müller) (Lep., Satyridae). I. Structures in correlation matrices. *Nota Lepidopterologica*, 7: 199–219.

Dennis, R.L.H., Porter, K. & Williams, W.R., 1986. Ocellation in *Coenonympha tullia* Müller) (Lep., Satyridae). II. Population differentiation and clinal variation in the context of climatically-induced antipredator defence strategies. *Entomologist's Gazette*, 37: 133–172.

Dennis, R.L.H. & Shreeve, T.G., 1988. Hostplant-habitat structure and the evolution of butterfly mate-location behaviour. *Zoological Journal of the Linnean Society*, 94: 301–318.
Dennis, R.L.H. & Shreeve, T.G., 1989. Butterfly wing morphology variation in the British Isles: the influence of climate, behavioural posture and the hostplant habitat. *Biological Journal of the Linnean Society*, 38: 323–348.
Dennis, R.L.H. & Shreeve, T.G., 1991. Climatic change and the British butterfly fauna: opportunities and constraints. *Biological Conservation*, 55: 1–16.
Dennis, R.L.H. & Williams, W.R., 1986. Butterfly 'diversity': regressing and a little latitude. *Antenna*, 10: 108–111.
Dennis, R.L.H. & Williams, W.R., 1987. Mate-location behaviour of the large skipper butterfly *Ochlodes venata*. Flexible strategies and spatial components. *Journal of the Lepidopterists' Society*, 41: 45–64.
Dennis, R.L.H., Williams, W.R. & Shreeve, T.G., 1991. A multivariate approach to the determination of faunal structures among European butterfly species (Lepidoptera: Rhopalocera). *Zoological Journal of the Linnean Society*, 101: 1–49.
Descimon, H. & Renon, C., 1975. Mélanisme et facteurs climatiques: II. Correlation entre melanisation et certains facteurs climatique chez *Melanargia galathea* (L.) (Lep., Satyridae) en France. *Archives de Zoologie Experimentale et Génerale*, 116: 437–468.
Denton, G.H. & Hughes, T.J., 1981. *The Last Great Ice Sheets*. New York: Wiley.
DeVries, P.J., 1987. *The Butterflies of Costa Rica and their Natural History*. Princeton: University Press.
Dickinson, R., 1989. Uncertainties of estimates of climatic change: a review. *Climatic Change*, 15: 5–13.
Digby, P.S.B., 1955. Factors affecting the temperature excess of insects in sunshine. *Journal of Experimental Biology*, 32: 279–298.
Dingle, H. 1982. Function of migration in the seasonal synchronization of insects. *Entomologia experimentalis et applicata*, 31: 36–48.
Dobson, A., Jolly, A. & Rubenstein, D., 1989. The greenhouse effect and biological diversity. *Trends in Ecology and Evolution*, 4: 64–68.
Dobzhansky, T., 1950. Evolution in the tropics. *American Scientist*, 38: 209–221.
Dohmen, G.P., 1988. Indirect effects of air pollutants: changes in plant/parasite interactions. *Environmental Pollution*, 53: 197–207.
Dolinger, P.M., Ehrlich, P.R., Fitch, W.L. & Breedlove, D.E., 1973. Alkaloid and predation patterns in Colorado lupine populations. *Oecologia*, 13: 191–204.
Douglas, M.M., 1978. *The Behavioural and Biophysical Strategies of Thermoregulation in Temperate Butterflies*. Ph.D. thesis, University of Kansas.
Douglas, M.M. & Grula, J.W., 1978. Thermoregulatory adaptations allowing ecological range expansion by the pierid butterfly *Nathalis iole* Boisduval. *Evolution*, 32: 776–783.
Douwes, P., 1976. Activity in *Heodes virgaureae* (Lep: Lycaenidae) in relation to air temperature, solar radiation, and time of day. *Oecologia*, 22: 287– 298.
Douwes, P., 1980. Periodical appearance of species in the genera *Oeneis* and *Erebia* in Fennoscandia (Lep: Satyridae). *Entomologia generalis*, 6: 151–157.
Dowdeswell, W.H., 1961. Experimental studies on natural selection in the butterfly, *Maniola jurtina*. Heredity, 16: 39–52.

Dowdeswell, W.H., 1962. A further study of the butterfly *Maniola jurtina* in relation to natural selection by *Apanteles tetricus*. *Heredity*, 17: 513–523.
Downes, J.A., 1965. Adaptations of insects in the arctic. *Annual Review of Entomology*, 10: 257–274.
Drake, V.A. & Farrow, R.A., 1988. The influence of atmospheric structure and motions on insect migration. *Annual Review of Entomology*, 33: 183–210.
Dudley, R., 1991. Thermoregulation in unpalatable danaine butterflies. *Functional Ecology*, 5: 503–506.
Duman, J.G. & Horwath, K.L., 1983. The role of haemolymph proteins in the cold tolerance of insects. *Annual Review of Physiology*, 45: 261–270.
Duman, J.G., Wu, D.W., Xu, L., Tursman, D. & Olsen, T.M., 1991. Adaptations of insects to subzero temperatures. *The Quarterly Review of Biology*, 66: 387–410.
Duncan, K., 1991. A comparison of the temperature records of Edinburgh and central England. *Weather*, 46: 169–173.
Dunn, T.C. & Parrack, J.D., 1986. *The Moths and Butterflies of Northumberland and Durham*. Northern Naturalists' Union: The Vasculum, Supplement No. 2. Part 1 Macroleidoptera.
Durden, C.J., & Rose, H., 1978. *Butterflies of the Middle Eocene: the Earliest Occurrence of Fossil Papilionidae (Lep.)*. Texas: Memorial Museum, Pearce-Sellards Series, 29: 1–25.
Eamus, D. & Jarvis, P.G., 1989. The direct effects of increases in the global and atmospheric CO_2 concentration on natural and commercial temperate trees and forests. *Advances in Ecological Research*, 19: 1–53.
Ehrlich, P.R., Breedlove, D.E., Brussard, P.F. & Sharp, M.A., 1972. Weather and the 'regulation' of subalpine populations. *Ecology*, 53: 243–247.
Ehrlich, P.R. & Gilbert, L.E., 1973. Population structure and dynamics of the tropical butterfly *Heliconius ethilla. Biotropica*, 5: 69–82.
Ehrlich, P.R. & Raven, P.H., 1965. Butterflies and plants: a study in coevolution. *Evolution*, 18: 586–608.
Ehrlich, P.R. & Wheye, D., 1986. Nonadaptive hilltopping behaviour in male checkerspot butterflies (*Euphydryas editha*). *American Naturalist*, 127: 477–483.
Ekholm, S., 1975. Fluctuations in butterfly frequency in central Nyland. *Notulae Entomologicae*, 55: 65–80.
Eliasson, C., 1991. Occurrence and biology of *Euphydryas maturna* (L.) (Lep., Nymphalidae) in central Sweden. *Entomologisk Tidskrift*, 112: 113–124.
Ellsaesser, H.W., 1990. A different view of the climatic effect of CO_2 – updated. *Atmosphera*, 3: 3–29.
Elsom, D. & Meaden, T., 1985. Devils and twisters screw things up. *New Scientist*, 27 June: 44–47.
Emmanuel, J.A., Shugart, H.H. & Stevenson, M.P., 1985. Climatic change and the broad-scale distribution of terrestrial ecosystem complexes. *Climatic Change*, 7: 26–43.
Emmel, T.C., 1989. The incredible butterfly diversity of the Rondonian rain forest in Brazil: a phenomenon soon to disappear. *News of the Lepidopterists' Society*, (4): 53–55.
Emmel, T.C. & Emmel, J.F., 1973. *The Butterflies of Southern California*. Los Angeles: County Museum of Natural History.
Emmel, T.C. & Leck, C.F., 1969. Seasonal changes in organization of tropical rain

forest butterfly populations in Panama. *Journal of Research on the Lepidoptera*, 8: 133–152.
Emmel, T.C. & Mattoon, S.O., 1990. *Butterflies of California*. Stanford: University Press.
Emmel, T.C., Minno, M.C. & Drummond, B.A., 1992. *Florissant Butterflies: a Guide to the Fossil and Present-Day Species of Central Colorado*. Stanford: University Press.
Emmet, A.M. & Heath, J., 1989. *The Moths and Butterflies of the British Isles. Volume 7, Part 1. The Butterflies*. Colchester: Harley Books.
Endler, J.A., 1978. A predator's view of animal colour patterns. *Evolutionary Biology*, 11: 319–364.
Endler, J.A., 1981. An overview of the relationships between mimicry and crypsis. *Biological Journal of the Linnean Society*, 16: 25–31.
Endo, K., 1972. Activation of the corpora allata in relation to ovarian maturation in the seasonal forms of the butterfly, *Polygonia c-aureum* L., *Development, Growth & Differentiation*, 14: 263–274.
Endo, K. & Funatsu, S., 1985. Hormonal control of seasonal morph determination in the swallowtail butterfly, *Papilio xuthus* L. (Lep: Papilionidae). *Journal of Insect Physiology*, 31: 669–674.
Endo, K. & Kamata, Y., 1985. Hormonal control of seasonal morph determination in the small copper butterfly, *Lycaena phlaeas daimio* Seitz. *Journal of Insect Physiology*, 31: 701–706.
Evans, J.G., 1975. *The Environment of Early Man in the British Isles*. London: Paul Elek.
Eyre, S.R., 1963. *Vegetation and Soils: A World Picture*. London: Edward Arnold.
Fairbanks, R.G., 1989. A 17,000-year glacio-eustatic sea level record: influence of glacial melting rates on the Younger Dryas event and deep ocean circulation. *Nature*, 342: 637–642.
Fajer, E.D., Bowers, M.D. & Bazzaz, F.A., 1989. The effects of enriched carbon dioxide atmospheres on plant/insect herbivore interactions. *Science*, 243: 1198 1200.
Fajer, E.D., Bowers, M.D. & Bazzaz, F.A., 1991. The effects of enriched CO_2 atmospheres on the buckeye butterfly, *Junonia Coenia. Ecology*, 72: 751–754.
Feeny, P., 1975. Biochemical coevolution between plants and their insect herbivores. In L.E. Gilbert & P.R. Raven (Eds) *Coevolution of Animals and Plants*. pp. 3–19. Austin: University of Texas Press.
Feltwell, J., 1990. Fluid re-cycling in a nymphalid. *Antenna*, 14: 52.
Findlay, J.A. & Findlay, R., 1983. Temperature-regulating strategies in adult butterflies. *News British Butterfly Conservation* Society, 31: 17–37.
Findlay, R., Young, M.R. & Findlay, J., 1983. Orientation behaviour in the Grayling butterfly: thermoregulation or crypsis? *Ecological Entomology*, 8: 145–153.
Flint, R.F., 1971. *Glacial and Quaternary Geology*. New York: Wiley.
Flitters, N.E., 1968. Insect body temperatures determined with microthermic thermometry. *Annals of the Entomological Society of America*, 61: 36–38.
Folland, C.K., Karl, T. & Vinnikov, K.YA., 1990. Observed Climate Variations and Change. In J.T. Houghton, G.J. Jenkins & J.J. Ephraums (Eds), *Climate Change. The IPCC Scientific Assessment*. pp. 195–238. Cambridge: University Press.

Follieri, M., Magri, D. & Sadori, L., 1988. 250,000-year pollen record from Valle di Castiglione (Roma). *Pollen et Spores*, 30: 329–356.

Ford, E.B., 1945. *Butterflies*. London: Collins.

Ford, E.B., 1975 (4th Edition). *Ecological Genetics*. London: Chapman & Hall.

France, R.L., 1991. Empirical methodology for predicting changes in species range extension and richness associated with climate warming. *International Journal of Biometeorology*, 34: 211–216.

Frenzel, B., 1979. Europe without forests. *Geographical Magazine*, 51: 756–761.

Fukuda, S. & Endo, K., 1966. Hormonal control of the development of seasonal forms in the butterfly, *Polygonia c-aureum* L. *Proceedings of the Japanese Academy*, 42: 1082–1087.

Fuller, R.J. & Warren, M.S., 1991. Conservation management in ancient and modern woodlands: responses of fauna to edges and rotations. In I.F. Spellerberg, F.B. Goldsmith & M.G. Morris (Eds), *The Scientific Management of Temperate Communities for Conservation*. 31st Symposium of the British Ecological Society Southampton 1989. pp. 445–471. Oxford: Blackwell Scientific Publications.

Futuyama, D.J., 1988. Selective factors in the evolution of host choice by phytophagous insects. In S. Ahmad (Ed.), *Herbivorous Insects*. pp. 227–244. London: Academic Press.

Garcia-Barros, E., 1987. Observaciones sobre la biologia de *Maniola jurtina* (L., 1758) en el centro de la peninsula Iberica: fenologia general del ciclo biologico, duracion del periodo de prepuesta y fecundidad potencial de las hembras (Lep. Nymphalidae). *Boletin Asociacion españa Entomologia*, 11: 235–247.

Garcia-Barros, E., 1988. Delayed ovarian maturation in the butterfly *Hipparchia semele* as a possible response to summer drought. *Ecological Entomology*, 13: 391–398.

Gardiner, B.O.C., 1980. The fatal effect of prolonged chilling on summer pupae of *Pieris brassicae* (L.) (Lep., Pieridae). *Entomologist's Gazette*, 31: 214–216.

Gardiner, B.O.C., 1988. Butterfly roosting posture: *Pieris* up, *Vanessa* down. *British Journal of Entomology and Natural History*, 1: 126–127.

Garland, S.P., 1981. *Butterflies of the Sheffield Area*. Sheffield: Sorby Natural History and Sheffield City Museum.

Gasse, F., Tehet, R., Durnad, A., Gibert, E. & Fontes, J.-C., 1990. The arid – humid transition in the Sahara and the Sahel during the last deglaciation. *Nature*, 346: 141–146.

Gaston, K.J. & Lawton, J.H., 1988. Patterns in the distribution and abundance of insect populations. *Nature*, 331: 709–711.

Gates, W.L., Rowntree, P.R. & Zeng, Q.-C., 1990. Validation of climate models. In J.T. Houghton, G.J. Jenkins & J.J. Ephraums (Eds), *Climate Change. The Scientific Assessment*. pp. 93–130. Cambridge: University Press.

Gee, J.A.R. & Giller, P.S., 1987. *Organization of Communities: Past & Present*. Oxford: Blackwell Scientific Publications.

Geiger, H., 1990. Enzyme electrophoretic methods in studies of systematics and evolutionary biology of butterflies. In O. Kudrna (Ed.), *Butterflies of Europe. Vol. 2, Introduction to Lepidopterology*. pp. 397–436. Wiesbaden: Aula-Verlag.

Geiger, H., Descimon, H. & Scholl, A., 1988. Evidence for speciation within nominal *Pontia daplidice* (Linnaeus, 1758) in southern Europe (Lepidoptera: Pieridae). *Nota Lepidopterologica*, 11: 7–20.

Geiger, R., 1965. *The Climate near the Ground.* Cambridge, Mass.: Harvard University Press.

Geiger, W., 1987. *Les Papillons de jour et leurs biotopes.* Bale: Ligue Suisse pour la Protection de la Nature.

Genthon, C., Barnola, J.M., Raynaud, O., Lorius, C., Jouzel, J., Barkov, N.I., Korotkevich, Y.S. & Kotlyakov, V.M., 1987. Vostok ice core: climatic responses to CO_2 and orbital forcing changes over the past climatic cycle. *Nature*, 329: 414–418.

Ghil, M. & Vautard, R., 1991. Interdecadal oscillations and the warming trend in global time series. *Nature*, 350: 324–327.

Gibo, D.L., 1981. Altitudes attained by migrating monarch butterflies, *Danaus plexippus* (Danaidae), as reported by glider pilots. *Canadian Journal of Zoology*, 59: 571–572.

Gibo, D.L. & Pallet, M.J., 1979. Soaring flight of monarch butterflies, *Danaus plexippus* (Lep. Danaidae), during the late summer migration in southern Ontario. *Canadian Journal of Zoology*, 57: 1393–1401.

Gilbert, L.E., 1984. The biology of butterfly communities. In R.I. Vane-Wright & P.R. Ackery (Eds), *The Biology of Butterflies.* Royal Entomological Society Symposium No. 11. pp. 41–54. London: Academic Press.

Gilbert, L.E. & Singer, M.C., 1973. Dispersal and gene flow in a butterfly species. *American Naturalist*, 107: 58–73.

Gilbert, L.E. & Singer, M.C., 1975. Butterfly ecology. *Annual Review of Ecology and Systematics*, 6: 365–397.

Gilbert, N., 1984a. Control of fecundity in *Pieris rapae.* I. The problem. *Journal of Animal Ecology*, 53: 581–588.

Gilbert, N., 1984b. Control of fecundity in *Pieris rapae.* II. Differential effects of temperature. *Journal of Animal Ecology*, 53: 589–597.

Gilbert, N., 1984c. Control of fecundity in *Pieris rapae.* III. Synthesis. *Journal of Animal Ecology*, 53: 599–609.

Gilbert, N., 1986. Control of fecundity in *Pieris rapae.* IV. Patterns of variation and their ecological consequences. *Journal of Animal Ecology*, 55: 317–329.

Gilchrist, G.W., 1990. The consequences of sexual dimorphism in body size for butterfly flight and thermoregulation. *Functional Ecology*, 4: 475–487.

Gilchrist, G.W. & Rutowski, R.L., 1986. Adaptive and incidental consequences of the *alba* polymorphism in an agricultural population of *Colias* butterflies: female size, fecundity, and differential dispersion. *Oecologia*, 68: 235–240.

Giller, P.S., 1984. *Community Structure and the Niche.* London: Chapman & Hall.

Glazier, D.S., 1986. Temporal variability of abundance and the distribution of species. *Oikos*, 47: 309–314.

Godwin, H., 1975. *The History of the British Flora.* Cambridge: University Press.

Goettel, M.S. & Philogene, B.J.R., 1978. Effects of photoperiod and temperature on the development of a univoltine population of the banded woollybear, *Pyrrharctia (Isia) isabella. Journal of Insect Physiology*, 24: 523–527.

Gorovaya, S.L. & Samersov, V.F., 1976. Razvitiye kapustnoy souki (*Barathra brassicae* L.) i kapustnoy belyanki (*Pieris brassicae* L.) u zavimosti ot soderzhaniya azotistykh veshchestu u korme. *Vestnik Akad. Nauk BSSR. Ser Biyal. Navuk*. 4: 126–127.

Gossard, T.W. & Jones, R.E., 1977. The effects of age and weather on egg-laying in *Pieris rapae* L. *Journal of Applied Ecology*, 14: 65–71.
Gould, S.J., 1985. The paradox of the first tier: an agenda for paleobiology. *Paleobiology*, 11: 2–12.
Gould, S.J., 1986. Evolution and the triumph of homology, or why history matters. *American Scientist*, 74: 60–69.
Gould, S.J. & Eldridge, N., 1977. Punctuated equilibria: the tempo and mode of evolution reconsidered. *Paleobiology*, 3: 115–151.
Gould, S.J. & Vrba, E.S., 1982. Exaptation – a missing term in the science of form. *Paleobiology*, 8: 4–15.
Graham, R.W. & Grimm, E.C., 1990. Effects of global climate change on the patterns of terrestrial biological communities. *Trends in Ecology and Evolution*, 5: 289–292.
Greatorix-Davies, J.N., Sparks, T.H., Hall, M.L. & Marrs, R.H., 1993. The influence of shade on butterflies in rides of coniferized lowland woods in southern England and implications for conservation management. *Biological Conservation*, 63: 31–41.
Green, F.H.W., 1964. A map of annual average potential water deficit in the British Isles. *Journal of Applied Ecology*, 1: 151–156.
Greenfield, M.D., 1982. The question of paternal investment in Lepidoptera: male-contributed proteins in *Plodia interpunctella. International Journal of Invertebrate Reproduction*, 5: 323–330.
Greensmith, J.T. & Tooley, M.J., 1982. Sea-level movements during the last deglacial hemicycle. *Proceedings of the Geologists' Association*, 93: 1–125.
Gregory, S., 1954. Accumulated temperature maps of the British Isles. *Transactions of the Institute of British Geographers*, 20: 59–73.
Gregory, S., 1988. *Recent Climatic Change: A Regional Approach*. London: Belhaven Press.
Grime, J.P., 1979. *Plant Strategies and Vegetation Processes*. New York: John Wiley & Sons.
Grime, J.P., 1989. The Stress debate: sympton of impending synthesis? *Biological Journal of the Linnean Society*, 37: 3–17.
Grime, J.P., Hodgson, J. & Hunt, R., 1988. *Comparative Plant Ecology: a Functional Approach to Common British Species*. London: Unwin Hyman.
Grossmueller, D.W. & Lederhouse, R.C., 1985. Oviposition site selection: an aid to rapid growth and development in the tiger swallowtail butterfly, *Papilio glaucus*. *Oecologia*, 66: 68–73.
Grubb, P.J., 1987. Global trends in species richness in terrestrial vegetation: a view from the northern hemisphere. In J.H.R. Gee & P.S. Giller (Eds), *Organization of Communities: Past and Present*. pp. 99–118. Oxford: Blackwell Scientific Publications.
Guetter, P.J. & Kutzbach, J.E., 1990. A modified Köppen classification applied to model simulations of glacial and interglacial climates. *Climatic Change*, 16: 193–215.
Guilford, T., 1988. The evolution of conspicuous colouration. *American Naturalist*, 131 Supplement: S7–S21.
Guilford, T., 1990. The secrets of aposematism: unlearned responses to specific colours and patterns. *Trends in Ecology and Evolution*, 10: 323.

Guillaumin, M. & Descimon, H., 1976. La notion d'espèce chez les Lépidoptères. In C. Bocquet, J. Genermont & M. Lamotte (Eds), *Les Problèmes de l'espèce dans le règne animal. Mémoires Société Zoologique de France*, 38: 129–201.

Guiot, J., Pons, A., de Beaulieu, J.L. & Reille, M., 1989. A 140,000-year continental climate reconstruction from two European pollen records. *Nature*, 338: 309–313.

Guppy, C.S., 1986. The adaptive significance of alpine melanism in the butterfly *Parnassius phoebus* F. (Lep: Papilionidae). *Oecologia*, 70: 205–213.

Guppy, J.C., 1969. Some effects of temperature on the immature stages of the armyworm, *Pseudaletia unipunctata* (Lep: Noctuidae), under controlled conditions. *Canadian Entomologist*, 101: 1320–1327.

Haffer, J., 1969. Speciation in Amazonian forest birds. *Science*, 165: 131–137.

Hagen, R.H. & Lederhouse, R.C., 1985. Polymodal emergence of the tiger swallowtail, *Papilio glaucus* (Lep: Papilionidae): source of a false second generation in central New York State. *Ecological Entomology*, 10: 19–28.

Haji-Mamat, H. & Tamashiro, M., 1988. The prevalence of infections of *Nosema meslini* (Microsporida: Nosematidae) in field populations of *Artogeia rapae* (Lep: Pieridae) adults in Hawaii. *Proceedings of the Hawaiian Entomological Society*, 28: 191–196.

Hall, M.L., 1981. *Butterfly Monitoring Scheme. Instructions for Independent Recorders*. Cambridge: Institute of Terrestrial Ecology (NERC).

Hamilton, W.J., 1973. *Life's Colour Code*. New York: McGraw Hill.

Hammen, van der, T., Wimjstra, T.A. & Zagwijn, W.H., 1972. The floral record of the late Cenozoic of Europe. In K.K. Turekian (Ed.), *The Late Cenozoic Glacial Ages*. pp. 391–424. New Haven: Yale University Press.

Hanninen, H., 1991. Does climatic warming increase the risk of frost damage in northern trees? *Plant, Cell and Environment*, 14: 449–454.

Hansen, J., Fung, I., Lacis, A., Rind, D., Lebedeff, S., Ruedy, R. & Russel, G., 1988. Global climate changes as forecast by GISS's three-dimensional model. *Journal of Geophyiscal Research*, 93: 9341–9364.

Hanski, I., 1982. Dynamics of regional distribution: the core and satellite hypothesis. *Oikos*, 38: 210–221.

Hanski, I. & Gilpin, M., 1991. Metapopulation dynamics: brief history and conceptual domain. *Biological Journal of the Linnean Society*, 42: 3–16.

Hanstock, S., 1991. Global warming. Accepting the implications. *Geography Review*, 4: 30–36.

Harland *et al.*, 1990. *A Geologic Time Scale 1978*. Cambridge: University Press.

Harrison, S., Murphy, D.D. & Ehrlich, P.R., 1988. Distribution of the bay checkerspot butterfly, *Euphydryas editha bayensis*: evidence for a metapopulation model. *American Naturalist*, 132: 360–382.

Harvey, L.D.D., 1991. Comments on 'the aerial fertilization effect of CO_2 and its implications for global carbon cycling and maximum greenhouse warming' by S.B. Idso. *Bulletin of the American Meteorological Society*, 72: 962–965.

Hassell, M.P., 1985. Insect natural enemies as regulating factors. *Journal of Animal Ecology*, 54: 323–334.

Hassell, M.P., Lawton, J.H. & May, R.M., 1976. Patterns of dynamical behaviour in single-species populations. *Journal of Animal Ecology*, 45: 471–486.

Hassell, M.P., Latto, J. & May, R.M., 1989. Seeing the wood for the trees: detecting

density dependence from existing life tables. *Journal of Animal Ecology*, 58: 883–892.
Hatch, D.J., 1973. British climate in generalized maps. *Weather*, 28: 509–516.
Hazel, W.D. & West, D.A., 1979. Environmental control of pupal colour in swallowtail butterflies (Lep: Papilioninae): *Battus philenor* (L.) and *Papilio polyxenes* Fabr. *Ecological Entomology*, 4: 393–400.
Hazel, W.N. & West, D.A., 1983. The effect of larval photoperiod on pupal colour and diapause in swallowtail butterflies. *Ecological Entomology*, 8: 37–42.
Heath, J., Pollard, E. & Thomas, J.A., 1984. *Atlas of Butterflies in Britain and Ireland*. London: Viking.
Heath, J.E. & Adams, P.A., 1967. Temperature regulation in the sphinx moth during flight. *Nature*, 205: 309–310.
Heinrich, B., 1970a. Thoracic temperature stabilization by blood circulation in a free-flying moth. *Science*, 168: 580–582.
Heinrich, B., 1970b. Nervous control of the heart during thoracic temperature regulation in a sphinx moth. *Science*, 169: 606–607.
Heinrich, B., 1971. Temperature regulation of the sphinx moth, *Manduca sexta*. II. regulation of heat loss by control of blood circulation. *Journal of Experimental Biology*, 54: 153–166.
Heinrich, B., 1972. Thoracic temperatures of butterflies in the field near the equator. *Comparative Biochemistry and Physiology*, 43: 459–467.
Heinrich, B., 1974. Thermoregulation in endothermic insects: body temperature is closely attuned to activity and energy supplies. *Science*, 185: 747–756.
Heinrich, B., 1979. Keeping a cool head: thermoregulation in honey bees. *Science*, 205: 1269–1271.
Heinrich, B., 1986. Thermoregulation and flight activity of a satyrine, *Coenonympha inornata* (Lep: Satyridae). *Ecology*, 67: 593–597.
Heinrich, B., 1989. Beating the heat in obligate insect endotherms: the environmental problem and the organismal solutions. *American Zoology*, 29: 1157–1168.
Heinrich, B., 1990. Is 'reflectance' basking real? *Journal of Experimental Biology*, 154: 31–43.
Hekstra, G., 1989. Sea level rise: regional consequences and responses. In N.J. Rosenberg, Wm. III Easterling, P.R. Crosson & J. Darmstadter (Eds), *Greenhouse Warming*. pp. 53–67. Washington DC Resources for the Future.
Henderson-Sellers, A., 1990. Modelling and monitoring 'greenhouse' warming. *Trends in Ecology and Evolution*, 5(9): 270–275.
Henderson-Sellers, A. & Robinson, P.J., 1986. *Contemporary Climatology*. Harlow: Longman.
Hendricksen, O.Q., 1985. Variation in the C:N ratio of substrate mineralised during forest humus decomposition. *Soil Biology and Biochemistry*, 17: 435–440.
Hengeveld, R., 1988. Mechanisms of biological invasions. *Journal of Biogeography*, 15: 819–828.
Hengeveld, R., 1990. *Dynamic Biogeography*. Cambridge: University Press.
Henriksen, H.J. & Kreutzer, I.B., 1982. *The Butterflies of Scandinavia*. Odense: Scandinavisk Bogforlag.
Heron, C., 1992. The networks of botanical creation. *New Scientist*, 8 February: 40–44.
Heslop-Harrison, J.W.H., 1947. The Pleistocene races of certain British insects and

distributional overlapping. *Entomologist's Record & Journal of Variation*, 59: 141–145.

Hester, A.J., Miles, J. & Gimingham, C.H., 1991a. Succession from heather moorland to birch woodland. I. Experimental alteration of specific environmental conditions in the field. *Journal of Ecology*, 79: 303–315.

Hester, A.J., Miles, J. & Gimingham, C.H., 1991b. Succession from heather moorland to birch woodland. II. Growth and competition between *Vaccinium myrtillus, Deschampsia flexuosa and Agrostis capillaris. Journal of Ecology*, 79: 317–328.

Hewitt, G.W., 1990. Divergence and speciation as viewed from an insect hybrid zone. *Canadian Journal of Zoology*, 68: 1710–1715.

Hidaka, T. & Takahashi, H., 1967. Temperature conditions and maternal effect as modifying factors in the photoperiodic control of the seasonal form in *Polygonia c-aureum* (Lep., Nymphalidae). *Annotationes Zoologicae Japonenses* 40: 200–204.

Higgins, L.G. & Hargreaves, B., 1983. *The Butterflies of Britain and Europe*. London: Collins.

Higgins, L.G. & Riley, N.D., 1983. *A Field Guide to the Butterflies of Britain and Europe*. London: Collins.

Higley, L.G., Pedigo, L.P. & Ostlie, K.R., 1986. DEGDAY: a program for calculating degree-days, and assumptions behing the day-degree approach. *Environmental Entomology*, 15: 999–1016.

Høegh-Guldberg, O., 1966. North European groups of *Aricia allous*: their variability and relationship to *Aricia agestis* (Denis & Schiff.). *Natura Jutlandica*, 13: 1–184.

Høegh-Guldberg, O., 1968. Evolutionary trends in the genus *Aricia*. (Lep.,). *Natura Jutlandica*, 14: 3–76.

Høegh-Guldberg, O. & Hansen, A.L., 1977. Phenotypic wing pattern modification by very brief periods of chilling of pupating *Aricia artaxerxes* (Lycaenidae). *Journal of the Lepidopterists' Society*, 31: 223–231.

Hoffman, R.J., 1978. Environmental uncertainty and evolution of physiological adaptation in *Colias* butterflies. *American Naturalist*, 112: 999–1015.

Holliday, N.J., 1985. Maintenance of the phenology of the winter moth (Lep: Geometridae). *Biological Journal of the Linnean Society*, 25: 221–234.

Holt, R.D., 1990. The microevolutionary consequences of climatic change. *Trends in Ecology and Evolution*, 5: 311–315.

Hooghiemstra, H., 1984. *Vegetational and Climatic History of the High Plains of Bogota, Columbia; a Continuous Record of the Last 3.5 Million Years*. Dissertationes Botanicae 79. Vaduz: J. Cramer Verlag.

Horn, H.S., 1978. Optimal tactics of reproduction and life-history. In J.R. Krebs & N.B. Davies (Eds), *Behavioural Ecology: An Evolutionary Approach*. pp. 411–429. Oxford: Blackwell Scientific Publications.

Horton, P., 1977. Local migrations of Lepidoptera from Salisbury plain in 1976. *Entomologist's Gazette*, 28: 281–283.

Hough, J.A. & Pimentel, D., 1978. Influence of host foliage on development, survival and fecundity of the gypsy moth. *Environmental Entomology*, 7: 97–102.

Houghton, J.T., Jenkins, G.J. & Ephraums, J.J. (Eds), 1990. *Climate Change. The IPCC Scientific Assessment*. Cambridge: University Press.

Houghton, R.A. & Skole, D.L., 1990. Changes in the global carbon between 1700 and 1985. In B.L. Turner (Ed.), *The Earth Transformed by Human Action*, Cambridge: University Press.

House, H.L., 1974. Nutrition. In M. Rockstein (Ed.), *The Physiology of Insecta*, Volume 5: 1–62. New York: Academic Press.

Hsu, K.J., 1977. Tectonic evolution of the Mediterranean basins. In A.E.M. Nairn, W.H. Kanes & F.G. Shehli (Eds), *The Ocean Basins and Margins*. Volume 4: 29–75. New York: Plenum Press.

Hsu, K.J., Montadert, L., Bernoulli, D., Cita, M.B., Erickson, A., Garrison, R.E., Kidd, R.B., Melieres, F., Muller, C. & Wright, R., 1977. History of the Mediterranean salinity crisis. *Nature*, 267: 399–403.

Hulme, M. & Jones, P.D., 1991. Temperatures and windiness over the United Kingdom during the winters of 1988/89 and 1989/90 compared with previous years. *Weather*, 46: 126–136.

Huntley, B., 1988. Europe. In B. Huntley & T. Webb (Eds), *Vegetation History*. pp. 341–383. Dordrecht Kluwer Academic Publishers.

Huntley, B., 1990. European vegetation history: palaeovegetation maps from pollen data – 13,000 yr BP to present. *Journal of Quaternary Science*, 5: 103–122.

Huntley, B. & Birks, H.J.B., 1983. *An Atlas of Past and Present Pollen Maps for Europe 0–13,000 Years Ago*. Cambridge: University Press.

Huntley, B. & Prentice, I.C., 1988. July temperatures in Europe from pollen data, 6000 years before the present. *Science*, 241: 687–690.

Huntley, B. & Webb, T. III (Eds). 1988. *Vegetation History*. Dordrecht: Kluwer Academic Publishers.

Huston, M., 1979. A general hypothesis of species diversity. *American Naturalist*, 113: 81–101.

Hutchinson, G.E., 1959. Homage to Santa Rosalia, or 'why are there so many kinds of animals?'. *American Naturalist*, 93: 145–159.

Idso, S.B., 1991. The aerial fertilization effect of CO_2 and its implications for global carbon cycling and maximum greenhouse warming. *Bulletin American Meteorological Society*, 72: 962–965.

Idso, S.B. & Balling, R.C., 1991. Evaluating the climatic effect of doubling atmospheric CO_2 via an analysis of Earth's historical temperature record. *The Science of the Total Environment*, 106: 239–242.

Imbrie, J. & Imbrie, K.P., 1979. *Ice Ages: Solving the Mystery*. London: Enslow Publishers.

Imbrie, J., Hays, J.D., Martinson, D.G., McIntyre, A., Mix, A.C., Morley, J.J., Pisias, N.G., Prell, W.L. & Shackleton, N.J., 1984. The orbital theory of Pleistocene climate: support from a revised chronology of the marine^{18}O record. In A. Berger, J. Imbrie, J. Hays, G. Kukla & B. Saltzma (Eds.), *Milankovitch and Climate*. pp. 269–305. Dordrecht: Reidel.

Ishizaki, H. & Kato, M., 1956. Environmental factors affecting the formation of orange pupa in *Papilio xuthus*. *Memoirs of the College Sciences, Kyoto University, Japan (B)*, 23: 11–18.

Jackson, C.S., Hodkinson, I.D. & Stanley, P., 1990. Cold-hardiness in the hawthorn psyllid *Cacopsylla melanoneura* (Forster) (Homoptera: Psylloidea). *Entomologist*, 109: 224–230.

Jacquemard, P., 1976. Relations entre la diapause de *Diparopsis watersi* (Roths.) (Lep. Noct.) et la diapause de son parasite *Eucarcelia* Spp (Dipt. Tachin.) dans le Nord du Caméroun. *Coton et Fibres Tropicales*, 31: 313–321.

Janzen, D.H., 1973. Sweep samples of foliage insects: effects of seasons, vegetation types, elevation, time of day and insularity. *Ecology*, 54: 687–708.
Janzen, D.H. (Ed.), 1983. *Costa Rican Natural History*. Chicago: Chicago Press.
Janzen, D.H., Ataroff, M., Farinas, M., Reyes, S., Rincon, N., Soler, A., Soviano, P. & Vera, M., 1976. Changes in the arthropod community along an elevation transect in the Venezuelan Andes. *Biotropica*, 8: 193–203.
Jarvis, F.V.L., 1966. The genus *Aricia* in Britain. *Proceedings of the South London Entomological and Natural History Society*, 1966: 37–60.
Jarzembowski, E.A., 1980. Fossil insects from the Bembridge Marsh. Palaeogene of the Isle of Wight, southern England. *Bulletin of the British Museum of Natural History* (Geology), 33: 237–243.
Jermy, T., 1984. Evolution of insect/hostplant relationships. *American Naturalist*, 124: 609–630.
Johnson, C.G., 1969. *Migration and Dispersal of Insects by Flight*. London: Methuen.
Jones, K.N., Odendaal, F.J. & Ehrlich, P.R., 1986. Evidence against the spermatorphore as paternal investment in the checkerspot butterflies (Euphydryas: Nymphalidae). *American Midland Naturalist*, 116: 1–6.
Jones, P.D., 1990. The climate of the past 1,000 years. *Endeavour*, 14: 129–136.
Jones, R.E., Hart, J.R. & Bull, G.D., 1982. Temperature, size and egg production in the cabbage butterfly, *Pieris rapae* L. *Australian Journal of Zoology*, 30: 223–232.
Jones, R.E., Nealis, V.G., Ives, P.M. & Scheermeyer, E., 1987. Seasonal and spatial variation in juvenile survival of the cabbage butterfly, *Pieris rapae*: evidence for patchy density-dependence. *Journal of Animal Ecology*, 56: 723–737.
Jones, R.E., Rienks, J.H. & Wilson, L., 1985. Seasonally and environmentally induced polyphenism in *Eurema laeta lineata* (Lep: Pieridae). *Journal of the Australian Entomological Society*, 24: 161–167.
Jong, R. de, 1972. Systematics and geographic history of the genus *Pyrgus* in the Palaearctic region (Lep. Hesperiidae). *Tijdschrift voor Entomologie*, 115: 1–60.
Joos, B., Casey, T.M., Fitzgerald, T.D. & Buttemer, W.A., 1988. Roles of the tent in behavioural thermoregulation of eastern tent caterpillars. *Ecology*, 69: 2004–2011.
Jordano, D., Retamosa, E.C. & Fernandez, H.J, 1991. Factors facilitating the continued presence of *Colias evagore* (Klug 1829) in southern Spain. *Journal of Biogeography*, 18: 637–646.
Jouzel, J., Lorius, C., Petit, J.R., Geuthon, C., Barkov, N.I., Kotlyakov, V.M. & Petrov, V.M., 1987. Vostok ice core: a continuous isotope temperature record over the last climatic cycle (160,000 years). *Nature*, 329: 403–406.
Kamm, J.A., 1972. Photoperiodic regulation of growth in an insect; response to progressive changes in daylength. *Journal of Insect Physiology*, 18: 1745–1749.
Kammer, A.E., 1968. Motor patterns during flight and warm-up in Lepidoptera. *Journal of Experimental Biology*, 48: 89–109.
Kammer, A.E., 1970. Thoracic temperatures, shivering and flight in the monarch butterfly, *Danaus plexippus* (L.). *Zeitschrift für vergleichende physiologie*, 68: 334–344.
Kareiva, P., 1987. Habitat fragmentation and the stability of predator-prey interactions. *Nature*, 326: 388–390.
Karlsson, B., 1987. Variation in egg weight, oviposition rate and reproductive

reserves with female age in a natural population of the speckled wood butterfly, *Pararge aegeria*. *Ecological Entomology*, 12: 473–476.

Karlsson, B. & Wickman, P.-O., 1989. The cost of prolonged life: an experiment on a nymphalid butterfly. *Functional Ecology*, 3: 399–405.

Karlsson, B. & Wickman, P.-O., 1990. Allometric increase in fecundity as explained by body size and resource allocation in the speckled wood butterfly, *Pararge aegeria*. Unpublished manuscript.

Karlsson, B. & Wiklund, C., 1985. Egg weight variation in relation to egg mortality and starvation endurance of newly hatched larvae in some satyrid butterflies. *Ecological Entomology*, 10: 205–211.

Kelly, P.M., Munro, M.A.R., Hughes, M.K. & Goodess, M., 1989. Climate and signature years in west European oaks. *Nature*, 340: 57–60.

Kelly, P.M. & Wigley, T.M.L., 1990. The influence of solar forcing trends on global mean temperature since 1861. *Nature*, 347: 460–462.

Kerr, R.A., 1992. A revisionist timetable for the ice ages. *Science*, 258: 220–221.

Kershaw, A.P., 1986. Climatic change and Aboriginal burning in north-east Australia during the last two glacial/interglacial cycles. *Nature*, 322: 47–49.

Kevan, P.G. & Baker, H.G., 1983. Insects as flower visitors and pollinators. *Annual Review of Entomology*, 28: 407–453.

Kevan, P.G., Jensen, T.S. & Shorthouse, J.D., 1982. Body temperatures and behavioural thermoregulation of high arctic woolly bear caterpillars and pupae (*Gynaephora rossi*, Lymantriiidae: Lep.) and the importance of sunshine. *Arctic and Alpine Research*, 14: 125–136.

Kevan, P.G. & Shorthouse, J.D., 1970. Behavioural thermoregulation by high arctic butterflies. *Arctic*, 23: 268–279.

Kimball, B.A., 1983a. Carbon dioxide and agricultural yield: An assemblage and analysis of 430 prior observations. *Agronomy Journal*, 75: 779–788.

Kimball, B.A., 1983b. *Carbon Dioxide and Agricultural Yield: An Analysis of 770 Prior Observations*. Washington, DC: US Water Conservation Laboratory.

Kingsolver, J.G., 1983. Ecological significance of flight activity in *Colias* butterflies: implications for reproductive strategy and population structure. *Ecology*, 64: 546–551.

Kingsolver, J.G., 1985a. Thermal ecology of *Pieris* butterflies (Lep: Pieridae): a new mechanism of behavioural thermoregulation. *Oecologia*, 66: 540–545.

Kingsolver, J.G., 1985b. Thermoregulatory significance of wing melanization in *Pieris* butterflies (Lep: Pieridae): physics, posture and pattern. *Oecologia*, 66: 546–553.

Kingsolver, J.G., 1985c. Butterfly thermoregulation: organismic mechanisms and population consequences. *Journal of Research on the Lepidoptera*, 24: 1–20.

Kingsolver, J.G., 1987a. Evolution and coadaptation of thermoregulatory behaviour and wing pigmentation pattern in pierid butterflies. *Evolution*, 41: 472–490.

Kingsolver, J.G., 1987b. Predation, thermoregulation and wing colour in pierid butterflies. *Oecologia*, 73: 301–306.

Kingsolver, J.G., 1988. Thermoregulation, flight, and the evolution of wing pattern in pierid butterflies: the topography of adaptive landscapes. *American Zoologist*, 28: 899–912.

Kingsolver, J.G. & Daniel, T.L., 1979. On the mechanics and energetics of nectar feeding in butterflies. *Journal of Theoretical Biology*, 76: 167–179.

Kingsolver, J.G. & Koehl, M.A.R., 1985. Aerodynamics, thermoregulation, and the evolution of insect wings: differential scaling and evolutionary change. *Evolution*, 39: 488–504.

Kingsolver, J.G. & Moffat, R.J., 1982. Thermoregulation and the determinants of heat transfer in *Colias* butterflies. *Oecologia*, 53: 27–33.

Kingsolver, J.G. & Watt, W.B., 1983. Thermoregulatory strategies in *Colias* butterflies: thermal stress and the limits to adaptation in temporally varying environments. *American Naturalist*, 121: 32–55.

Kingsolver, J.G. & Wiernasz, D.C., 1987. Dissecting correlated characters: adaptive aspects of phenotypic co-variation in melanization pattern of *Pieris* butterflies. *Evolution*, 41: 491–503.

Kingsolver, J.G. & Wirnasz, D.C., 1991a. Development, function, and the quantitative genetics of wing-melanism pattern in *Pieris* butterflies. *Evolution*, 45: 1480–1492.

Kingsolver, J.G. & Wiernasz, D.C., 1991b. Seasonal polyphenism in wing-melanism pattern and thermoregulatory adaptation in *Pieris* butterflies. *The American Scientist*, 137: 816–830.

Kitano, H., 1978. Studies on the ovipositing activity and host-searching behaviour of *Apanteles glomeratus* L. (Hym. Braconidae), a parasitoid of the cabbage butterfly. *Kontyu*, Tokyo, 46: 152–161.

Kolp, O., 1976. Submarine Uferterrasen der südlichen Ost- und Nordsei als Marken des Holozänen Meeransteigs und der Überflutungsphasen der Ostsee. *Petermanns Geographische Mitteilungen*, 120: 1–23.

Krassilov, V.A., 1991. The origin of Angiosperms: new and old problems. *Trends in Ecology and Evolution*, 6: 215–220.

Krogh, A., 1948. Determination of temperature and and heat production in insects. *Zeitschrift für vergleichende Physiologie*, 31: 274–280.

Krogh, J., 1955. Notes on temperature measurements indicative of special organization in arctic and subarctic plants for utilization of radiated heat from the sun. *Physiologia Plantarum*, 8: 836–839.

Kudrna, O., 1986. *Aspects of the Conservation of Butterflies of Europe*. In Butterflies of Europe Series, Volume 8. Wiesbaden: Aula-Verlag.

Kuhn, A., 1926. Uber die Anderung des Zeichnungsmeisters von Schmetterlingen durch Temperaturreize und das Grundschema der Nymphalidenzeichnung. *Nachrichten van der Gesellschaft der Wissenschaften zu Goettingen. Mathematisch-physikalishe Klasse*: 120–141.

Kukal, O. & Duman, J.G., 1989. Switch in the overwintering strategy of two insects and latitudinal differences in cold hardiness. *Canadian Journal of Zoology*, 67: 825–827.

Kuo, C., Lindberg, C. & Thomson, D.J., 1990. Coherence established between atmospheric CO2 and global temperature. *Nature*, 343: 709–714.

Kutzbach, J.E. & Gallimore, R.G., 1988. Sensitivity of a coupled atmosphere/mixed ocean model to changes in orbital forcing at 9000 years BP. *Journal of Geophysical Research*, 93D: 803–821.

Kutzbach, J.E. & Guetter, P.J., 1986. The influence of changing orbital parameters and surface boundary conditions on climatic simulations for the past 18,000 years. *Journal of Atmospheric Science*, 41: 1726–1759.

Kutzbach, J.E. & Wright, H.E. Jr, 1985. Simulation of the climate of 18.000 years

BP – Results for the North American North-Atlantic European sector and a comparison with the geologic record of North America. *Quaternary Science Reviews*, 4: 147–187.

Lacis, A.A., Wuebbles, D.J. & Logan, J.A., 1990. Radiative forcing of climates by changes in the vertical distribution of ozone. *Journal of Geophysics Research*, 95: 9971–9981.

Laing, J.E. & Levin, D.B., 1982. A review of the biology and a bibliography of *Apanteles glomeratus* (L.) (Hym: Braconidae). *Biocontrol News Information*, 3: 7–23.

Lamas, G., 1983. How many species in your backyard? *News of the Lepidopterists' Society*, (4): 53–55.

Lamb, H.H., 1966. *The Changing Climate*. London: Methuen.

Lamb, H.H., 1977. *Climate: Present, Past and Future*. London: Methuen.

Lamb, H.H., 1982. *Climate, History and the Modern World*. London: Methuen.

Lamb, H.H. & Johnson, A.I., 1966. *Secular Variations of the Atmospheric Circulation since 1750*. Geophysical Memoirs No. 110. London: HMSO.

Landing, B.H., 1984. *Factors in the Distribution of Butterfly Colour and Behaviour Patterns – Selected Aspects*. Los Angeles: Landing.

Landsberg, H.E., 1981. *The Urban Climate*. New York: Academic Press.

Larsen, T.B., 1974. *Butterflies of the Lebanon*. Beirut: National Council for Scientific Research.

Larsen, T.B., 1976. The importance of migration to the butterfly faunas of Lebanon, East Jordan, and Egypt (Lep., Rhopalocera). *Notulae Entomologicae*, 56: 73–83.

Larsen, T.B., 1984a. *Butterflies of Saudi Arabia and its Neighbours*. London: Stacey International.

Larsen, T.B., 1984b. The zoogeographical composition and distribution of the Arabian butterflies (Lepidoptera: Rhopalocera). *Journal of Biogeography*, 11: 119–158.

Larsen, T.B., 1986. Tropical butterflies of the Mediterranean. *Nota Lepidopterologica*, 9: 63–77.

Larsen, T.B. & Larsen, K., 1980. *Butterflies of Oman*. Edinburgh: Bartholomew.

Lattin, G. de, 1967. *Grundriss der Zoogeographie*. Stuttgart: G. Fischer Verlag.

Law, R., 1979. Ecological determinants in the evolution of life histories. In R.M. Anderson, B.D. Turner & L.R. Taylor (Eds), *Population Dynamics*. pp. 81–103. Oxford: Blackwell.

Lawton, J.H., 1986. Surface availability and insect community structure: the effects of architecture and fractal dimension of plants. In B.E. Juniper & T.R.E. Southwood (Eds), *Insects and the Plant Surface*. pp. 317–331. London: Edward Arnold.

Lawton, J.H., MacGarvin, M. & Heuds, P.A., 1987. Effects of altitude on the abundance and species richness of insect herbivores on bracken. *Journal of Animal Ecology*, 56: 147–160.

Leather, S.R., 1988. Size, reproductive potential and fecundity in insects: things aren't as simple as they seem. *Oikos*, 51: 386–389.

Lederhouse, R.C., 1982. Territorial defence and lek behaviour of the black swallowtail butterfly, *Papilio polyxenes*. *Behavioural Ecology and Sociobiology*, 10: 109–118.

Lederhouse, R.C., 1983. Population structure, residency and weather related mortality in the black swallowtail butterfly, *Papilio polyxenes*. *Oecologia*, 59: 307–311.
Lee, R.E., 1989. Insect cold-hardiness: to freeze or not to freeze. *BioScience*, 39: 308–313.
Lee, R.E., Chen, C.P. & Denlinger, D.L., 1987. A rapid cold-hardening process in insects. *Science*, 238: 1414–1417.
Lee, R.E. & Denlinger, D.L. (Eds), 1991. *Insects at Low Temperature*. London: Chapman & Hall.
Lees, A.D., 1968. Photoperiodism in insects. In A.C. Giese (Ed.), *Photophysiology*. pp. 47–137. New York: Academic Press.
Lees, E., 1962a. Factors determining the distribution of the speckled wood butterfly (*Pararge aegeria*) in Gt. Britain. *Entomologist's Gazette*, 132: 101–103.
Lees, E., 1962b. On the voltinism of *Coenonympha pamphilus* L., (Lep: Satyridae). *Entomologist*, 95: 5–6.
Lees, E., 1965. Further observations on the voltinism of *Coenonympha pamphilus* (L.) (Lep., Satyridae). *Entomologist*, 98: 43–45.
Lees, E., 1969. Voltinism of *Polyommatus icarus* Rott. (Lep., Lycaenidae) in Britain. *Entomologist*, 102: 194.
Lees, E., 1970. A univoltine race of *Pieris napi* L. (Lep: Pieridae). *Entomologist*, 103: 260–262.
Lees, E. & Archer, D.M., 1980. Diapause in various populations of *Pieris napi* L. from different parts of the British Isles. *Journal of Research on the Lepidoptera*, 19: 96–100.
Leestmans, R., 1975. Etude biogéographique et écologique des Lépidoptères des îles Canaries. (Insecta, Lepidoptera). *La Laguna, Vieraea*, 4: 9–116.
Legg, G., 1978. A note on the diversity of world Lepidoptera (Rhopalocera). *Biological Journal of the Linnean Society*, 10: 343–347.
Leibenstein, H., 1976. Beyond Economic Man. Cambridge, Mass.: Harvard University Press.
Leigh, T.F. & Smith, R.F., 1959. Flight acivity of *Colias philodice eurytheme* Boisduval in response to its physical environment. *Hilgardia*, 28: 569–624.
Lelieveld, J. & Heintzenberg, J., 1992. Sulphate cooling effect on climate through in-cloud oxidation of anthropogenic SO_2. *Science*, 258: 117–120.
Levins, R., 1968. *Evolution in Changing Environments*. Princeton: University Press.
Lewin, R., 1989. Widely spread species avoid mass extinctions. *New Scientist*, 124 (25 November): 34.
Lewontin, R.C., 1965. Selection for colonizing ability. In H.G. Baker & G.L. Stebbins (Eds), *The Genetics of Colonizing Species*. New York: Academic Press.
Lewontin, R.C., 1972. Testing the theory of natural selection. *Nature*, 236: 181–182.
Liljequist, G.H., 1970. *Klimatologi*. Stockholm: GLA.
Lincoln, D.E. & Couvet, D., 1989. The effect of carbon supply on allocation to allelochemicals and caterpillar consumption of peppermint. *Oecologia*, 78: 112–114.
Lincoln, D.E., Couvet, D. & Sionit, N., 1986. Response of an insect herbivore to hostplants grown in enriched carbon dioxide atmospheres. *Oecologia*, 69: 566–570.

Loader, C. & Damman, H., 1991. Nitrogen content of foodplants and vulnerability of *Pieris rapae* to natural enemies. *Ecology*, 72: 1586–1590.

Lockwood, J.G., 1985. *World Climatic Systems*. London: Edward Arnold.

Long, A.G., 1980. The mystery of the orange tip butterfly. *Entomologist's Record and Journal of Variation*, 92: 292–293.

Longstaff, G.B., 1912. *Butterfly Hunting in Many Lands*. London.

Lorkovic, Z., 1958. Some peculiarities of spatially and sexually restricted gene exchange in the *Erebia tyndarus* group. *Cold Spring Harbour Symposium on Quantitative Biology*, 23: 319–325.

Lorkovic, Z., 1962. The genetics and reproductive isolating mechanisms of the *Pieris napi – bryoniae* group. *Journal of the Lepidopterists' Society*, 16: 5–19; 105–127.

Lorkovic, Z., 1978. Types of hybrid sterility in diurnal Lepidoptera; speciation and taxonomy. *Acta Entomologica Jugoslavica*, 14: 13–26.

Lorkovic, Z., 1990. The butterfly chromosomes and their application in systematics and phylogeny. In O. Kudrna (Ed.), *Butterflies of Europe, Volume 2, Introduction to Lepidopterology*. pp. 332–396. Wiesbaden: Aula-Verlag.

Lovelock, J.E., 1988. *The Ages of Gaia. A Biography of Our Living Earth*. London: W.W. Norton & Co.

Ludwig, K.R., Simmons, K.R., Szabo, B.J., Winograd, I.J., Landwehr, J.M., Riggs, A.C. & Hoffman, R.J., 1992. Mass spectrometric ^{230}Th–^{234}U–^{238}U dating of the Devil's Hole calcite vein. *Science*, 258: 284–287.

Lundgren, L., 1977. The role of intra- and interspecific male:male interactions in *Polyommatus icarus* Rott., and some other species of blues. *Journal of Research on the Lepidoptera*, 16: 249–264.

MacArthur, R.H., 1962. Some generalized theorems of natural selection. *Proceedings of the National Academy of Sciences, U.S.*, 48: 1893–1897.

MacArthur, R.H. & Wilson, E.O., 1967. *The Theory of Island Biogeography*. Princeton: University Press.

McCoy, E.D., 1990. The distribution of insects along elevation gradients. *Oikos*, 58: 313–322.

McKay, H.V.M., 1988. *Plant Vulnerability and Egg-laying by Butterflies*. Ph.D. thesis (CNAA), Oxford Polytechnic.

McLeod, L., 1968. Controlled environment experiments with *Precis octavia* Cram. (Nymphalidae). *Journal of Research on the Lepidoptera*, 7: 1–18.

McLeod, L., 1984. Seasonal polyphenism in African *Precis* butterflies. In R.I. Vane-Wright & P.R. Ackery (Eds), *The Biology of Butterflies*. Royal Entomological Society Symposium No. 11. pp. 313–315. London: Academic Press.

McManus, M.L., 1988. Weather, behaviour and insect dispersal. *Memoirs of the Entomological Society of Canada*, 146: 71–94.

McNeil, J.N. & Fields, P.G., 1985. Seasonal diapause development and diapause termination in the European skipper, *Thymelicus lineola*. *Journal of Insect Physiology*, 31: 467–470.

McWhirter, K.G., 1967. Quantum genetics of human blood groups and phoneme-preferences. *Heredity*, 22: 162–163.

McWhirter, K.G., 1985. Intensive natural selection of spot genotypes in stable populations of *Maniola jurtina cassiteridum*. *Heredity*, 20: 160.

Magnus, D.B.E., 1958. Experimental analysis of some 'overoptimal' sign-stimuli in the mating behaviour of the fritillary butterfly *Argynnis paphia*. *Proceedings of the 10th International Congress of Entomology*, 2: 405–418.

Majerus, M.E.N., 1979. The status of *Anthocharis cardamines hibernica* Williams (Lep., Pieridae), with special reference to the Isle of Man. *Entomologist's Gazette*, 30: 245–248.

Major, J., 1988. Endemism: a botanical perspective. In A.A. Myers & P.S. Giller (Eds), *Analytical Biogeography*. pp. 117–146. London: Chapman & Hall.

Malcolm, S.B., Cockrell, B.J. & Brower, L.P., 1987. Monarch butterfly voltinism: effects of temperature constraints at different latitudes. *Oikos*, 49: 77–82.

Manley, G., 1974. Central England temperatures: monthly means 1659–1973. *Quarterly Journal of the Royal Meteorological Society*, 100: 389–405.

Marland, G., 1989. Fossil fuels CO2 emissions: Three countries account for 50 per cent in 1988. CDIAC Communications, Winter 1989, 1–4, Carbon dioxide Information Analysis Center, Oak Ridge National Laboratory, USA.

Marshall, L.D., 1980. *Paternal Investment in Colias philodice-eurytheme butterflies (Lep: Pieridae)*. M.Sc. thesis Arizona State University.

Marshall, L.D., 1982. Male nutrient investment in the Lepidoptera: what nutrients should males invest? *American Naturalist*, 120: 273–279.

Marshall, L.D., 1985. Protein and lipid composition of *Colias philodice* and *C. eurytheme* spermatophores and their changes over time (Pieridae). *Journal of Research on the Lepidoptera*, 24: 21–30.

Marshall, L.G., 1988. Extinction. In A.A. Myers & P.S. Giller (Eds), *Analytical Biogeography*. pp. 219–254. London: Chapman & Hall.

Masaki, S., 1973. Climatic adaptation and photoperiodic response in the band-legged ground cricket. *Evolution*, 26: 587–600.

Masaki, S., 1978. Seasonal and latitudinal adaptations in the life-cycles of crickets. In H. Dingle (Ed.), *Evolution of Insect Migration and Diapause*. pp. 72–100. New York: Springer-Verlag.

Masaki, S., 1980. Summer diapause. *Annual Review of Entomology*, 251: 1–25.

Masó, A., 1992. *Danaus chrysippus* (L.) en Mediterranée occidentale. Migrations dès 1979 jusqu'à 1985. *Nota Lepidopterologica*, Supplement No. 3: 53–60.

Mathavan, S. & Muthukrishnan, J., 1976. Effects of ration levels and restriction of feeding durations on food utilization in *Danaus chrysippus*. *Entomologia Experimentalis et Applicata*, 19: 155–162.

Mathavan, S. & Pandian, T.J., 1975. Effects of temperature on food utilization in the monarch butterfly *Danaus chrysippus*. *Oikos*, 26: 60–64.

Matheson, J.W. and Decker, G.C., 1965. Development of the European corn borer at controlled constant and variable temperatures. *Journal of Economic Entomology*, 58: 344–349.

Mattson, W.J., 1980. Herbivory in relation to plant nitrogen content. *Annual Review of Ecology and Systematics*, 11: 119–161.

May, M.L., 1979. Insect thermoregulation. *Annual Review of Entomology*, 24: 313–349.

May, P.G., 1985. Nectar uptake rates and optimal nectar concentrations of two butterfly species. *Oecologia*, 66: 381–386.

May, R.M., 1981. (2nd Edition) *Theoretical Ecology: Principles and Application*. Oxford: Blackwell Scientific Publications.

Mayes, J.C., 1991. Recent trends in summer rainfall in the United Kingdom. *Weather*, 46: 190–196.
Mayr, E., 1942. *Systematics and the Origin of Species*. New York: Columbia University Press.
Mayr, E., 1965. What is a fauna? *Zoologisches Jabrbuch der Systematik*, 92: 473–486.
Mazur, P., 1984. Freezing of living cells: mechanisms and implications. *American Journal of Physiology*, 247: C125–C142 (Cell Physiology, 16).
Meier, M.F., 1990. Greenhouse effect. Reduced rise in sea level. *Nature*, 343: 115–116.
Melillo, T.V., Callaghan, F.I., Woodward, F.I., Salati, E. & Sinha, S.K., 1990. Effects on ecosystems. In J.T. Houghton, G.J. Jenkins & J.J. Ephraums (Eds), *Climate Change. The IPCC Scientific Assessment*. pp. 283–310. Cambridge: University Press.
Mell, R., 1931. Die Trockenzeitform als Hemmungserscheinung (*Diagora nigrivena* Leech) als Trockenzeitform von *Hestina assimilis* (L.). *Biologisches Zentralblatt.*, 51: 187–194.
Mellars, P., 1974. The Palaeolithic and Mesolithic. In C. Renfrew (Ed.), *British Prehistory: A New Outline*. pp. 41–99. London: Duckworth.
Melling, T., 1984. Extraordinary survival abilities of larvae of the large heath: *Coenonympha tullia* (Müll). *Entomologist's Record and Journal of Variation*, 96: 236.
Melling, T., 1987. *The Ecology and Population Structure of a Butterfly Cline*. Ph.D., thesis, University of Newcastle-upon-Tyne.
Meteorological Office, 1973. Climatological Memorandum No. 51A. *Averages of Mean Sea Level Barometric Pressure at 9h for the United Kingdom 1941–1970*. Bracknell, Berks.: Climatological Services.
Meteorological Office, 1975a. Climatological Memorandum No. 72. *Maps of Average Bright Sunshine over the United Kingdom 1941–1970*. Bracknell, Berks.: Climatological Services.
Meteorological Office, 1975b. Climatological Memorandum No. 73. *Maps of Mean and Extreme Temperatures over the United Kingdom 1941–1970*. Bracknell, Berks.: Climatological Services.
Meteorological Office, 1980–83. *Tables of Temperature, Relative Humidity, Precipitation and Sunshine for the World*. London: HMSO. Pt 1–4.
Michaels, P., 1991. Global pollution's silver lining. *New Scientist*, 23 November: 40–49.
Mikkola, K., 1976. On the pairing and number of generations in Finland of *Aglais urticae* (Lepidoptera: Nymphalidae). *Notulae Entomoligicae*, 56: 99–102.
Mikkola, K. & Kononenko, V.S., 1989. Flight year of the alternate-year *Xestia* moths (Lep, Noctuidae) in north-eastern Siberia. A character from the ice ages? *Nota Lepidopterologica*, 12: 144–152.
Miles, J., 1981. *Effect of Birch on Moorlands*. Cambridge: Institute of Terrestrial Ecology.
Miles, P.W., Aspinall, D. & Correll, A.T., 1982. The performance of two chewing insects on water-stressed foodplants in relation to changes in their chemical composition. *Australian Journal Zoology*, 30: 347–355.

Miller, K., 1982. Cold-hardiness strategies of some adult and immature insects overwintering in interior Alaska. *Comparative Biochemistry and Physiology*, 73A: 595–604.
Miller, K.F., 1985. *Windthrow Hazard Classification*. Forestry Commission Leaflet 85.
Miller, K.G., Fairbanks, R.G. & Mountain, G.S., 1987. Tertiary isotope synthesis, sea level history and continental margin erosion. *Paleoceanography*, 2: 1–19.
Milne, A., 1957. The natural control of insect populations. *Canadian Entomologist*, 89: 193–213.
Mitchell, J.F.B., Manabe, S., Tokioka, T. & Meleshko, V., 1990. Equilibrium climate change. In J.T. Houghton, G.J. Jenkins & J.J. Ephraums (Eds), *Climate Change. The IPCC Scientific Assessment*. pp. 131–172. Cambridge: University Press.
Mitchell, J.F.B., Senior, C.A. & Ingram, W.J., 1989. CO_2 and climate: a missing feedback? *Nature*, 341: 132–134.
Mitchell, J.F.B. & Zeng, Q.-C., 1991. Climatic change prediction. *The Meteorological Magazine*, 120: 153–163.
Mitter, C., Farrell, B. & Wiegmann, B., 1988. The phylogenetic study of adaptive zones: has phytophagy promoted insect diversification? *American Naturalist*, 132: 107–128.
Monteith, G.B., 1982. Dry season aggregations of insects in Australian monsoon forests. *Memoirs of the Queensland Museum*, 20: 533–543.
Montesinos, J.L.V., 1985. Diversity and species richness of butterflies and skippers in central Spain habitats. *Journal of Research on the Lepidoptera*, 24: 364–371.
Mooney, H.A., Drake, B.G., Luxmoore, R.J., Oechel, W.C. & Pitelka, L.F., 1991. Predicting Ecosystem responses to elevated CO_2 concentrations. *BioScience*, 41: 96–104.
Moore, P.D., 1989. No smoke without fire. *Nature*, 342: 226–227.
Moore, R.A. & Singer, M.C., 1987. Effects of maternal age and adult diet on egg weight in the butterfly *Euphydryas editha*. *Ecological Entomology*, 12: 401–408.
Moore, S.D., 1987. Male-biased mortality in the butterfly *Euphydryas editha*: a novel cost of mate-acquisition. *American Naturalist*, 130: 306–309.
Morison, J.I.L., 1990. Plant and ecosystem responses to increasing atmospheric CO_2. *Trends in Ecology and Evolution*, 5: 69–70.
Morison, J.I.L. & Gifford, R.M., 1984. Plant growth and water use with limited water supply in high CO_2 concentrations. I. Leaf area, water use and transpiration. *Australian Journal of Plant Physiology*, 11: 361–374.
Morse, D.R., Lawton, J.H., Dodson, M.M. & Williamson, M.H., 1985. Fractal dimension of vegetation and the distribution of arthropod body lengths. *Nature*, 314: 731–733.
Morton, A.C.G., 1985. *The Population Biology of an Insect with a Restricted Distribution: Cupido minimus Fuessley* (Lep. Lycaenidae), Ph.D. thesis, University of Southampton.
Mossman, R., 1896. The meteorology of Edinburgh. *Transactions of the Royal Society of Edinburgh*, 1: 63–207.
Murphy, D.D., 1982. *The Role of the Adult Resources in the Population Biology of Checkerspot Butterflies of the Genus Euphydryas*. Ph.D. Thesis, Stanford University, Palo Alto, CA.

Murphy, D.D., 1983. Nectar sources as constraints on the distribution of egg masses by the checkerspot butterfly, *Euphydryas chalcedona* (Lep: Nymphalidae). *Environmental Entomology*, 12: 463–466.
Murphy, D.D., 1986. Factors to the distribution of butterfly colour and behaviour patterns – selected aspects. (Review.) *Journal of Research on the Lepidoptera*, 24: 376–379.
Murphy, D.D., Launer, A.E. & Ehrlich, P.R., 1983. The role of adult feeding in egg production and population dynamics of the checkerspot butterfly *Euphydryas editha*. *Oecologia, Berlin*, 56: 257–263.
Murray, M.B., Cannell, M.G. & Smith, R.I., 1989. Date of budburst of fifteen tree species in Britain following climatic warming. *Journal of Applied Ecology*, 26: 693–700.
Musk, L.F., 1985. Glacial and post-glacial climatic conditions in North-west England. In R.H. Johnson, *The Geomorphology of North-west England*. pp. 59–79. Manchester: University Press.
Musk, L.F., 1988. *Weather Systems*. Cambridge: University Press.
Muyshondt, A. & Muyshondt, A. Jr, 1974. Gregarious seasonal roosting of *Smyrna Kaewinskii* adults in El Salvador (Nymphalidae). *Journal of the Lepidopterists' Society*, 28: 224–229.
Myers, A.A. & Giller, P.S., 1988. *Analytical Biogeography*. London: Chapman & Hall.
Myers, J.H., 1985. Effect of physiological condition of the host plant on the ovipositional choice of the cabbage white butterfly, *Pieris rapae*. *Journal of Animal Ecology*, 54: 193–204.
Nakasuji, F. & Kimura, M., 1984. Seasonal polyphenism of egg size in a migrant skipper, *Parnara guttata guttata* (Lep: Hesperiidae). *Kontyu*, Tokyo, 52: 253–259.
Neilson, R.P., 1986. High resolution climatic analysis and Southwest biogeography. *Science*, 232: 27–34.
Neilson, R.P., 1987. Biotic regionalization and climatic controls in western North America. *Vegetatio*, 70: 135–147.
Nekrutenko, Y.P., 1965. Tertiary nymphalid butterflies and some phylogenetic aspects of systematic Lepidopterology. *Journal of Research on the Lepidoptera*, 4: 149–158.
Neville, A.C. & Weis-Fogh, T., 1963. The effect of temperatures on locust flight muscle. *Journal of Experimental Biology*, 40: 111–121.
Nicolai, V., 1986. The bark of trees: thermal properties, micro-climate and fauna. *Oecologia*, 69: 148–160.
Nie, N.H., Hull, C.H., Jenkins, J.G., Steinbrenner, K. & Bent, D.H., 1975. *Statistical Package for the Social Sciences*. (2nd Edition.) London: McGraw-Hill.
Nielsen, E.T. & Nielsen, H.T., 1959. Temperature preferred by the pierid *Ascia monuste L*. *Ecology*, 40: 181–185.
Nijhout, H.F., 1981. The color patterns of butterflies and moths. *Scientific American*, 245: 104–115.
Nijhout, H.F., 1984. Colour pattern modification by coldshock in Lepidoptera. *Journal of Embryology and Experimental Morphology*, 81: 287–305.
Nijhout, H.F., 1985. The developmental physiology of colour patterns in Lepidoptera. *Advances in Insect Physiology*, 18: 181–247.

Nijhout, H.F., 1991. *The Development and Evolution of Butterfly Wing Patterns.* Washington DC: Smithsonian Institute Press.

Nijhout, H.F. & Wray, G.A., 1986. Homologies in the colour patterns of the genus *Charaxes* (Lep: Nymphalidae). *Biological Journal of the Linnean Society*, 28: 387–410.

Nijhout, H.F., Wray, G.A. & Gilbert, L.E., 1990. An analysis of the phenotypic effects of certain colour genes in *Heliconius* (Lep: Nymphalidae). *Biological Journal of the Linnean Society*, 40: 357–372.

Nishida, T. 1955. The phenomenon of arrested insect development in the Hawaiian islands. *Proceedings of the Hawaiian Entomological Society*, 15: 575–582.

Norby, R.J., Pastor, J. & Melillo, J.M., 1986. Carbon–nitrogen interactions in CO_2 enriched white oak: physiological and long-term perspectives. *Tree Physiology*, 2: 223–241.

Nordtug, T. & Melo, T.B., 1988. Diurnal variations in natural light conditions at summertime in arctic and subarctic areas in relation to light detection in insects. *Holarctic Ecology*, 11: 202–209.

Nylin, S., 1988. Hostplant specialization and seasonality in a polyphagous butterfly, *Polygonia c-album* (Nymphalidae). *Oikos*, 53: 381–386.

Nylin, S., 1989. Effects of changing photoperiod in the life-cycle regulation of the comma butterfly, *Polygonia c-album* (Nymphalidae). *Ecological Entomology*, 14: 209–218.

Nylin, S., 1992. Seasonal plasticity in life history traits: growth and development in *Polygonia c-album* (Lep., Nymphalidae). *Biological Journal of the Linnean Society*, 47: 301–323.

Nylin, S. & Svard, L., 1991. Latitudinal patterns in the size of European butterflies. *Holarctic Ecology*, 14: 192–202.

Nylin, S., Wickman, P.-O. & Wiklund, C., 1989. Seasonal plasticity in growth and development of the speckled wood butterfly, *Pararge aegeria* (Satyrinae). *Biological Journal of the Linnean Society*, 38: 155–171.

Oechel, W.C. & Riechers, G.H., 1986. Impacts of increasing CO_2 on natural vegetation, particularly the tundra. In *Proceedings of Climate-Vegetation Workshop*, 36–42. NASA/GSFC, Greenbelt, MD, 27–29 January 1986.

Oechel, W.C. & Strain, B.R., 1985. Native species responses to increased carbon dioxide concentration. In B.R. Strain & J.D. Cure (Eds), *Direct Effects of Increasing Carbon Dioxide on Vegetation.* pp. 117–154. Washington DC: US Department of Energy.

Oerlemans, J. & Fortvin, J.P.F., 1992. Sensitivity of glaciers and small ice caps to greenhouse warming. *Science*, 258: 115–117.

Officer, C.B., Hallam, A., Drake, C.L. & Devine J.D., 1987. Late Cretaceous and paroxysmal Cretaceous/ Tertiary extinctions. *Nature*, 326: 143–149.

Ohsaki, N., 1982. Comparative population studies of three *Pieris* butterflies, *P. rapae, P. melete* and *P. napi*, living in the same area. III. Differences in the annual generation numbers in relation to habitat selection by adults. *Research in Population Ecology*, 24, 193–210.

Oke, T.R., 1987. *Boundary Layer Climates*. London: Methuen.

Osbrink, W.E., Trumble, J.T. & Wagner, R.E., 1987. Host suitability of *Phaseolus lunata* for *Trichoplusia ni* (Lepidoptera: Noctuidae) in a controlled carbon dioxide atmosphere. *Environmental Entomology*, 16: 639–644.

Overpeck, J.T., Rind, D. & Goldberg, R., 1990. Climate-induced changes in forest disturbance and vegetation. *Nature*, 343: 51–53.

Owen, D.F., 1971. *Tropical Butterflies*. Oxford: Clarendon Press.

Owen, D.F., 1980. *Camouflage and mimicry*. Oxford: University Press.

Owen, D.F. & Chanter, D.O., 1972. Species diversity and seasonal abundance in *Charaxes butterflies* (Nymphalidae). *Journal of Entomology* (A), 46: 135–143.

Oxford Dictionary for Scientific Writers and Editors, 1991. Oxford: University Press.

Papageorgis, C., 1975. Mimicry in Neotropical butterflies – Why are there so many different wing-coloration complexes in one place? *American Scientist*, 63: 522–532.

Parker, D.E., 1988. The climate of England and Wales during the past 150 years. *Journal of the Royal Agricultural Society*, 149: 90–101.

Parker, L., 1984. The ecological genetics of the speckled wood butterfly, *Pararge aegeria* L. – a preliminary study. *Heredity*, 52: 179–188.

Parry, M.L., 1978. *Climatic Change, Agriculture and Settlement*. Folkestone, Kent: Wm Dawson & Sons Ltd., Cannon House.

Parry, M.L., 1981. Evaluating the impact of climatic change. In C. Delano-Smith & M.L. Parry (Eds), *Consequences of Climatic Change*. pp. 3–16. Nottingham: University Press.

Parry, M.L., Porter, J.H. & Carter, T.R., 1990. Agriculture: climatic change and its implications. *Trends in Ecology & Evolution*, 5: 318–322.

Parsons, P.A., 1988. Evolutionary rates: effects of stress upon recombination. *Biological Journal of the Linnean Society*, 35: 49–58.

Partridge, L. & Harvey, P.H., 1988. The ecological context of life history evolution. *Science*, 241: 1449–1455.

Pastor, J. & Post, W.M., 1988. Response of northern forests to CO_2-induced climate change. *Nature*, 334: 55–58.

Patterson, C., 1983. Aims and methods in biogeography. *Systematics Association* Special Volume, 23: 1–28.

Pears, N., 1977. *Basic Biogeography*: London: Longman.

Perry, D.A., Borchers, J.G., Turner, D.P., Gregory, S.V., Perry, C.R., Dixon, R.K., Hart, S.C., Kauffman, B., Neilson, R.P. & Sollins, P. 1991. Biological feedbacks to climatic change. Terrestrial ecosystems as sinks and sources of carbon and nitrogen. *The Northwest Environmental Journal*, 7: 203–232.

Peterken, G.F. & Game, M., 1984. Historical factors affecting the number and distribution of vascular plant species in the woodlands of central Lincolnshire. *Journal of Ecology*, 72: 155–182.

Peters, R.H., 1991. *A Critique for Ecology*. Cambridge: University Press.

Peters, R.L. & Darling, D.J.S., 1985. The greenhouse effect and nature reserves. *BioScience*, 35: 707–717.

Petersen, B., 1947. Die geographische Variation einiger Fenno-Skandischer Lepidopteren. *Zoologiska Bidragfran Uppsala*, 26: 329–531.

Petersen, B., 1952. The relations between *Pieris napi and Pieris bryoniae. Transactions of the International Congress of Entomology* (9), 1: 83.

Pettier, W.R. & Tushingham, A.M., 1989. Global sea level rise and the greenhouse effect. Might they be connected?. *Science, New York*, 244: 806–810.

Pianka, E.R., 1966. Latitudinal gradients in species diversity; a review of concepts. *American Naturalist*, 100: 33–46.

Pianka, E.R., 1972. r- and K-selection or b and d selection? *American Naturalist*, 106: 581–588.
Pianka, E.R., 1974. *Evolutionary Ecology*. New York: Harper & Row.
Pickett, S.T.A. & White, P.S. (Eds), 1985. *The Ecology of Natural Disturbance and Patch Dynamics*. New York: Academic Press.
Pimm, S.L. & Redfearn, A., 1988. The variability of population densities. *Nature*, 334: 613–614.
Pivnick, K.A. & McNeil, J.N., 1985. Effects of nectar concentration on butterfly feeding: measured feeding rates for *Thymelicus lineola* (Lep: Hesperiidae) and a general feeding model for adult Lepidoptera. *Oecologia, Berlin*, 66: 226–237.
Pivnick, K.A. & McNeil, J.N., 1987. Diel patterns of activity of *Thymelicus lineola* adults (Lep: Hesperiidae) in relation to weather. *Ecological Entomology*, 12: 197–207.
Pivnick, K.A. & McNeil, J.N., 1988. Puddling in butterflies. *Physiological Ecology*, 12: 461–472.
Plant, C.W., 1987. *The Butterflies of the London Area*. London: Natural History Museum, c/o Passmore Edwards Museum.
Platt, C.M.R., 1989. The role of cloud microphysics in high-cloud feedback effects on climate change. *Nature*, 341: 428–429.
Pollard, E., 1977. A method for assessing changes in the abundances of butterflies. *Biological Conservation*, 12: 115–134.
Pollard, E., 1979. Population ecology and change in range of the white admiral butterfly *Ladoga camilla* L. in England. *Ecological Entomology*, 4: 61–74.
Pollard, E., 1982a. Monitoring the abundance of butterflies in relation to the management of a nature reserve. *Biological Conservation*, 24: 317–328.
Pollard, E., 1982b. Observations on the migratory behaviour of the painted lady butterfly, *Vanessa cardui* (L.) (Lep., Nymphalidae). *Entomologist's Gazette*, 33: 99–103.
Pollard, E., 1984. Fluctuations in the abundance of butterflies, 1976–1982. *Ecological Entomology*, 9: 179–188.
Pollard, E., 1988. Temperature, rainfall and butterfly numbers. *Journal of Applied Ecology*, 25: 819–828.
Pollard, E., 1991a. Changes in the flight period of the hedge brown butterfly *Pyronia tithonus* during range expansion. *Journal of Animal Ecology*, 60: 737–748.
Pollard, E., 1991b. Synchrony of population fluctuations: the dominant influence of widespread factors on local butterfly populations. *Oikos*, 60: 7–10.
Pollard, E., Hall, M.L. & Bibby, T.J., 1986. *Monitoring the Abundance of Butterflies*, 1976–1985. Peterborough: Nature Conservancy Council.
Pollard, E. & Lakhani, K.H., 1985. *Butterfly Monitoring Scheme: Effects of Weather on Abundance*. pp. 54–56. Abbots Ripton, Monks Wood Experimental Station: Institute of Terrestrial Ecology, Annual Report 1984.
Pollard, E. & Yates, T.J., 1992. The extinction and foundation of local butterfly populations in relation to population variability and other factors. *Ecological Entomology*, 17: 249–254.
Polson, A. & Tripconey, D., 1970. A virus disease of the lucerne caterpillar, *Colias electo* L. *Phytophylactica*, 2: 17–20.
Porter, A.H. & Geiger, H., 1988. Genetic and phenotypic population structure of the *Coenonympha tullia* complex (Lepidoptera: Nymphalidae: Satyrinae) in

California: no evidence for species boundaries. *Canadian Journal of Zoology*, 66: 2751–2765.

Porter, A.H. & Shapiro, A.M., 1990. Lock-and-key hypothesis of mechanical isolation in a butterfly (Lepidoptera: Pieridae) hybrid zone. *Annals of the Entomological Society of America*, 83: 107–114.

Porter, K., 1982. Basking behaviour in larvae of the butterfly *Euphydryas aurinia. Oikos*, 38: 308–312.

Porter, K., 1983. Multivoltinism in *Apanteles bignellii* and the influence of weather on synchronization with its host *Euphydryas aurinia. Entomologia experimentalis et applicata*, 34: 155–162.

Porter, K., 1992. Eggs and egglaying. In R.L.H. Dennis (Ed.), *The Ecology of Butterflies in Britain*. pp. 46–72. Oxford: University Press.

Porter, W.P. & Gates, D.M., 1969. Thermodynamic equilibria of animals. *Ecology Monographs*, 39: 227–244.

Poston, F.L., Hammond, R.B. & Pedigo, L.P., 1977. Growth and development of the painted lady on soybeans (Lep: Nymphalidae). *Journal of the Kansas Entomological Society*, 54: 387–408.

Pratt, C., 1986–87. A history and investigation into the fluctuations of *Polygonia c-album* L. The comma butterfly. *Entomologist's Record and Journal of Variation*, 98: 197–203, 244–250; 99: 21–27, 69–80.

Preece, R.C., Coxon, P. & Robinson, J.E., 1986. New biostratigraphic evidence of the Post-glacial colonization of Ireland and for Mesolithic forest disturbance. *Journal of Biogeography*, 13: 487–509.

Price, P.W., Bouton, C.E., Gross, P., McPheron, B.A., Thompson, J.N. & Weis, A.E., 1980. Interaction among three trophic levels: influence of plants on interactions between insect herbivores and natural enemies. *Annual Review of Ecology and Systematics*, 11: 41–65.

Prosser, C.L., 1973. *Comparative Animal Physiology*. Philadelphia: W.B. Saunders.

Pullin, A.S., 1986a. Influence of the foodplant, *Urtica dioica*, on larval development, feeding efficiencies, and voltinism of a specialist insect, *Inachis io. Holarctic Ecology*, 9: 72–78.

Pullin, A.S., 1986b. Effect of photoperiod and temperature on the life-cycle of different populations of the peacock butterfly *Inachis io. Entomologia experimentalis et applicata*, 41: 237–242.

Pullin, A.S., 1986c. *Life-history Strategies of the Butterflies Inachis io and Aglais urticae Feeding on Nettle Urtica dioica*, Ph.D. thesis (CNAA), Oxford Polytechnic.

Pullin, A.S., 1987a. Changes in leaf quality following clipping and regrowth of *Urtica dioica*, and consequences for a specialist insect herbivore, *Aglais urticae. Oikos*, 49: 39–45.

Pullin, A.S., 1987b. Adult feeding time, lipid accumulation, and overwintering in *Aglais urticae* and *Inachis io* (Lep: Nymphalidae). *Journal of Zoology*, London, 211: 631–641.

Pullin, A.S., 1988. Environmental cues and variable voltinism patterns in *Aglais urticae* (L.) (Lep., Nymphalidae). *Entomologist's Gazette*, 39: 101–112.

Pullin, A.S., 1992. Butterfly overwintering strategies. *Butterfly Conservation News*, 52: 43–47.

Pullin, A.S. & Bale, J.S., 1989a. Effects of low temperature on diapausing *Aglais urticae* and *Inachis io* (Lep: Nymphalidae): cold hardiness and overwintering survival. *Journal of Insect Physiology*, 35: 277–281.

Pullin, A.S. & Bale, J.S., 1989b. Effects of low temperature on diapausing *Aglais urticae* and *Inachis io* (Lep: Nymphalidae): overwintering physiology. *Journal of Insect Physiology*, 35: 283–290.

Pullin, A.S. & Bale, J.S., 1989c. Influence of diapuase and temperature on cryoprotectant synthesis and cold hardiness in pupae of *Pieris brassicae*. *Comparative Biochemistry and Physiology*, 94A: 499–503.

Pullin, A.S. & Bale, J.S., 1989d. Effects of ecdysone, juvenile hormone and haemolymph transfer on cryoprotectant metabolism in diapausing and non-diapausing pupae of *Pieris brassicae*. *Journal of Insect Physiology*, 35: 911–918.

Pullin, A.S., Bale, J.S. & Fontaine, X.L.R., 1991. Physiological aspects of diapause and cold tolerance during overwintering in *Pieris brassicae*. *Physiological Entomology*, 16: 447–456.

Pullin, A.S. & Knight, T.M., 1992. Induction and termination of reproductive diapause in a neotropical beetle, *Chelymorpha alternans* (Coleoptera: Chrysomelidae). *Journal of Zoology*, 227: 509–516.

Rackham, O. 1974. Temperatures of plant communities as measured by pyrometer and other methods. In G.C. Bainbridge, R. Bainbridge & O. Rackham (Eds), *Light as an Ecological Factor II*. pp. 423–449. Oxford: Blackwell.

Rackham, O., 1986. *The History of the Countryside*. London: J.M. Dent & Sons Ltd.

Rahman, M., 1969. Effects of different foods on the development of *Pieris rapae* L. larvae (Lep: Pieridae). *Pakistan Journal of Zoology*, 1: 35–40.

Ramanathan, V., 1988. The greenhouse theory of climate change: a test by inadvertent global experiment. *Science*, 240: 293–299.

Rapport, D.J., 1991. Myths in the foundations of economics and ecology. *Biological Journal of the Linnean Society*, 44: 185–202.

Raup, D.M. & Sepkoski, J.J., 1984. Periodicity of extinctions in the geologic past. *Proceedings of the National Academy of Sciences (USA)*, 81: 801–805.

Raup, D.M. & Sepkoski, J.J., 1988. Testing for periodicity of extinction. *Science*, 241: 94–96.

Rawlins, J.E., 1980. Thermoregulation by the black swallowtail butterfly, *Papilio polyxenes* (Lep: Papilionidae). *Ecology*, 61: 345–357.

Rawlins, J.E. & Lederhouse, R.C., 1978. The influence of environmental factors on roosting in the black swallowtail, *Papilio polyxenes asterius* Stoll (Papilioidae). *Journal of the Lepidopterist's Society*, 32: 145–159.

Rawlins, J.E. & Lederhouse, R.C., 1981. Developmental influences of thermal behaviour on monarch caterpillars (*Danaus plexippus*): an adaptation for migration (Lep: Nymphalidae: Danainae). *Journal of the Kansas Entomological Society*, 54: 387–408.

Raymo, M.E., Ruddiman, W.F. & Froelich, P.N., 1988. Influence of Late Cenozoic mountain building on ocean geochemical cycles. *Geology*, 16: 649–653.

Reznick, D., 1985. Cost of reproduction: an evaluation of the empirical evidence. *Oikos*, 44: 257–267.

Ricklefs, R.E., 1979. *Ecology*. Newton, MA: Chiron Press.

Rienks, J.H., 1985. Phenotypic response to photoperiod and temperature in a tropical pierid butterfly. *Australian Journal of Zoology*, 33: 837–847.

Riley, D. & Spolton, L., 1974. *World Weather & Climate*. Cambridge: University Press.

Rind, D., 1990. Puzzles from the tropics. *Nature*, 346: 317–318.

Rind, D., Chiou, E. W., Chu, W., Larsen, J., Oltmans, S., Lerner, J., McCormick, M.P. & McMaster, L., 1991. Positive water vapour feedback in climate models confirmed by satellite data. *Nature*, 349: 500–503.

Ritland, D.B., 1986. The effect of temperature on expression of the dark phenotype in female *Papilio glaucus* (Papilionidae). *Journal of Research on the Lepidoptera*, 25: 179–187.

Ritland, D.B. & Scriber, J.M., 1985. Larval development rates of three putative subspecies of tiger swallowtail butterflies, *Papilio glaucus*, and their hybrids in relation to temperature. *Oecologia*, 65: 185–193.

Robbins, R.K., 1980. The lycaenid 'false head' hypothesis: historical review and historical analysis. *Journal of the Lepidopterists' Society*, 34: 194–208.

Robbins, R.K., 1981. The 'false head' hypothesis: predation and wing pattern variation of lycaenid butterflies. *American Naturalist*, 118: 770–775.

Roberston, F.W., 1960a. The ecological genetics of growth in *Drosophila* I. Body size and development time on different diets. *Genetical Research, Cambridge*, 1: 288–304.

Roberston, F.W., 1960b. The ecological genetics of growth in *Drosophila* II. Selection for large body size on different diets. *Genetical Research, Cambridge*, 1: 305–318.

Robertson, T.S., 1980a. Weather and variation in *Argynnis paphia* (L.). *Entomologist's Gazette*, 31: 93–102.

Robertson, T.S., 1980b. Seasonal variation in *Pararge* (L.). (Lep., Satyridae). *Entomologist's Gazette*, 31: 151–156.

Robinson, R., 1971. *Lepidoptera Genetics*. Oxford: Pergamon Press.

Roff, D., 1980. Optimizing development time in a seasonal environment: the 'ups and downs' of clinal variation. *Oecologia*, 45: 202–208.

Roff, D., 1983. Phenological adaptation in a seasonal environment: a theoretical perspective. In V.K. Brown & I. Hodek (Eds), *Diapause and Life-cycle Strategies in Insects*. pp. 253–270. The Hague: Junk.

Roff, D.A., 1990. Understanding the evolution of insect life-cycles: the role of genetic analysis. In F. Gilbert (Ed.), *Insect Life Cycles. Genetics, Evolution and Co-ordination*. pp. 3–27, London: Springer-Verlag.

Roland, J., 1982. Melanism and diel activity of alpine *Colias* (Lep: Pieridae). *Oecologia*, 53: 214–221.

Rosen, B.R., 1988. Biogeographical patterns: a perceptual overview. In A.A. Myers & P.S. Giller (Eds), *Analytical Biogeography*. pp. 23–55. London: Chapman & Hall.

Rosenberg, N.J., 1974. *Microclimate: The Biological Environment*. New York: Wiley.

Rosenberg, N.J., 1981. The increasing CO_2 concentration in the atmosphere and its implication on agricultural productivity. I Effects on photosynthesis, transpiration and water use efficiency. *Climatic Change*, 3: 265–279.

Rossignol-Strick, M. & Planchais, N., 1989. Climate patterns revealed by pollen and oxygen isotope records of a Tyrrhenian sea core. *Nature*, 342: 413–416.

Rothschild, M., 1981. The mimicrats must move with the times. *Biological Journal of the Linnean Society*, 16: 21–23.

Rousseau, D.-D. & Puisségur, J.-J., 1990. A 350,000-year climatic record from the loses sequence of Achenheim, Alsace, France. *Boreas*, 19: 203–216.

Routledge, R.D., 1983. Evenness indices: are any admissable? *Oikos*, 40: 149–151.

Ruddiman, W.F., 1990. Changes in climate and biota on geologic time scales. *Trends in Ecology and Evolution*, 5: 285–288.

Ruddiman, W.F. & Kutzbach, J.C., 1991. Plateau uplift and climatic change. *Scientific American*, 264: 42–51.

Ruddiman, W.F. & McIntyre, A., 1981. The north Atlantic ocean during the last deglaciation. *Palaeo*, 35: 145–214.

Ruddiman, W.F. & Raymo, M.E., 1988. Northern hemisphere climatic regimes during the past 3ma: possible tectonic connections. *Philosophical Transactions of the Royal Society of London*, 318B: 411–430.

Ruddiman, W.F., Raymo, M.E. & McIntyre, A., 1986. Matuyama 41,000-year cycles. North Atlantic ocean and northern hemisphere ice sheets. *Earth and Planetary Science Letters*, 80: 117–129.

Runkle, J.R., 1989. Synchrony of regeneration, gaps and latitudinal differences in tree species diversity. *Ecology*, 70: 546–547.

Ruszczyk, A., 1987. Distribution and abundance of butterflies in the urbanization zones of Porto Alegre, Brazil. *Journal of Research on the Lepidoptera*, 25: 157–178.

Rutowski, R.L., 1980. Courtship solicitation by females of the checkered white butterfly, *Pieris protodice. Behavioural Ecology and Sociobiology*, 7: 113–117.

Rutowski, R.L., 1981. Courtship behaviour of the dainty sulfur butterfly, *Nathalis iole* with a description of a new, facultative male display (Pieridae). *Journal of Research on the Lepidoptera*, 20: 161–169.

Rutowski, R.L., 1982a. Epigamic selection by males as evidenced by courtship partner preferences in the checkered white butterfly (*Pieris protodice*). *Animal Behaviour*, 30: 108–112.

Rutowski, R.L., 1982b. Mate choice and Lepidopteran behaviour. *The Florida Entomologist*, 65: 72–82.

Rutowski, R.L., Gilchrist, G.W. & Terkanian, B., 1987. Female butterflies mated with recently mated males show reduced reproductive output. *Behavioural Ecology and Sociobiology*, 20: 319–322.

Rutowski, R.L., Long, C.E., Marshall, L.D. & Vetter, R.S., 1981. Courtship solicitation by *Colias* females (Lep: Pieridae). *American Midland Naturalist*, 105: 334–340.

Ryrholm, N., 1988. An extralimital population in a warm climatic outpost. The case of the moth *Idaea dilutaria* in Scandinavia. *International Journal of Biometeorology*, 32: 205–216.

Sage, R.F. & Sharkey, T.D., 1987. The effect of temperature on the occurrence of O_2 and CO_2 insensitive photosynthesis in field grown plants. *Plant Physiology*, 84: 658–664.

Salt, R.W., 1936. Studies on the freezing process in insects. *Technical Bulletin of the Minnesota Agricultural Experimental Station*, 116: 1–41.

Salt, R.W., 1956. Influence of moisture content and temperature on cold-hardiness of hibernating insects. *Canadian Journal of Zoology*, 34: 283–294.

Salt, R.W., 1966a. Factors influencing nucleation in supercooled insects. *Canadian Journal of Zoology*, 44: 117–133.

Salt, R.W., 1966b. Effect of cooling rate on the frezing temperature of supercooled insects. *Canadian Journal of Zoology*, 44: 655–659.
Salt, R.W., 1968. Location and quantitative aspects of ice nucleators in insects. *Canadian Journal of Zoology*, 46: 329–333.
Sato, Y., 1979. Experimental studies on parasitization by *Apanteles glomeratus* L. (Hym: Braconidae). IV. Factors leading a female to the host. *Physiological Entomology*, 4: 63–70.
Sauer, J.D., 1988. *Plant Migration. The Dynamics of Geographic Patterning in Plant Species*. Berkeley: University of California Press.
Saunders, D.S., 1981. Insect photoperiodism – the clock and the counter: a review. *Physiological Entomology*, 6: 99–116.
Saunders, D.S., 1982. *Insect Clocks*. Oxford: Pergamon Press.
Sbordoni, V. & Forestiero, S., 1984. *The World of Butterflies*. Poole, Dorset: Blandford Press.
Scali, V., 1971. Imaginal diapause and gonadal maturation of *Maniola jurtina* (Lepidoptera, Satyrinae) from Tuscany. *Journal of Animal Ecology*, 40: 467–472.
Scali, V., 1972. Spot-distribution in *Maniola jurtina*: Tuscan archipelago, 1968–70. *Heredity*, 29: 25–36.
Scali, V. & Masetti, M., 1975. Variazioni intrastagionali dello spotting e selezione in *Maniola jurtina. Atti Accademia Nazionale dei Lincei Rendiconti*, 58: 244–257.
Schaffer, W.M., 1974. Optimal reproductive effort in fluctuating environments. *American Naturalist*, 108: 783–790.
Schneider, S.H., 1989. The greenhouse effect: science and policy. *Science, New York*, 243: 771–781.
Schoener, T.W., 1988. Ecological interactions. In A.A. Myers & P.S. Giller (Eds), *Analytical Biogeography*. pp. 255–297. London: Chapman & Hall.
Schwanwitsch, B.N., 1924. On the ground plan of wing-pattern in Nymphalidae and certain other families of the rhopalocerous Lepidoptera. *Proceedings of the Zoological Society*, London, 1924: 509–528.
Scott, J.A., 1968. Hilltopping as a mating mechanism to aid the survival of a low density species. *Journal of Research on the Lepidoptera*, 7: 191–204.
Scott, J.A., 1972. Biogeography of Antillean butterflies. *Biotropica*, 4: 32–45.
Scott, J.A., 1973. Lifespan of butterflies. *Journal of Research on the Lepidoptera*, 12: 225–230.
Scott, J.A., 1979. Hibernal diapause on North American Papilionoidea and Hesperioidea. *Journal of Research on the Lepidoptera*, 18: 171–200.
Scott, J.A. & Epstein, M.E., 1986. Factors affecting phenology in a temperate insect community. *The American Midland Naturalist*, 117: 103–118.
Scott, M.J., Rosenberg, N.J., Edmonds, J.A., Cushman, R.M., Darwin, R.F., Yohe, G.W., Liebetrau, A.M., Hunsaker, C.T., Bruns, D.A., DeAngelis, D.L., Hales, J.M., 1990. Consequences of climatic change for the human environment. *Climate Research*, 1: 63–79.
Scriber, J.M., 1973. Latitudinal gradients in larval feeding specialization of the world Papilionidae (Lepidoptera). *Psyche*, 80: 355–373.
Scriber, J.M., 1984. Larval foodplant utilization by the world Papilionidae (Lepidoptera): latitudinal gradients reappraised. *Tokurana*, 6/7: 1–50.
Scriber, J.M. & Lederhouse, R.C., 1983. Temperature as a factor in the development

and feeding ecology of tiger swallowtail caterpillars, *Papilio glaucus* (Lep.). *Oikos*, 40: 95–102.

Scriber, J.M. & Slansky, F. [Jr], 1981. The nutritional ecology of immature insects. *Annual Review of Entomology*, 26: 183–211.

Sedjo, R.A. & Solomon, A.M., 1989. Climate and forests. In N.J. Rosenberg, Wm III Easterling, P.R. Crosson & J. Darmstadter (Eds), *Greenhouse Warming: Abatement and Adaptation*. pp. 105–119. Washington DC: Resources for the Future.

Sepkoski, J.J., 1989. Periodicity in extinctions and the problem of catastrophism in the history of life. *Journal of the Geological Society*, London, 146: 7–19.

Shackleton, N.J., 1989. Deep trouble for climatic change. *Nature*, 342: 616–617.

Shackleton, N.J., Imbrie, J. & Pisias, N.J., 1988. The evolution of oceanic oxygen-isotope variability in the North Atlantic over the past three million years. *Philosophical Transactions of the Royal Society of London*, 318B: 679–688.

Shackleton, N.J. & Pisias, N.J., 1985. Ocean carbon isotope constraints on oxygen and carbon in the Cenozoic atmosphere. The Carbon Cycle and Atmospheric CO_2: Natural Variations Archean to Present. *Geophysical Monograph*, 32: 412–417.

Shapiro, A.M., 1973. Altitudinal migration of butterflies in the central Sierra Nevada. *Journal of Research on the Lepidoptera*, 12: 231–235.

Shapiro, A.M., 1975a. The temporal component of butterfly species diversity. In M.L. Cody & J.M. Diamond (Eds), *Ecology and Evolution of Communities*. pp. 181–195. Cambridge, Mass.: Harvard University Press.

Shapiro, A.M., 1975b. Why do California tortoiseshells migrate? *Journal of Research on the Lepidoptera*, 14: 93–97.

Shapiro, A.M., 1976. Seasonal polyphenisms. *Evolutionary Biology*, 9: 259–333.

Shapiro, A.M., 1977. Autumnal false broods of multivoltine butterflies. *Journal of Research on the Lepidoptera*, 16: 83–86.

Shapiro, A.M., 1978. Developmental and phenotypic responses to photoperiod and temperature in an equatorial montane butterfly, *Tatochila xanthodice*. *Biotropica*, 10: 297–301.

Shapiro, A.M., 1979. The phenology of *Pieris napi microstriata* (Lep: Pieridae) during, and after the 1975–77 California drought, and its evolutionary significance. *Psyche*, 86: 1–10.

Shapiro, A.M., 1980a. Egg-load assessment and carryover diapause in *Anthocharis* (Pieridae). *Journal of the Lepidopterists' Society*, 34: 307–315.

Shapiro, A.M., 1980b. Physiological and developmental responses to photoperiod and temperature as data in phylogenetic and biogeographic inference. *Systematic Zoology*, 29: 335–341.

Shapiro, A.M., 1980c. Convergence in pierine polyphenisms (Lepidoptera). *Journal of Natural History*, 14: 781–802.

Shapiro, A.M., 1980d. Mediterranean climate and butterfly migration. An overview of the California fauna. *Atalanta*, 11: 181–188.

Shapiro, A.M., 1982a. Survival of refrigerated *Tatochila* butterflies (Lep: Pieridae) as an indicator of male nutrient investment in reproduction. *Oecologia*, 53: 139–140.

Shapiro, A.M., 1982b. Redundancy in pierid polyphenisms: pupal chilling induces vernal phenotype in *Pieris occidentalis* (Pieridae). *Journal of the Lepidopterists' Society*, 36: 174–177.

Shapiro, A.M., 1984a. Non-diapause overwintering by *Pieris rapae* (Lep: Pieridae) and *Papilio zelicaon* (Lep: Papilionidae) in California: adaptiveness of type III diapause-induction curves. *Psyche*, 91: 161–169.
Shapiro, A.M., 1984b. The genetics of seasonal polyphenisms and the evolution of 'general purpose genotypes' in butterflies. In K. Wohrmann & V. Loeschcke (Eds), *Population Biology and Evolution*. pp. 16–38. Berlin: Springer-Verlag.
Shapiro, A.M., 1984c. Experimental studies on the evolution of seasonal polyphenism. In R. Vane-Wright & P. Ackery (Eds), *The Biology of Butterflies*. Royal Entomological Symposium No. 11. pp. 297–307. London: Academic Press.
Shapiro, A.M., 1989. The zoogeography and systematics of the Argentine Andean and Patagonian pierid fauna. *Journal of Research on the Lepidoptera*, 28: 137–238.
Shaw, M.R., 1981. Parasitism by Hymenoptera of larvae of the white admiral butterfly, *Ladoga camilla* (L.) in England. *Ecological Entomology*, 6: 333–335.
Shaw, R.H., 1967. *Groundlevel Climatology*. Symposium 1965 Berkeley Publication No. 86. Washington, DC: ASA.
Shelly, T.E. & Ludwig, D., 1985. Thermoregulatory behaviour of the butterfly *Calisto nubila* (Satyridae) in a Puerto Rican forest. *Oikos*, 44: 229–233.
Shepherd, R.F., 1958. Factors controlling the internal temperature of spruce budworm larvae, *Choristoneura fumiterana* (Clem.). *Canadian Journal of Zoology*, 36: 779–786.
Sheppard, P.M., Turner, J.R.G., Brown, K.S., Benson, W.W. & Singer, M.C., 1985. Genetics and the evolution of Müllerian mimicry in *Heliconius* butterflies. *Philosophical Transactions of the Royal Society* (B), 308: 433–613.
Sherman, P.W. & Watt, W.B., 1973. The thermal ecology of some *Colias* butterfly larvae. *Journal of Comparative Physiology*, 83: 25–40.
Shields, O., 1967. Hilltopping. *Journal of Research on the Lepidoptera*, 6: 69–178.
Shields, O., 1974. Toward a theory of butterfly migration. *Journal of Research on the Lepidoptera*, 13: 217–238.
Shields, O., 1976. Fossil butterflies and the evolution of Lepidoptera. *Journal of Research on the Lepidoptera*, 15: 132–143.
Shields, O., 1989. World numbers of butterflies. *Journal of the Lepidopterists' Society*, 43: 178–183.
Shields, O. & Dvorak, S.K., 1979. Butterfly distribution and continental drift between the Americas, the Caribbean and Africa. *Journal of Natural History*, 13: 221–250.
Shreeve, T.G., 1984. Habitat selection, mate location, and microclimatic constraints on the activity of the speckled wood butterfly *Pararge aegeria*. *Oikos*, 42: 371–377.
Shreeve, T.G., 1986a. Egg-laying by the speckled wood butterfly (*Pararge aegeria*): the role of female behaviour, host plant abundance and temperature. *Ecological Entomology*, 11: 229–236.
Shreeve, T.G., 1986b. The effect of weather on the life cycle of the speckled wood butterfly *Pararge aegeria*. *Ecological Entomology*, 11: 325–332.
Shreeve, T.G., 1987. The mate location behaviour of the male speckled wood butterfly, *Pararge aegeria*, and the effect of phenotypic differences in hind-wing spotting. *Animal Behaviour*, 35: 682–690.
Shreeve, T.G., 1989. The extended flight period of *Maniola jurtina* on chalk

downland: seasonal changes of the adult phenotype and evidence for a population of mixed origins. *Entomologist*, 108: 202–215.
Shreeve, T.G., 1990. Microhabitat use and hindwing phenotype in *Hipparchia semele* (Lepidoptera, Satyrinae): thermoregulation and background matching. *Ecological Entomology*, 15: 201–213.
Shreeve, T.G., 1992. Adult behaviour, in R.L.H. Dennis (Ed.), *The Ecology of Butterflies in Britain*. pp. 22–45. Oxford: University Press.
Shreeve, T.G. & Dennis, R.L.H., 1991. How mobile is the British butterfly fauna? *BBCS News*, 48: 20–24.
Shreeve, T.G. & Dennis, R.L.H., 1992. The development of butterfly settling posture: the role of predators, climate, hostplant-habitat and phylogeny. *Biological Journal of the Linnean Society*, 45: 57–69.
Shugart, H.H., 1990. Using ecosystem models to assess potential consequences of global climatic change. *Trends in Ecology and Evolution*, 5: 303–307.
Sibatani, A., 1983. A compilation of data on wing homeosis in Lepidoptera. *Journal of Research on the Lepidoptera*, 22: 1–46.
Sibly, R.M. & Calow, P., 1989. A life-cycle theory of responses to stress. *Biological Journal of the Linnean Society*, 37: 101–116.
Sibly, R.M. & Monk, K.A., 1987. A theory of grasshopper life-cycles. *Oikos*, 48: 186–194.
Silberglied, R.E., 1984. Visual communication and sexual selection among butterflies. In R.I. Vane-Wright & P.R. Ackery (Eds), *The Biology of Butterflies*. Royal Entomological Society Symposium No. 11. pp. 207–223. London: Academic Press.
Simmons, I.G., 1990. The mid-Holocene ecological history of the moorlands of England and Wales and its relevance for conservation. *Environmental Conservation*, 17: 61–69.
Simmons, I.G. & Tooley, M.J. (Eds), 1981. *The Environment in British Prehistory*. London: Duckworth.
Sims, S.R. & Shapiro, A.M., 1983. Pupal diapause in *Battus philenor* (Lep.: Papilionidae). *Annals of the Entomological Society of America*, 76: 407–412.
Singer, M.C., 1972. Complex components of habitat suitability within a butterfly colony. *Science*, 176: 75–77.
Singer, M.C., 1982. Sexual selection for small size in male butterflies. *American Naturalist*, 119: 440–443.
Singer, M.C. & Ehrlich, P.R., 1979. Population dynamics of the checkerspot butterfly, *Euphydryas editha*. *Fortschritte der Zoologie*, 25: 53–60.
Slansky, F., 1974. Relationship of larval foodplants and voltinism patterns in temperate butterflies. *Psyche*, 81: 243–253.
Slansky, F. & Feeny, P., 1977. Stabilization of the rate of nitrogen accumulation by larvae of the cabbage butterfly on wild and cultivated food plants. *Ecological Monographs*, 47: 209–228.
Slobodkin, L.B. & Sanders, H.L., 1969. On the contribution of environmental predictability to species diversity. In G.M. Woodwell & H.H. Smith (Eds), *Diversity and Stability in Ecological Systems*. Volume 22: 82–93. Brookhaven Symposium on Biology.
Smart, P., 1985. *The Illustrated Encyclopaedia of the Butterfly World*. London: Salamander Books.
Smith, A.G., 1978. Environmental factors influencing pupal colour determination

in Lepidoptera. I. Experiments with *Papilio polytes, Papilio demoleus* and *Papilio polyxenes. Proceedings of the Royal Society of London*, B, 200: 295–329.

Smith, A.G., 1980. Environmental factors influencing pupal colour determination in Lepidoptera. II. Experiments with *Pieris rapae, Pieris napi* and *Pieris brassicae. Proceedings of the Royal Society of London*, B, 207: 163–186.

Smith, A.G., Hurley, A.M. & Briden, J.C., 1981. *Phanerozoic Palaeocontinental World Maps*. Cambridge: University Press.

Smith, D.A.S., Shoesmith, E.A. & Smith, A.G., 1988. Pupal polymorphism in the butterfly *Danaus chrysippus* (L.): environmental, seasonal and genetic influences. *Biological Journal of the Linnean Society*, 33: 17–50.

Smith, F.B., 1991. Regional air pollution, with special emphasis on Europe. *Quarterly Journal of the Meteorological Society*, 117: 657–683.

Smith, J.M., 1989. The causes of extinction. *Philosophical Transactions of the Royal Society*, London B, 325: 241–252.

Smith, R.H., Sibly, R.M. & Moller, H., 1987. Control of size and fecundity in *Pieris rapae*: towards a theory of butterfly life-cycles. *Journal of Animal Ecology*, 56: 341–350.

Sneath, P.H.A. & Sokal, R.R., 1973. *Numerical Taxonomy. The Principles and Practice of Numerical Classification*. San Francisco: W.H. Freeman & Company.

Somme, L., 1982. Supercooling and winter survival in terrestrial arthropods. *Comparative Biochemistry and Physiology*, 73A: 519–543.

Southwood, T.R.E., 1977. Habitat, the templet for ecological strategies? *Journal of Animal Ecology*, 46: 337–365.

Springer, P. & Boggs, C.L., 1987. Resource allocation to oocytes: heritable variation with altitude in *Colias philodice eriphyle* (Lep: Pieridae). *American Naturalist*, 127: 252–256.

Srygley, R.B. & Chai, P., 1990a. Predation and the elevation of thoracic temperature in brightly coloured Neotropical butterflies. *American Naturalist*, 135: 766–787.

Srygley, R.B. & Chai, P., 1990b. Flight morphology of Neotropical butterflies: palatability and distribution of mass to the thorax and abdomen. *Oecologia*, 84: 491–499.

Stamp, N.E. & Bowers, M.D., 1990. Body temperature, behaviour and growth of early spring caterpillars (*Hemileuca lusina*, Saturnidae). *Journal of the Lepidopterists' Society*, 44: 143–155.

Stanley, S.M., 1985. Rates of evolution. *Palaeobiology*, 11: 13–26.

Starkel, L., 1977. The palaeogeography of mid and east Europe during the last cold stages with west European comparisons. *Philosophical Transactions of the Royal Society*, London B, 280: 351–372.

Stearns, S.C., 1976. Life-History tactics: a review of the ideas. *The Quarterly Review of Biology*. 51: 3–47.

Stearns, S.C., 1977. The evolution of life-history traits. *Annual Review of Ecology & Systematics*. 8: 145–171.

Stearns, S.C., 1989. Trade-offs in life-history evolution. *Functional Ecology*, 3: 259–268.

Stehli, F.G., Douglas, R.G. & Newell, N.D., 1969. Generation and maintenance of gradients in taxonomic diversity. *Science*, 164: 947–949.

Stephenson, N.L., 1990. Climatic control of vegetation distribution: the role of the water balance. *American Naturalist*, 135: 649–670.

Stern, V.G. & Smith, R.F., 1960. Factors affecting egg production and oviposition in populations of *Colias eurytheme* Boisduval (Lep: Pieridae). *Hilgardia*, 29: 411–454.

Stevens, G.C., 1989. The latitudinal gradient in geographical range: how so many species coexist in the tropics. *American Naturalist*, 133: 240–256.

Stewart-Oaten, A. & Murdock, W.W., 1990. Temporal consequences of spatial density dependence. *Journal of Animal Ecology*, 59: 1027–1045.

Stiling, P., 1988. Density dependent processes and key factors in insect populations. *Journal of Animal Ecology*, 57: 581–593.

Stirling, R., 1982. *The Weather of Britain*. London: Faber & Faber.

Stone, G.N., Amos, J.N., Stone, T.F., Knight, R.L., Gay, H. & Parrott, F., 1988. Thermal effects on activity patterns and behavioural switching in a concurse of foragers on *Stachytarpheta mutabilis* (Verbenaceae) in Papua New Guinea. *Oecologia*, 77: 56–63.

Storey, K.B., 1990. Biochemical adaptation for cold-hardiness in insects. *Philosophical Transactions of the Royal Society of London*, B. 326: 635–654.

Storey, K.B. & Storey, J.M., 1988. Freeze tolerance in animals. *Physiological Review*, 68: 27–84.

Stork, N.E., 1988. Insect diversity: facts, fiction and speculation. *Biological Journal of the Linnean Society*, 35: 321–337.

Stradling, D.J., 1976. The nature of the mimetic pattens of the brassolid genera, *Caligo* and *Eryphanis*. Ecological Entomology, 1: 135–138.

Strain, B.R., 1987. Direct and indirect effects of increasing atmospheric CO_2 on plants and ecosystems. *Trends in Ecology and Evolution*, 2: 18–21.

Strangeways, I., 1991. Gaps in the climate map. *New Scientist*, 23 November: 46.

Strong, D.R. Jr, Lawton, J.H. & Southwood, T.R.E., 1984. *Insects on Plants. Community Patterns and Mechanisms*. Oxford: Blackwell Scientific Publications.

Stuart, A.J., 1974. Pleistocene history of the British vertebrate fauna. *Biological Review*, 49: 225–266.

Suc, J.P., 1984. Origin and evolution of the Mediterranean vegetation and climate in Europe. *Nature*, 307: 429–432.

Suffert, F., 1927. Zur vergleichende analyse der Schmetterlingspuppe. *Zeitschrift für Morphologie der Tiere*, 14: 338–359.

Sullivan, C.R. & Wellington, W.G., 1953. The light reactions of larvae of tent caterpillars, *Malacosoma disstria* Hbn., *M. americanum* (Fab.) and *M. pluviale* (Dyar). *Canadian Entomologist*, 85: 297–310.

Svard, L. & Wiklund, C., 1988. Fecundity, egg weight and longevity in relation to multiple matings in females of the monarch butterfly. *Behavioural Ecology and Sociobiology*, 23: 39–43.

Svard, L. & Wiklund, C., 1989. Mass and production rate of ejaculates in relation to monandry/polyandry in butterflies. *Behavioural Ecology and Sociobiology*, 24: 395–402.

Sway, C.A.M. van, 1990. An assessment of the changes in butterfly abundance in the Netherlands during the 20th century. *Biological Conservation*, 52: 287–302.

Tabashnik, B.E., 1980. Population structure of pierid butterflies. III. Pest population of *Colias philodice eriphyle*. *Oecologia*, 47: 175–183.

Takata, N., 1961. Studies on the host preference of the common cabbage butterfly

Pieris rapae crucivora (Bois). XII. Successive rearing of the cabbage butterfly larva with certain host plants and its effect on the oviposition preference of the adult. *Japanese Journal of Ecology*, 11: 147–154.
Tallis, J.H., 1991a. *Plant Community History*. London: Chapman & Hall.
Tallis, J.H., 1991b. Forest and moorland in the south Pennine uplands in the mid-Flandrian period. III. The spread of moorland – local, regional and national. *Journal of Ecology*, 79: 401–415.
Tanaka, S., Denlinger, D.L. & Wolda, H., 1987. Daylength and humidity as environmental cues for diapause termination in a tropical beetle. *Physiological Entomology*, 12: 213–224.
Tauber, M.J., Tauber, C.A. & Masaki, S., 1986. *Seasonal Adaptations of Insects*. Oxford: University Press.
Taylor, J.A., 1980. Environmental changes in Wales during the Holocene period. In J.A. Taylor (Ed.), *Culture and Environment in Prehistoric Wales*. pp. 101–130. British Archaeological Report Series 76. Oxford.
Taylor, L.R. & Taylor, R.A.J., 1977. Aggregation, migration and population mechanics. *Nature*, 265: 415–421.
Taylor, M.F.J., 1988. Field measurements of the dependence of life history on plant nitrogen and temperature for a herbivorous moth. *Journal of Animal Ecology*, 57: 873–891.
Tenow, O. & Nilssen, A., 1990. Egg cold-hardiness and topoclimatic limitations to outbreaks of *Epirrata autumnata* in northern Fennoscandia. *Journal of Applied Ecology*, 27: 723–734.
Terborgh, J., 1985. Vertical component of plant species diversity in temperate and tropical forests. *American Naturalist*, 126: 760–776.
Thomas, C.D., 1984. Oviposition and egg load assessment by *Anthocharis cardamines* (L.) (Lep: Pieridae). *Entomologist's Gazette*, 35: 145–148.
Thomas, C.D., 1985. Specializations and polyphagy of *Plebejus argus* (Lep., Lycaeidae) in North Wales. *Ecological Entomology*, 10: 325–340.
Thomas, C.D., 1991. Habitat use and geographical ranges of butterflies from the wet lowlands of Costa Rica. *Biological Conservation*, 55: 269–281.
Thomas, J.A., 1974. *Factors Influencing the Numbers and Distribution of the Brown Hairstreak Thecla betulae L., and the Black Hairstreak Strymonidia pruni L. (Lep., Lycaenidae)*. Ph.D. thesis, University of Leicester.
Thomas, J.A., 1983. The ecology and conservation of Lysandra bellargus (Lep: Lycaenidae) in Britain. *Journal of Applied Ecology*, 20: 59–83.
Thomas, J.A., 1984. The conservation of butterflies in temperate countries: past efforts and lessons for the future. In R.I. Vane-Wright & P.R. Ackery (Eds), *The Biology of Butterflies*. Royal Entomological Society Symposium 11. pp. 333–353. London: Academic Press.
Thomas, J.A., 1986. *R.S.N.C. Guide to Butterflies of the British Isles*. London: Country Life Books (Newnes Books).
Thomas, J.A., 1989. *Maculinea arion*. In A.M. Emmet & J. Heath (Eds). *The Butterflies of Great Britain and Ireland*. Volume 7, Part 1. (Hesperiidae to Nymphalidae). 171–175. Colchester: Harley Books.
Thomas, J.A. & Merrett, P., 1980. Observations of butterflies in the Purbeck Hills in 1976 and 1977. *Proceedings of the Dorset Natural History and Archaeological Society*, 99: 112–119.

Thomas, J.A. & Simcox, D.J., 1982. A quick method for estimating larval populations of *Melitaea cinxia* during surveys. *Biological Conservation*, 22: 315–322.

Thomas, J.A., Thomas, C.D., Simcox, D.J. & Clarke, R.T., 1986. Ecology and declining status of the silver-spotted skipper butterfly (*Hesperia comma*) in Britain. *Journal of Applied Ecology*, 23: 365–380.

Thompson, J.D., 1991. Pheotypic plasticity as a component of evolutionary change. *Trends in Ecology and Evolution*, 6: 246–249.

Thomsen, R., 1990. Effect of climate variability and change in groundwater in Europe. *Nordic Hydrology*, 21: 185–194.

Thomson, G., 1970a. The distribution and nature of *Pieris napi thomsoni* Warren (Lep., Pieridae). *Entomologist's Record and Journal of Variation*, 82: 255–261.

Thomson, G., 1970b. On the nature of *Maniola jurtina splendida* B. White (Lep., Satyridae). *Entomologist's Record and Journal of Variation*, 82: 261–268.

Thomson, G., 1973. Temperature effects on *Maniola jurtina* (L.) (Lep., Satyridae). *Entomologist's Record and Journal of Variation*, 85: 109–114.

Thomson, G., 1980. *The Butterflies of Scotland. A Natural History*. London: Croom Helm.

Thomson, G., 1987a. *Maniola chia* – a new satyrid from the Greek island of Chios (Lep., Nymphalidae: Satyrinae. *Phegea*, 15: 13–22.

Thomson, G., 1987b. *Enzyme Variation at Morphological Boundaries in Maniola and Related Genera*, Ph.D. thesis, University of Stirling.

Thomson, G., 1990. *Maniola halicarnassus* – a new satyrid from south-west Turkey (Lep., Nymphalidae: Satyrinae). *Phegea*, 18: 149–155.

Tindale, N.B., 1980. Origin of the Lepidoptera, with description of a new mid-Triassic species and notes on the origin of the butterfly stem. *Journal of the Lepidopterists' Society*, 34: 263–285.

Tooley, M.J., 1989. The flood behind the embankment. *Geographical Magazine*, 61 (11): 32–36.

Tooley, M.J., 1991. Future sea-level rise: a working paper. *Proceedings of the Institution of Civil Engineers*, 90: 1101–1105.

Tooley, M.J. & Sheail, G.M., 1985. *The Climatic Scene* (Essays in Honour of G. Manley). London: Allen & Unwin.

Topham, P.B. & Alphey, T.J.W., 1985. Faunistic analysis of Lonidorid nematodes in Europe. *Journal of Biogeography*, 12: 165–174.

Trivers, R.L., 1972. Parental investment and sexual selection. In B. Campbell (Ed.), *Sexual Selection and the Descent of Man 1871–1971*. pp. 136–179. Chicago: Aldine.

Tsuji, J.S., Kingsolver, J.G. & Watt, W.B., 1986. Thermal physiological ecology of *Colias* butterflies in flight. *Oecologia*, 69: 161–170.

Turchin, P., 1990. Rarity of density dependence or population with lags? *Nature*, 344: 660–663.

Turner, J., 1984. Pollen diagrams from Cross Fell and their implications for former tree-lines. In E.Y. Haworth & J.W.G. Lund (Eds), *Lake Sediments and Environmental History*. pp. 317–357. Leicester: University Press.

Turner, J. & Hodgson, J., 1979. Studies in the vegetational history of the Northern Pennines. I. Variations in the composition of the early Flandrian forests. *Journal of Ecology*, 67: 629–646.

Turner, J.R.G., 1963. A quantitative study of a Welsh colony of the large heath butterfly, *Coenonympha tullia* Müller (Lepidoptera). *Proceedings of the Royal Entomological Society*, London (A) 38: 101–112.

Turner, J.R.G., 1971. Two thousand generations of hybridization in a *Helicolius* butterfly. *Evolution*, 25: 471–482.

Turner, J.R.G., 1984. Mimicry: The palatability spectrum and its consequences. In R.I. Vane-Wright & P.R. Ackery (Eds), *The Biology of Butterflies*. Royal Entomological Society Symposium No. 11. pp. 141–161. London: Academic Press.

Turner, J.R.G., 1986. Why are there so few butterflies in Liverpool? Homage to Alfred Russel Wallace. *Antenna*, 10: 18–24.

Turner, J.R.G., 1987. The evolutionary dynamics of Batesian and Muellerian mimicry: similarities and differences. *Ecological Entomology*, 12: 81–95.

Turner, J.R.G., Gatehouse, C.M. & Corey, C.A., 1987. Does solar energy control organic diversity? Butterflies, moths and the British climate. *Oikos*, 48: 195–205.

Turner, J.R.G., Lennon, J.J. & Lawrenson, J.A., 1988. British bird species distributions and the energy theory. *Nature*, 335: 539–541.

Tyndale-Biscoe, M. & Hughes, R.D., 1968. Changes in the female reproductive system as age indicators in the bushfly *Musca retustissima* Wlk. *Bulletin of Entomological Research*, 59: 129–141.

Underwood, A.J., 1989. The analysis of stress in natural populations. *Biological Journal of the Linnean Society*, 37: 51–78.

Unwin, D.M., 1980. *Microclimate Measurement for Ecologists*. London: Academic Press.

Urquhart, F.A., 1960. *The Monarch Butterfly*. Toronto: University Press.

Utida, S., 1957. Developmental zero temperatures in insects. *Japanese Journal of Applied Entomological Zoology*, 1: 46–53.

Vane-Wright, R.I., 1976. A unified classification of mimetic resemblances. *Biological Journal of the Linnean Society*, 8: 25–56.

Vane-Wright, R.I., 1979. Towards a theory of the evolution of butterfly colour patterns under directional and disruptive selection. *Biological Journal of the Linnean Society*, 11: 141–152.

Vane-Wright, R.I., 1980. A classification of sexual interactions, and the evolution of species-specific coloration in butterflies. Nota *Lepidopterologica*, 3: 91–93.

Vane-Wright, R.I., 1984. The role of pseudosexual selection in the evolution of butterfly colour patterns. In R.I.Vane-Wright & P.R. Ackery (Eds), *The Biology of Butterflies*. Royal Entomological Society Symposium No. 11. pp. 251–253. London: Academic Press.

Varley, G.C., Gradwell, G.R. & Hassell, M.P., 1973. *Insect Population Ecology*. Oxford: Blackwell Scientific Publications.

Viejo, J.L., Galiano, E.F. & Sterling, A., 1992. The importance of riparian forests in the conservation of butterflies in central Spain. *Nota Lepidopterologica*, Supplement No. 3: 29–42.

Vielmetter, W., 1958. Physiologie des Verhaltens zur Sonnenstrahlung bei dem tagfalter *Argynnis paphia* L. – I Untersuchungen im Freiland. *Journal of Insect Physiology*, 2: 13–37.

Vincent, P., 1981. From theory into practice – a cautionary tale of island biogeography. *Area*, 13: 115–118.

Wagener, P.S., 1988. What are the valid names for the two genetically different

taxa currently included within *Pontia daplidice* (Linnaeus, 1758)? (Lep., Pieridae) *Nota Lepidopterologica*, 11: 21–38.
Waldbauer, G.P., 1978. Phenological adaptation and the polymodal emergence petterns of insects. In H. Dingle (Ed.), *Evolution of Insect Migration and Diapause.* 127–144. New York: Springer-Verlag.
Walker, M.J.C. & Lowe, J.J., 1990. Reconstructing the environmental history of the last glacial-interglacial transition: evidence from the Isle of Skye, Inner Hebrides, Scotland. *Quaternary Science Review*, 9: 15–49.
Wallace, A.R., 1876. *The Geographical Distribution of Animals*. London: MacMillan.
Wallace, A.R., 1889. *Darwinism*. London.
Wallén, C.C., 1970. *The Climates of Northern and Western Europe*. Amsterdam: Elsevier Scientific Publications.
Warren, M.S., 1984a. The biology and status of the wood white butterfly, *Leptidea sinapis* L. in the British Isles. *Entomologist's Gazette*, 35: 207–223.
Warren, M.S., 1984b. The status of the heath fritillary butterfly *Mellicta athalia* Rott. in Britain. *Biological Conservation*, 29: 287–305.
Warren, M.S., 1987. The ecology and conservation of the heath fritillary butterfly, *Mellicta athalia*. II. Adult population structure and mobility. *Journal of Applied Ecology*, 24: 483–498.
Warren, M.S., 1989. Pheasants and fritillaries: is there really any evidence that pheasant rearing may have caused butterfly declines? *British Journal of Entomology and Natural History*, 2: 169–175.
Warren, M.S., 1992a. Butterfly populations. In R.L.H. Dennis (Ed.), The *Ecology of Butterflies in Britain*. pp. 73–92. Oxford: University Press.
Warren, M.S., 1992b. The conservation of British butterflies. In R.L.H. Dennis (Ed.), *The Ecology of Butterflies in Britain*. pp. 246–274. Oxford: University Press.
Warren, M.S. & Key, R.S., 1991. Woodlands: past, present and potential for insects. In N.M. Collins & J.A. Thomas (Eds), *The Conservation of Insects and their Habitats*, Proceedings of the XV Symposium of the Royal Entomological Society of London. London: Academic Press.
Warren, M.S., Thomas, C.D. & Thomas, J.A., 1984. The status of the Heath Fritillary butterfly *Mellicta athalia* Rott. in Britain. *Biological Conservation*, 29: 287–305.
Warren, M.S., Pollard, E. & Bibby, T.J., 1986. Annual and long-term changes in a population of the wood white butterfly *Leptidea sinapis*. *Journal of Animal Ecology*, 55: 707–719.
Warren, R.G., 1984. *Atlas of the Lepidoptera of Staffordshire. Part 1: Butterflies*, Hanley: Stoke-on-Trent City Museum & Art Gallery.
Warrick, R.A. & Barrow, E.M., 1991. Climate change scenarios for the UK *Transactions of the Institute of British Geographers N.S.*, 16: 387–399.
Warrick, R.A. & Oerlemans, H., 1990. Sea level rise. In J.T. Houghton, G.J. Jenkins & J.J. Ephraums (Eds), *Climate Change. The IPCC Scientific Assessment*. pp. 257–282. Cambridge: University Press.
Wasserthal, L.T., 1975. The role of butterfly wings in regulation of body temperatures. *Journal of Insect Physiology*, 21: 1921–1930.
Wasserthal, L.T., 1983. Haemolymph flows in the wings of pierid butterflies visualized by vital staining (Insecta, Lepidoptera). *Zoomorphology*, 103: 177–192.
Watanabe, M., 1988. Multiple matings increase the fecundity of the yellow

swallowtail butterfly, *Papilio xuthus* L., in summer generations. *Journal of Insect Behaviour*, 1: 17–30.
Watson, E., 1977. The periglacial environment of Great Britain during the Devensian. *Philosophical Transactions of the Royal Society of London* B, 280: 183–197.
Watson, R.T., Rodhe, H., Oeschger, H. & Siegenthaler, U., 1990. Greenhouse gases and aerosols. In J.T. Houghton, G.J. Jenkins & J.J. Ephraums (Eds), *Climate Change. The IPCC Scientific Assessment*. pp. 1–40. Cambridge: University Press.
Watt, W.B., 1968. Adaptive significance of pigment polymorphisms in *Colias* butterflies. I. Variation of melanin pigment in relation to thermoregulation. *Evolution*, 22: 437–458.
Watt, W.B., 1969. Adaptive significance of pigment polymorphism in *Colias* butterflies. II Thermoregulation in *Colias eurytheme*. *Proceedings of the National Academy of Sciences*, USA, 63: 767–774.
Watt, W.B., 1973. Adaptive significance of pigment polymorphisms in *Colias* butterflies. III Progress in the study of the '*alba*' variant. *Evolution*, 27: 537–548.
Watt, W.B., 1977. Adaptation at specific loci. I Natural selection on phosphoglucose isomerase of *Colias* butterflies: biochemical and population aspects. *Genetics*, 87: 177–194.
Watt, W.B., 1983. Adaptation at specific loci. II Demographic and biochemical elements in the maintenance of the *Colias* PGI polymorphism. *Genetics*, 103: 691–724.
Watt, W.B., 1985. Bioenergetics and evolutionary genetics: opportunities for new synthesis. *American Naturalist*, 125: 118–143.
Watt, W.B. & Bowden, S.R., 1966. Chemical phenotypes of pteridine colour forms in *Pieris* butterflies. *Nature*, 210: 304–306.
Watt, W.B., Carter, P.A. & Blower, S.M., 1985. Adaptation at specific loci. IV Differential mating success among glycolytic allozyme genotypes of *Colias* butterflies. *Genetics*, 109: 157–175.
Watt, W.B., Cassin, R.C. & Swan, M.S., 1983. Adaptation at specific loci. III Field behaviour and survivorship differences among *Colias* PGI genotypes are predictable from in vitro biochemistry. *Genetics*, 103: 725–739.
Watt, W.B., Hoch, P.C. & Mills, S.G., 1974. Nectar resource use by *Colias* butterflies: chemical and visual aspects. *Oecologia*, 14: 353–374.
Wcislo, W.T., 1989. Behavioural environments and evolutionary change. *Annual Review of Ecology & Systematics*, 20: 137–169.
Weiss, S.B., Murphy, D.D. & White, R.R., 1988. Sun, slope and butterflies: topographic determinants of habitat quality for *Euphydryas editha*. *Ecology*, 69: 1486–1496.
West, D.A., Snellings, W.M. & Herbek, T.A., 1972. Pupal colour dimorphism and its environmental control in *Papilio polyxenes asterius* Stoll (Lep: Papilionidae). New *York Entomological Society*, 80: 205–211.
West, D.A. & Hazel, W.N., 1985. Pupal colour dimorphism in swallowtail butterflies: timing of the sensitive period and environmental control. *Physiological Entomology*, 10: 113–119.
West, R.G., 1977. 2nd Edition. *Pleistocene Geology and Biology*. London: Longman.
West-Eberhard, M.J., 1989. Phenotypic plasticity and the origins of diversity. *Annual Review of Ecology and Systematics*, 20: 249–278.

Whalley, P., 1986. A review of the current fossil evidence of Lepidoptera in the Mesozoic. *Biological Journal of the Linnean Society*, 28: 253–271.

White, E.J. & Perry, A.H., 1989. Classification of the climate of England and Wales based on agroclimatic data. *International Journal of Climatology*, 9: 271–291.

White, R.R., 1974. Food plant defoliation and larval starvation of *Euphydryas editha. Oecologia* 14, 307–315.

Wickman, P.-O., 1985a. The influence of temperature on the territorial and mate-locating behaviour of the small heath butterfly, *Coenonympha pamphilus* (L.) (Lep: Satyridae). *Behavioural Ecology and Sociobiology*, 16: 233–238.

Wickman, P.-O., 1985b. Teritorial defence and mating success in males of the small heath butterfly, *Coenonympha pamphilus* L. (Lep., Satyridae). *Animal Behaviour*, 33: 1162–1168.

Wickman, P.-O., 1988. Dynamics of mate-searching behaviour in a hilltopping butterfly, *Lasiommata megera* (L.): the effects of weather and male density. *Zoological Journal of the Linnean Society*, 93: 357–377.

Wickman, P.-O. & Karlsson, B., 1987. Changes in egg colour, egg weight and oviposition rate with the number of eggs laid by wild females of the small heath butterfly, *Coenonympha pamphilus. Ecological Entomology*, 12: 109–114.

Wickman, P.-O. & Karlsson, B., 1989. Abdomen size, body size and the reproductive effort of insects. *Oikos*, 56: 209–214.

Wickman, P.-O. & Wiklund, C., 1983. Territorial defence and its seasonal decline in the speckled wood butterfly (*Pararge aegeria*). *Animal Behaviour*, 31: 1206–1216.

Wickman, P.-O., Wiklund, C. & Karlsson, B., 1990. Comparative phenology of four satyrine butterflies inhabiting dry grasslands in Sweden. *Holarctic Ecology*, 13: 238–246.

Wigglesworth, V., 1972. The *Principles of Insect Physiology*. London: Chapman & Hall.

Wigley, T.M.L., 1991. Could reducing fossil-fuel emissions cause global warming? *Nature*, 349: 503–506.

Wigley, T.M.L., Lough, J.M. & Jones, P.D., 1984. Spatial patterns of precipitation in England and Wales and a revised, homogeneous England and Wales precipitation series. *Journal of Climatology*, 4: 1–25.

Wigley, T.M.L. & Raper, S.C.B., 1990. Natural variability of the climatic system and detection of the greenhouse effect. *Nature*, 344: 324–327.

Wiklund, C., 1975. Pupal colour polymorphism in *Papilio machaon* and the survival in the field of cryptic versus non-cryptic pupae. *Transactions of the Royal Entomological Society of London*, 127: 73–84.

Wiklund, C., 1981. Generalist vs. specialist oviposition behaviour in *Papilio machaon* (Lepidoptera) and functional aspects on the hierarchy of oviposition preferences. *Oikos*, 36: 163–170.

Wiklund, C., 1984. Egg-laying patterns in butterflies in relation to their phenology and the visual apparency and abundance of their hostplants. *Oecologia*, 63: 23–29.

Wiklund, C. & Åhrberg, C., 1978. Host plants, nectar source plants, and habitat selection of males and females of *Anthocharis cardamines* (Lep). *Oikos*, 31: 169–183.

Wiklund, C. & Persson, A., 1983. Fecundity, egg weight variation and its relation

to offspring fitness in the speckled wood butterfly, *Pararge aegeria*, or why don't butterfly females lay more eggs? *Oikos*, 40: 53–63.
Williams, C.B., 1958. *Insect Migration*. London: Collins.
Williams, E.H., 1981. Thermal influences on oviposition in the montane butterfly *Euphydryas gillettii. Oecologia,* 50: 342–346.
Williams, R.B.G., 1975. The British climate during the last glaciation; an interpretation based on periglacial phenomena. In A.E. Wright & F. Moseley (Eds), *Ice Ages; Ancient and Modern*. pp. 95–117. Geological Journal Special Issue No. 6. Liverpool: Seel House Press.
Willmer, P.G., 1980. The effects of a fluctuating environment on the water relations of larval lepidoptera. *Ecological Entomology*, 5: 271–292.
Willmer, P.G., 1982. Microclimate and the environmental physiology of insects. *Advances in Insect Physiology*, 16: 1–57.
Willmer, P.G., 1983. Thermal constraints on activity patterns in nectar-feeding insects. *Ecological Entomology*, 8: 455–469.
Willmer, P.G. & Unwin, D.M., 1981. Field analysis of insect heat budgets: reflectance, size and heating rates. *Oecologia*, 50: 250–255.
Willmot, K.J., 1985. A survey of Glanville fritillary roosting sites. *Butterfly Conservation News*, 35: 35–36.
Wiltshire, E.P., 1958. The natural history of *Papilio machaon* in Bagdad. *Transactions of the Royal Entomological Society of London*, 110: 221–244.
Winograd, I.J., Coplen, T.B., Landwehr, J.M., Riggs, A.C., Ludwig, K.R., Szabo, B.J., Kolesar, P.T. & Revesz, K.M., 1992. Continuous 500,000-yr climate record from vein calcite in Devil's Hole, Nevada. *Science*, 258: 255–260.
Wolda, H., 1987. Altitude, habitat and tropical insect diversity. *Biological Journal of the Linnean Society*, 30: 313–323.
Wolda, H., 1988. Insect seasonality: why? *Annual Review of Ecology and Systematics*, 19: 1–18.
Wolfe, J.A., 1987. Late Cretaceous–Cenozoic history of deciduousness and the terminal Cretaceous event. *Palaeobiology*, 13: 215–226.
Wolfe, J.A., 1990. Palaeobotanical evidence for a marked temperature increase following the Cretaceous/Tertiary boundary. *Nature*, 343: 153–156.
Wolfe, J.A., 1991. Palaeobotanical evidence for a June 'impact winter' at the Cretaceous/Tertiary boundary. *Nature*, 352: 420–423.
Wolpert, L., 1969. Positional information and the spatial pattern of cellular differentiation. *Journal of Theoretical Biology*, 25: 1–47..
Woodin, S.J., 1989. Environmental effects of air pollution in Britain. *Journal of Applied Ecology*, 26: 749–761.
Woodward, I., 1989. Plants in the greenhouse world. *New Scientist*, 166: 1–4.
Woodwell, G.M., 1989. The warming of the industrialized middle latitudes 1985–2050: causes and consequences. *Climatic Change*, 15: 31–50.
Wourms, M.K. & Wasserman, F.E., 1985. Bird predation on Lepidoptera and the reliability of beak-marks in determining predation pressure. *Journal of the Lepidopterists' Society*, 39: 239–261.
Wright, D.H., 1983. Species-energy theory: an extension of species-area theory. *Oikos*, 41: 496–506.
Wright, S.J., 1982. The shifting balance theory and micromutation. *Annual Review of Genetics*, 16: 1–20.

Wynther-Blyth, M.A., 1957. *Butterflies of the Indian Region*. Bombay: Bombay Natural History Society.
Young, A.M., 1980. The interaction of predators and 'eyespot butterflies' feeding on rotting fruits and soupy fungi in tropical forests: variations on a theme developed by the Muyshondts and Arthur Shapiro. *Entomologist's Record and Journal of Variation*, 92: 63–69.
Zachariassen, K.E., 1985. Physiology of cold tolerance in insects. *Physiological Review*, 65: 799–832.
Zeuner, F.E., 1962. Notes on the evolution of the Rhopalocera (Lep.). *Proceedings of the 11th Congress of Entomology*, Vienna, 1: 310–313.
Zwolfer, H., 1978. Mechanismen und Ergebnisse der Co-evolution von phytophagen und entomophagen Insekten und höheren Pflanzen. *Sonderdruck aus Naturwissenschaften Verein Hamburg*, 2: 7–50.

Index

References to figures are in *italic type*. References to tables are in **bold type**.